The Effects of Fishing on Marine Ecosystems and Communities

The Effects of Fishing on Marine Ecosystems and Communities

Stephen J. Hall

Professor of Marine Biology
The Flinders University of South Australia

Blackwell
Science

Editorial Offices:
Osney Mead, Oxford OX2 0EL
25 John Street, London WC1N 2BL
23 Ainslie Place, Edinburgh EH3 6AJ
350 Main Street, Malden
MA 02148 5018, USA
54 University Street, Carlton
Victoria 3053, Australia
10, rue Casimir Delavigne
75006 Paris, France

Other Editorial Offices:

Blackwell Wissenschafts-Verlag GmbH
Kurfürstendamm 57
10707 Berlin, Germany

Blackwell Science KK
MG Kodenmacho Building
7–10 Kodenmacho Nihombashi
Chuo-ku, Tokyo 104, Japan

First published 1999

Set in 10/13pt Times
by DP Photosetting, Aylesbury, Bucks
Printed and bound in Great Britain by
MPG Books Ltd, Bodmin, Cornwall

DISTRIBUTORS

Marston Book Services Ltd
PO Box 269
Abingdon
Oxon OX14 4YN
(*Orders:* Tel: 01235 465500
Fax: 01235 465555)

USA
Blackwell Science, Inc.
Commerce Place
350 Main Street
Malden, MA 02148 5018
(*Orders:* Tel: 800 759 6102
781 388 8250
Fax: 781 388 8255)

Canada
Login Brothers Book Company
324 Saulteaux Crescent
Winnipeg, Manitoba R3J 3T2
(*Orders:* Tel: 204 837 2987
Fax: 204 837 3116)

Australia
Blackwell Science Pty Ltd
54 University Street
Carlton, Victoria 3053
(*Orders:* Tel: 03 9347 0300
Fax: 02 9347 5001)

A catalogue record for this title is available from the British Library

ISBN 0-632-04112-9

Library of Congress
Cataloging-in-Publication Data
Hall, Stephen, J.
The effects of fisheries on ecosystems and communities/Stephen J. Hall.
p. cm.
Includes bibliographical references and index.
ISBN 0-632-04112-9
1. Fisheries – Environmental aspects.
2. Marine ecology. 3. Fish communities.
I. Title.
QH545.F33H35 1998
577.7'27 – dc21 98-28325
CIP

To Lynda, Patrick and Daniel for making me keep a healthy balance
between work and play

Contents

Series Foreword

Fish biologists (a.k.a. fish freaks) like to explain, to any innocent bystander, how fish have evolved an astonishing array of adaptations, so much so that it seems difficult for them to comprehend why anyone would study anything else. Yet, at the same time, fish are among the last wild creatures to be hunted by humans for food. As a consequence, few fish biologists today would not recognise the need to reconcile exploitation with conservation as the major challenge to our science, and so to encourage the study of fish biology, ecology and physiology to probe new frontiers. The Blackwell Science *Fish Biology and Aquatic Resources Series* is an initiative aimed at providing key peer-reviewed books in this fast-moving field.

While bony fish stem from a great radiation that followed the invention of the swimbladder in the Cretaceous period 200 million years ago, some fish groups, such as the sharks, lungfish, and sturgeons, are more ancient beasts. Survivors from earlier eras may be more widespread than we think: the deep sea coelacanths, known only from the Indian Ocean, have recently turned up in Indonesia. In addition, ancient fish may be more effectively adapted to specialised niches than their ancient body plans would suggest. For example, rays and angel sharks have perfected the art of the ambush predator and most cartilaginous fishes can detect electric discharges from the nerves of their prey.

Bony fish have evolved into an amazing array of habitats and niches. As well as the open sea, there are fish in lakes, ponds, rivers, rock pools, deserts, in the great deeps of the sea, in the extreme cold of the Antarctic, in warm waters of high alkalinity or of low oxygen, and in habitats like estuaries where their physiology is challenged by continuous change. Air-breathing climbing perch, which are regularly found up trees, walking catfish and mangrove mudskippers are repeating the land invasion of their Carboniferous ancestors. We can marvel at high-speed swimming adaptations in the fins, tails, gills and muscles of marlins, sailfish and warm blooded tunas; gliding and flapping flight in several groups of fish; swinging, protrusible jaws providing suction-assisted feeding that have evolved in parallel in groupers, cods and carps; parental care in mouth brooding cichlids; the birth of live young in toothcarps, rockfish and blennies; immense migrations in salmon, shads and tunas; and even the so-called four-eye fish with eyes divided into upper air and lower water-adapted sections.

Sadly, large sectors of this stunning biodiversity are at risk from human activities. In fresh water, for example, the largest mass extinction since the end of the dinosaurs has occurred as the introduced Nile perch in Lake Victoria eliminated over 100 species of endemic haplochromine fish. In the oceans, we have barely begun to understand the profound changes that have accompanied the vast expansion of human fishing over the past 100 years. Stephen Hall's *The Effects of Fishing on Marine Ecosystems and Communities*, the first book in the new Blackwell Science *Fish Biology and Aquatic Resources Series*, addresses this issue head on. This is the first book to review what we know of the impact of fishing on marine ecosystems, and bridges the author's move from the ICES arena in Europe to southern hemisphere fisheries in Australia.

After defining the problem – the previously unrealized impact of fishing on ecosystems – the author describes direct effects of fishing on fish population mortality rates through bycatch, discards, the impacts of trawling on the structure and ecology of benthos, and human effects on rocky shores, mudflats and coral reefs. The next section reviews ecotrophic effects such as removal of predators, competitors and forage fish. Three case studies are then described in detail: Georges Bank, the North Sea and the Gulf of Thailand. Hall discusses useful measures that may be employed for biodiversity and ecosystem health. The book closes with a discussion of mitigation through rational fisheries management; Hall's final prescription is to reduce effort, reduce effort, and reduce effort.

For many years, fisheries scientists were in denial that ecosystems could be adversely affected by fishing; indeed, the idea would be greeted by hostility and derision. Although the remnants of this attitude are still far from dead, the pendulum has swung pretty much to the other side, and we see many scrambling to get onto the bandwagon, often with little comprehension of our current state of knowledge. This exciting book redresses the balance by drawing a clear map of what is known and what remains unknown, and hence supplies a very appropriate theme with which to kick off the new Blackwell Science *Fish Biology and Aquatic Resources Series*. I am happy to begin my work as Series Editor with Stephen Hall's book and I am confident that it will be widely cited and will provide a valuable source of reference for some time to come.

Professor Tony J. Pitcher
Editor, Blackwell Science Fish Biology and Aquatic Resources Series
Director, Fisheries Centre, University of British Columbia,
Vancouver, Canada

Preface

Purpose

When I last looked there were 125 188 articles in a prominent scientific abstracts database and over 1200 textbooks available to buy on fisheries and fisheries related subjects. Given this level of coverage, one might reasonably ask why another book on marine fisheries is necessary. The simple answer is that, despite all that has been written, understanding the wider implications of fishing activity is still in its infancy. In comparison with conventional fisheries biology, there have been relatively few research programs which consider the direct mortalities that fishing imposes on non-target populations, the consequences of physical damage to habitats that some fishing gears cause, or the possible knock-on effects of reducing the populations of target species for other parts of the marine system. Yet, with growing recognition of the need to conduct and manage our activities within a wider, more environmentally sensitive framework, these and related issues are increasingly being debated by scientists and policy makers around the world.

To be fair, the relative neglect of the wider effects of fishing by fisheries biologists is understandable – as a scientific discipline fisheries biology was born from the need to manage populations of the harvested species and maintain sustainable fisheries. Given the human and financial resources required to simply monitor exploited populations and measure the effects of fishing on them, it is little wonder that a wider perspective has been accorded low priority up to now. However, with the seemingly unstoppable rise in the levels of exploitation and increase in the range of harvested species, and as populations (particularly in more developed countries) adopt progressively more environmentally aware attitudes, it was inevitable that the massive human intervention that fishing represents would come under greater scrutiny. As with many other activities such as waste disposal, chemical usage or energy policies, scientists and politicians are increasingly being asked whether they fully understand the consequences of society's actions.

This book has two broad objectives. The first is to draw together and synthesise a rather disparate body of literature that ought, in my view, be the cornerstone of a debate about the wider effects of fisheries. I hope that I have organised that literature into a framework that helps to focus the debate. The second objective is of course to contribute my own views and conclusions about the relative importance of the various topics that fall within the broad scope of fisheries effects. One referee for an early draft of this book commented that it seemed to contain more questions than answers. Sadly on many levels this is true. There are many aspects of this issue where our data and understanding are fragmentary and it would be foolish to suggest that this were not so. Indeed, part of the reason for writing this book was to help identify the most pressing questions to be included on the current research agenda. This is not to say, however, that another decade of research is required before anything can be done. On the contrary, there are a number of approaches to mitigating the environmental effects of fisheries that could be more actively pursued. Some of these are noted in the conclusion sections for each chapter or in Chapter 9 where some of the promising approaches for mitigating fisheries effects are discussed in more detail.

As far as my qualifications to write this book are concerned, I claim no particular authority, other than an active interest in the subject and a fortuitous exposure to many aspects of the problem. It is through the activities of the International Council for the Exploration of the Seas (ICES) Working Group on the Ecosystem Effects of Fishing Activities, that I first became aware of these issues. More recently, a move to Australia has provided me with an invaluable additional perspective, as has membership of a Scientific Committee on Oceanic Research Working Group which has been working to understand the impact of fisheries on the dynamics and stability of marine ecosystems.

Scope and structure

What am I really talking about when I refer to the wider implications of fishing activity? Put simply, the entirety of effects, not simply those on target species. I define a target species as one for which there is a directed fishery; in other words, fishermen go out and deliberately catch them. Non-target species include those of commercial value which, if caught, would be landed, as well as those which are either not caught, or are of no interest and would always be discarded.

As a foundation it is also useful to distinguish between the direct and indirect effects of fishing. Direct effects can be summarised as follows:

(1) Fishing mortality on species populations, either by catching them (and either landing or throwing them back) or by killing them during the fishing process without actually retaining them in the gear, and/or exposing or damaging them and making them vulnerable to scavengers and other predators.
(2) Increasing food available to other species in the system by discarding unwanted fish, fish offal and benthos.
(3) Disturbing and/or destroying habitats by the action of some fishing gears.

These topics form the basis for Part 1, and it is in many ways the least problematic part of this book. This is because the causal chain to be elucidated is short and relatively unambiguous – documenting effects is relatively easy. Less easy, however, is quantifying the overall scale of direct effects and its importance for species populations and communities. Even here, however, progress is being made.

Part 2, deals with indirect effects i.e. the knock-on effects that follow from a direct effect. As we will see later, fishing has led to density reductions of immense proportions for target species. These reductions, along with the direct effects noted above might in themselves be expected to have consequences for the predators, prey and competitors of a target species. Thus, an important element of Part 2 is the fisheries biologists' own examination of how fish populations interact with one another. There is something to be learnt about how marine systems react, and the consequences of fishing for the system as a whole, from examining the findings of multi-species fisheries research and from exploring the relative changes in the abundance of commercially exploited species.

Examining what is known of these species relationships will of course take up a substantial part of this book, but using the words ecosystem and community in the title also serves to emphasise a somewhat broader perspective. In essence, it signifies an intention to consider if and how the harvesting of one or a suite of species effects changes, not only in the component populations of the many other species in the system, but also to the structure and function of the community and ecosystem as a whole.

One might, perhaps, take the view that the effects of fisheries on an ecosystem are best expressed with reference to summed effects on constituent populations. In other words, the subject of Parts 1 and 2 of this book would be sufficient. To an extent, I agree with this sentiment, but I also believe that ecosystems have, to use a well-worn phrase 'emergent properties' which cannot be deduced by considering the responses of the individual components. How might fishing affect the distribution of individuals among species, the size-distribution of communities, or the flow of materials and energy through ecosystems? Could existing balances between, or the character of, the processes of primary production, primary consumption and secondary production be altered? These, questions along with the question of the most appropriate way to think about ecosystem status seem to me to be important – they form the basis of Part 3.

The final part of the book (Part 4) deals with the issue of fisheries management and asks how it might be broadened to take account of these wider issues. In particular it focuses on some of the strategies that seem most promising for mitigating fishing effects.

One comment on ecological terms should be made at this point. In common with Paine (1994) I take the view that one shouldn't worry too much if the meaning of words such as ecosystem and community are somewhat vague. For me a community is simply the collection of all species that occur in a given place. I reserve the related term assemblage for any selected subset of the species in a community. More often than not the boundaries between communities will be indistinct but we recognise communities when we see them. An estuary has a community that is distinct from that in adjacent fully marine sub-tidal areas, but it would be difficult to say where one ends and the other begins. Similarly, the meaning of the word ecosystem is problematic if one thinks too hard about it – we all feel we know what one is, but articulating a satisfactory definition and identifying a firm boundary is difficult.

With respect to the marine realm, perhaps the best definition of an ecosystem centres around the concept of the Large Marine Ecosystem (LME) (see, for example, Sherman , 1991). LMEs have been defined as relatively large regions of the global Exclusive Economic Zones (EEZs), generally of the order of $>200\,000\,km^2$ and characterised by unique bathymetry, hydrography and productivity. Within an LME marine populations are considered to have adapted reproductive, growth and feeding strategies and although there are obviously exchanges across the boundaries, LMEs can be viewed as relatively discrete units. It has been argued (legitimately in my view) that these LMEs should be the focus of attention as regional units for the implementation of management actions leading to sustained and predictable yields of living resources. This is particularly important for the LMEs on the coastal margins. On a global basis, nearly 95% of the biomass yield from the oceans is produced within the 200 mile limits of the EEZs of coastal states. Moreover, it is these areas that are subject to the most intensive over-exploitation of fisheries, and which are experiencing increased stress from pollution and other environmental perturbations.

Biases and uncertainties

Philosophers of science and sociologists have largely debunked the myth that science, as it is actually practised by scientists, is an entirely objective pursuit in which the acquisition and analysis of data is unbiased. On the contrary, it has become quite clear that scientific research and the conclusions drawn from it is socially constructed. In other words, the way data are collected, the weight that is given to various types of knowledge and the way that data are interpreted are all affected by the social and political context in which the scientist operates (for a lucid account of these issues in a fisheries context see Finlayson, 1994). Like many scientists I am rather uncomfortable with this conclusion, but I have no doubt that it is true. It seems reasonable, therefore, to try examine at the outset the potential cultural biases that I am likely to impose in writing a book of this kind.

Perhaps the first point to make concerns geographic bias. Almost all my career has been spent studying Northern European temperate systems. My experience of polar, boreal and tropical habitats is superficial at best. The second bias relates to topic area since, although I have been well exposed to issues of relevance to this book, through meetings, workshops and (most of all) my own reading, a large part of my practical research activity has centred on soft-sediment benthic communities. With these points in mind, an old adage is relevant – the more you look at a problem the less you seem to know. My suspicion is that I am less likely to be cautious and to stress the uncertainty of my conclusions when the subject under discussion is some distance from the geographic and subject areas in which I have an active research involvement. To counteract this likely bias, I have tried to make a clear distinction between the data on which conclusions are drawn and the conclusions themselves, thereby providing the reader with his/her own opportunity to make a judgement. Of course, deciding what data to include in analyses offers opportunities for bias also and the only defence I can offer is that I have tried to be as exhaustive as possible, within the space limits imposed by a book of sensible proportions. In taking a global view, I am acutely aware of likely deficiencies in coverage, but I hope that the examples and issues explored here will provide a sound basis for discussion.

Stephen J. Hall

Acknowledgements

The first part of this book was written while I was working for the Scottish Office Agriculture Fisheries and Environment Department, Marine Laboratory in Aberdeen, Scotland. I cannot express adequately my gratitude for the resources, support and encouragement that SOAFED provided and for the stimulating environment created by my colleagues and friends. In particular, I thank Simon Greenstreet, John Hislop, Peter Wright, Phil Kunzlik and Robin Cook for clarifying my thoughts at various stages and Sarah, Elizabeth and Sandra in the library for always coming up with the goods. Thanks also to Tony Hawkins, the Director for recognising the importance of strategic research and supporting efforts to do it.

On a wider front, this book would not have been written, were it not for my involvement with the ICES Working Group on the Ecological Effects of Fisheries Activity. It was working with this group that introduced me to the many facets of the fisheries effects problem and led me to consider more deeply issues that lay beyond my area of immediate expertise. I especially wish to thank (in no particular order) Niels Daan, John Pope, Jake Rice, Henrik Gislason, Han Lindeboom, Jeremy Collie, Chris Frid and Mark Tasker who made lively and thought provoking contributions to our various debates. Later work with members of a SCOR Working Group broadened my experience and understanding further. Replies to specific questions by George Hunt and others, and comment on drafts by Steve Hawkins, Nick Polunin, Steve Blaber, Mike Kaiser, Jeremy Collie, Simon Greenstreet, John Hislop and two anonymous reviewers were also a great help.

The second half of this book was written after moving to Australia (that's why it was late). This change of scene also brought with it valuable new perspectives and a number of new friends, who have helped at various stages, both with settling into a new life and this project. With regard to the latter, a special mention goes to Ian Poiner, Keith Sainsbury and Paul McShane. Contributions by Steve Eayres, David McDonald and Chris Glass to a workshop on by-catch reduction have also been drawn upon. I also owe a debt of thanks to Susan Gray for keeping the lab running while I disappeared to write, and helping out with the boring stuff when I was desperate.

Much of the material presented here is drawn from comprehensive syntheses by others. I have tried to ensure that these efforts are fully acknowledged. In preparing something of this kind, however, omissions and errors are always possible and I apologise in advance for any deficiencies in these respects. Thanks are especially due to Mike Kaiser and Simon Jennings for providing me with a final version of their forthcoming review (Jennings & Kaiser, 1998) – despite the almost complete independence of the two works, the similarity of many of the conclusions is striking.

Various people have commented on draft chapters, provided information or simply given a second opinion. I think I have mentioned them all above, but ommissions are always possible. If you are not included I apologise profusely and will buy you a drink next time I see you. I am sincerely grateful to everyone who helped for taking up the challenge. Thanks also to the staff at Blackwells (Ian Sherman, Richard Miles and Shahzia Chaudhri) for quiet support and understanding when deadlines slipped.

Part 1
Direct Effects on Populations

Introduction

In 1992 the world catch of marine fish was approximately 85 million tonnes, representing a seafood harvest which provides around 13% of the total animal protein for human consumption (Idyll, 1978). However, the importance of seafood for the overall food supply is much greater than indicated by this statistic because more than half the world's population depends on fish for the majority of the animal protein it consumes. For many poor countries it is often the only source. As a resource, therefore, food harvested from the sea is an essential element for the well-being of a very large proportion of the human population. By 2010 it is estimated that the world population will reach 7200 million, by which time, at unchanged relative prices, total demand for fish will reach 140 million tonnes of which 110 million will be for human consumption and 30 million for fish meal (Garcia & Newton, 1994). With such demand in prospect it seems almost certain that fisheries management and concerns about the impact of fishing on the environment will continue to be a focus for attention.

Chapter 1, provides an overview of the status of global fisheries and some background to post-war fisheries development. Since the most profound effects of fishing are on target stocks, this is ground that must be covered. However, this treatment is provided more as background than anything else, since many other volumes have been devoted to this topic. In this chapter an overview of some of the management approaches that have been adopted and the difficulties that have arisen is also provided. The material in Chapter 1 is offered to provide an introduction and context for the rest of the book, but for those who are familiar with the history of fisheries and fisheries management it can be safely skipped.

Chapter 2, reviews our current knowledge of by-catch and discards in world fisheries and considers the implications of this activity. In many areas of the world a wide variety of fishing gears are used, each focusing on one or a few species. Unfortunately, this focus does not mean that non-target species, sexes or size-classes are excluded from catches. On the contrary, incidental catches can be the source of bitter dispute between fishermen, a monumental problem for fisheries managers and a seemingly unnecessary source of mortality and waste. Disagreements over levels of by-catch have resulted in many regulations seeking to protect fisheries resources and the interests of specific fisheries sectors. The history of these efforts shows that the management problems surrounding this issue are truly formidable. In the last decade, however, discussion of by-catch and discarding has become prominent for other reasons. In particular, heightened interest can be attributed to the increased awareness of marine conservation issues and the effects of fishing activities on populations of marine mammals, birds and turtles. The interest probably also reflects growing recognition of how poorly we seem to be able to manage many fisheries and how wasteful discarding is. My purpose here is not to deal specifically with the management issues concerning the by-catch and discards of target fish and shellfish species in a region, but with the effects on the taxa, which none of the fisheries are interested in landing. As we will see, when effects are large enough to be detectable they usually show reductions in population sizes. In some cases, however, positive effects on populations can also be observed.

Chapter 3 focuses on sub-tidal benthic habitats and considers the consequences of trawling activity for infaunal and epifaunal taxa. This is a topic that has received considerable attention, yet there remain key deficiencies in the available data. In particular, uncertainty concerning the fine-scale distribution of trawling activity and the distribution of benthic epifaunal communities that are most likely to be affected by trawling is a major impediment to both assessment of the problem and the implementation of any mitigation strategies that might be desirable.

Chapter 4 examines inter-tidal and shallow sub-tidal habitats. These habitats have been treated separately from those in Chapter 3, because the topic of trawling impacts is sufficiently large to warrant a chapter of its own. However, many of the issues for the inter-tidal systems examined in Chapter 4 are similar to those examined previously. In addition, however, effects on hard substrata are considered, including coral reefs and rocky shores, where the pressure of increasing human populations along the coastal margins is taking its toll.

Chapter 1
Trends in World Fisheries and the Essence of the Problem

1.1 A brief history

The most accessible assessments of the extent and state of world fisheries are provided by FAO who collate available data in biennial reports. It is convenient to start with the trend in world landings (Fig. 1.1), which shows a relatively steady increase into the 1990s. Although this trend looks impressive, it obscures some important features of the world fishery that should be appreciated – drawing heavily on both the FAO assessments and Garcia & Newton (1994), a brief history of world fisheries is in order.

The period immediately after the Second World War saw intense fisheries development, which continued until 1958. During this period fish production increased from 17.7 to 28.4 million tonnes and the main policy objective of the FAO was the reconstruction of European economies, including the fisheries sector. In

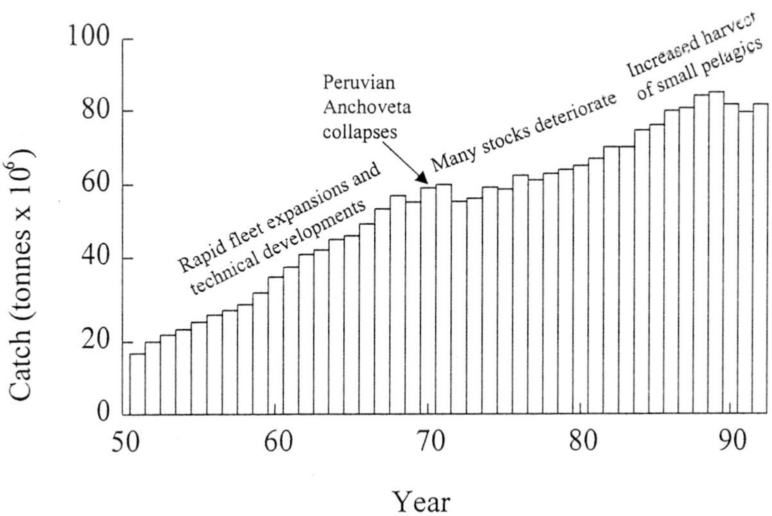

Fig. 1.1 Growth in global marine fisheries catch from 1948–1992. (Adapted from Figure 1, FAO, 1994. Reproduced with permission of the Food and Agriculture Organization of the United Nations.)

1945, the major problems affecting fisheries in the northern hemisphere were identified by the first FAO Technical Committee. They included resource depletion (including whales), discarding practices and the lack of data for high seas management (at this stage the term high seas denoted waters beyond 3 miles from the coast). These problems have a familiar ring today, but the committee also noted that resources in the southern hemisphere were under-utilised, signalling opportunities for long-range fisheries expansion. In the early 1950s important fishery resources such as the Hokkaido sardine, the North Sea and Atlanto-Scandian herring and the Californian pilchard all decreased or collapsed. Although natural environmental fluctuations almost certainly played their part in these events, it seems likely that high levels of exploitation were a contributory factor.

Between 1959 and 1972 there was a rapid geographic expansion of fisheries and world catch rose from 30 to 60 million tonnes. For developing countries and financial institutions the focus was on the assessment of world resources in support of this expansion. The FAO undertook surveys in West Africa, the Bay of Bengal and off the Atlantic coast of South America, and long-distance fleets from many nations expanded their operations supported by subsidy schemes. This increase in effort coupled with the introduction of technological developments into the fisheries of developing nations (e.g. synthetic fibres, better engines) put increasing pressure on traditional fishing grounds. An industrial fishing sector also emerged in some developing countries during this period and valuable shrimp fisheries developed in Asia and India, Kuwait and Pakistan. These fisheries were major sources of foreign exchange for developing nations. During this period severe overfishing occurred in the coastal areas of many developing countries and the period saw a series of dramatic fishery collapses (e.g. the anchoveta in Peru, which collapsed from 12 million to 2 million tonnes in the early 1970s). Significantly, the growth of world fisheries between 1959 and 1972 was accompanied by a growing public concern for the mismanagement of fisheries for marine mammals. This concern heralded the beginning of the era in which those not directly involved in the exploitation of fisheries began to play a role in the development of fisheries policy.

Between 1972 and 1982 fisheries production increased from 60 to 68 million tonnes and the status of stocks deteriorated except in a few areas. During this period, the expansion of fisheries was constrained by rising fuel prices and the growing need to negotiate conditions of access. Increasing numbers of coastal states extended their jurisdiction to 200 miles in anticipation of international legislation, negotiations for which began in 1973. Despite the deterioration in many stocks, the early 1980s saw a rise in catches. This increase occurred, partly because the stocks of low-value small pelagic species were at a peak in their natural fluctuations, but it was also due to the development of the fishery for Alaskan pollock. The period also saw the final expansion of distant water fleets into the Indian Ocean, the South Pacific and the south-west Atlantic in the search for high value species such as tunas, shrimps and cephalopods. However, the increase in overall production during this period belied the evident over-exploitation of more traditional demersal fish resources. Moreover, for the first time, there was discussion of more fundamental changes to ecosystems in the Gulf of Thailand, the North Sea and off West Africa. In these areas large long-lived species were apparently being replaced with smaller short-lived ones. Those areas of the continental shelf which were still freely accessible (e.g. off Western Sahara, Namibia and in the southwest Atlantic) became the targets for increased pressure from high seas fleets which also intensified their efforts to exploit oceanic species. In developing countries the growth of new fishing industries was often accompanied by an overall increase in the level of fishing effort as the number of artisinal fishers exploiting nearshore resources increased. This increase led in turn to conflicts between the artisinal and commercial sector.

Deficiences in fisheries management were by this time well recognised and, in particular the concept of Maximum Sustainable Yield (MSY), hitherto a touchstone for fisheries biologists, came under criticism (e.g. Larkin, 1977). Many states began to consider the adoption of lower targets for fishing mortality than those implied by MSY in the belief that these would improve yields and the economic performance of their fisheries.

Between 1983 and 1992 catches increased from 68 to about 85 million tonnes and global issues of sustainabilty and the environmental implications of fisheries began to dominate. States began to consider the need for greater protection of non-target species which appeared to be under threat and conflicts arising from the effects of high seas drift netting for tuna on dolphin populations was a particularly prominent issue. In 1992 the culmination of this concern and the continuing problems of over-exploitation of fishery resources led to the International Conference on Responsible Fishing, held in Cancun, Mexico, which elaborated a declaration containing a series of principles. These principles emphasised the problems of the environment and the need to protect critical habitats, reduce by-catch and discards and other environmental impacts of fishing and aquaculture. There followed a more complete specification of these principles with the FAO Code of Conduct for Responsible Fisheries (FAO, 1996).

1.2 The current status of world fisheries

So how would one characterise the current status of world fisheries? One perspective comes from considering the FAO assessments of the various world stocks. This assessment classified stocks as occupying states ranging from 'under-exploited' to 'overfished'. Although convenient, it is worth pointing out here the difficulty of a classification system that uses this latter term because it really only has meaning with reference to a specific objective – only if you state your objective can you say that you've overfished.

Overfishing (in a biological sense) can be divided into two classes, recruitment overfishing and growth overfishing. Recruitment overfishing occurs when a stock is depleted to a level where there is an unacceptable risk that the remaining adults will be insufficient to produce enough offspring to maintain the stock. This situation is most likely to occur in pelagic species where the individuals often form dense aggregations that can be easily detected so that catches and catch rates can remain high even when the stock is severely depleted. Additionally, many pelagic species are prone to dramatic natural fluctuations in recruitment success without any help from fishermen (e.g. Soutar & Isaacs, 1974), often due to changes in the hydrographic conditions (e.g. the strength of upwelling). Importantly, collapses can occur with little or no warning, the spectacular collapse of Peru's anchoveta fishery and the herring fisheries in the North Atlantic were almost complete surprises. Although there have been relatively few dramatic collapses due to recruitment failure in demersal fish there are notable exceptions, e.g. Arctic cod, Pacific cod, George's Bank haddock (Beddington & Rettig, 1984) and northern cod (Myers *et al.*, 1997).

The other group that is particularly prone to recruitment overfishing comprises those marine species that are economically valuable but have low reproductive capacities. Turtles, marine mammals and elasmobranch species clearly fall into this category and, in contrast to pelagic or demersal fish species, the additional concern beyond a collapse in the yield is the possibility of species extinctions. It is difficult to ignore the prospect that fishing might lead to extinctions, particularly of such charismatic creatures as marine mammals. One of the most graphic examples of the effects on target stocks must surely be the exploitation of whales in the Southern Ocean. Seven species or subspecies of baleen whale occur in this region and all were subject to very intensive exploitation. Of the eight species of toothed whale only the sperm whale was taken in any numbers. The total reported catch of whales from 1904 to the moratorium at the beginning of the 1986–7 season was 1.483×10^6 individuals (Kock & Shimadzu, 1994). Examination of the catch data show that species were in effect mined out sequentially – as one stock was depleted to uneconomic levels the fishery moved on to the next. Most whale populations remain only a fraction of their initial sizes.

The alternative form of overfishing is termed growth overfishing which describes a state where fish are harvested too early in their life. At one extreme one can remove a few large old fish, at the other, many small young fish – in between there is an optimal age at which the product of numbers and body size is maximal. Growth overfishing probably occurs most often for demersal species. Most of the North Sea gadoid stocks are in this condition at present.

Although talk of overfishing ought to be with reference to a specific objective (biological or otherwise), in many publications, including those from the FAO, the term is used in a much less precise way. In these cases, the best one can do is assume that the stock is in a particularly poor state. Notwithstanding difficulties of definition, of collecting data on a global scale, and the undoubted biases and uncertainties, the conclusion that the majority of world fishery resources are either fully or over-exploited is robust (Fig. 1.2). A strong case can be made that we are reaching the limits of production for wild marine resources. This limit was predicted to be

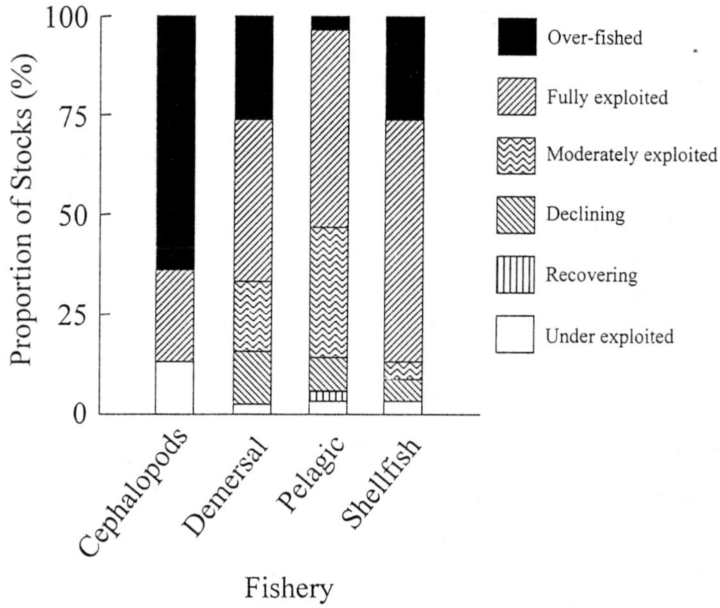

Fig. 1.2 The proportion of stocks in states ranging from under-exploited to overfished for four categories of fishery. (Compiled from data in FAO, 1994. Reproduced with permission of the Food and Agriculture Organization of the United Nations.)

approximately 100 million tonnes in the 1970s and may not be too far wrong. Garcia & Newton (1994), for example, state that annual fish production is unlikely to rise beyond 115–120 million tonnes of which about 100 million will be from capture fisheries and 15–20 million from aquaculture. By 2010, these figures imply a deficit in production of 20–25 million tonnes. As Garcia & Newton (1994) point out, it is difficult to predict how the world fisheries system will react, but what seems certain is that unless present policies are drastically changed the conditions are set for an acceleration in the degradation in the resource base. It is difficult to imagine that this will not have wider implications for the structure and function of ecosystems and communities.

1.3 The essence of the fisheries management problem

Despite the importance, and the widespread recognition of the need to manage fishery resources to sustain and preferably optimise yields, it would be hard to argue that from a global perspective current management practices are satisfactory. Perhaps the most fundamental reason for this is that fish stocks which are treated as a common resource for all to exploit will rarely be harvested optimally. Beddington & Rettig (1984) summarise the classical picture for the development of a fishery, which explains this principle.

(1) Fishers first enter the fishery, obtain high catches and make large profits.
(2) Seeing the early entrant's success, other fishers are attracted in, which prompts those already there to increase investment in vessels, gear etc. to maintain their profitability.
(3) With the increased fishing effort the size of the stock falls, leading to reduced catches per unit of fishing effort (CPUE) and cut-throat competition for a limited resource.
(4) Catch rates and profits fall to the point where, at best, most fishers can only break even. At this point, further increase in effort leads to financial losses.

This last condition is termed the bioeconomic equilibrium which is often viewed as the inevitable end point for an open-access fishery. For high value fish resources this benign-sounding term often describes a parlous state. Fisheries in this condition are usually in an undesirable condition, both ecologically and environmentally.

It has been widely argued that the pattern of development described above is an almost inevitable feature for the uncontrolled exploitation of a common property resource, a view expounded most often with reference to Hardin's evocative phrase 'The Tragedy of the Commons' (Hardin, 1968). Some see 'The Tragedy' as an argument for strict external regulation of a fishery, while others argue that it can best be avoided if common property fisheries can be converted to private fisheries. Neither approach appears to have any logical primacy and Clark (1973) concludes that over-exploitation can result from either common property competitive exploitation or private property maximisation of profits. A particular danger arises for populations that are economically valuable but have low reproductive capacities, especially when high rates of discount are applied. Setting a high discount means that a high premium is placed on extracting an income from the resource now compared to some time in the future. The choice of discount rate is crucial for determining how a fishery behaves. Indeed, contrary to most people's intuition, with high discount rates it can make economic sense to invest heavily in a new fishery and increase the fishing effort to a point where the stock collapses. This logic holds if the initial returns on investment are very high and that money can be re-invested with similar or greater financial yields in another venture (Clark, 1976). This kind of logic can be applied, for example, to some deep-sea fish stocks which can be viewed as mining rather than sustainable fishing operations.

The above features go a long way to explaining why so many of the world's fisheries resources are over-exploited – over-exploitation is the natural outcome of un- (or under-) restrained access to common property. If we want to exploit fish stocks sensibly they have to be managed through legal and social instruments which restrain access in some way. Sadly, many fisheries still do not have such restraints imposed – they are doomed to failure. But it is legitimate to ask why, when efforts are made to manage fisheries by restraining access in some way, they are not more successful? Two factors might contribute to the failure: (1) flawed objectives, or (2) inappropriate or ineffective action to achieve the required changes. Let us consider the objectives first.

1.4 Biological objectives for fisheries

The fundamental biological objective for managing fisheries on a sustainable basis is remarkably simple – on average, catch should be balanced by recruitment. The problem is that for most stocks, recruitment cannot be predicted. In an ideal world there would be a clear relationship between the size of the stock this year and the number of new fish entering the fishery (the recruits) next year. In most cases, however, this relationship seems to be very weak indeed. The problem of predicting recruitment is by far the most difficult one facing fisheries' biologists and is the one which adds most uncertainty to predictions. Fishing itself adds to this uncertainty because it usually acts to remove the older age groups from a stock, thereby reducing the period into the future over which you can make predictions. If you only have one or two age groups, making predictions beyond one or two years will depend on unknown levels of future recruitment. More formally, the variance in the stock biomass approaches the variance in recruitment the harder you fish the stock.

Because of the difficulties of predicting or measuring recruitment, much of the early work on fisheries' models avoided absolute catch predictions and focussed instead on yield per recruit. These models are used to examine the effects on the yield of varying the pattern of exploitation (factors such as the size and age at which fish are first caught) and the level of fishing mortality imposed. Analysis of yield per recruit models shows that, for a given stock structure (age composition, growth rate), there is an optimal combination of exploitation pattern and fishing effort which maximises the yield. From such analyses, coupled with assumptions of average recruitment, the concept of Maximum Sustainable Yield (MSY) emerged. Fishing for MSY is essentially an attempt to obtain the maximum average weight of catch over the long term. In a sense the term is rather misleading since the word sustainable has connotations of predictability and constancy which belie the, often high, short-term variability in recruitment that affects immediate yields. If one chooses to set MSY as an objective (and it often was, at least implicitly) then a rate of fishing mortality must be specified to achieve it for any given pattern of exploitation. The tools for doing this are well developed and, in the past, much of the advice from biologists was given in terms of rates of fishing mortality that would lead to MSY over the long term. This reference level of fishing mortality is usually termed F_{max}. In the past there was considerable debate about whether MSY can be said to exist at all, given that fish populations are rarely, if ever, at equilibrium (e.g. Larkin, 1977). There was also much hand wringing over whether F_{max} was the most suitable goal.

The outcome of these debates was that other reference points for lower levels of exploitation are now often advocated. One widely used alternative is the $F_{0.1}$

strategy, in which the slope of the yield per recruit versus fishing mortality is 0.1 times (10%) of the initial slope. The basis for choosing this level of fishing mortality is that it should have little effect on overall yield but will reduce the probability of a population crash. As Hilborn & Walters (1992) point out, this 10% value is entirely arbitrary, but it is probably in the right ballpark. If we know the expected number of recruits into the fishery (from pre-recruit surveys for example) the fish harvesting rate can be set for the short term using $F_{0.1}$ or something like it. This strategy imposes a constant rate of harvest on the stock and seems fairly reasonable, although even this may not be sufficiently conservative. Walters & Pearse (1996), for example, calculated that the mortality rate calculated for the Northern cod fishery that corresponded to $F_{0.1}$ would have to be reduced by 50% to give even a moderate level of protection from risk of stock collapse. This 50% reduction would mean that only 10% of the stock could be taken each year.

I do not propose to debate the relative merits of constant harvest rates or other strategies, but it should be acknowledged that the approach has been strongly criticised in some quarters. It can be argued, however, that it is rather difficult to evaluate the success of $F_{0.1}$ or one of its variants as a management strategy because the levels of fishing effort required for obtaining it have rarely, if ever, been achieved. Indeed, despite the telling criticisms of MSY by Larkin (1977) he also stresses that it provides a valuable rough index of production potential for major commercial species and is a reasonable first cut management policy. It must be recognised, however, that once obtained it should be expected that it will not be sustained. While one might debate whether $F_{0.1}$ or some other more stringent regime is sufficiently conservative in most cases, for single stocks, there is in principle nothing particularly unreasonable about aiming for levels of fishing mortality that ought to give high average long-term yields. A more important difficulty is restraining fishing effort so that the required level of mortality can be achieved.

Others would dissent from this view, among them Wilson *et al.* (1994) who argue that the entire basis for fisheries management is flawed and that a complete shift in approach is warranted. The basic premise of Wilson's argument is that fish population dynamics are chaotic and cannot, therefore, be predicted in the long term. Thus, attempts to regulate fisheries through control of catch or fishing mortality, achieved by, for example, catch quotas are futile because we have a massive and economically intractable information problem. Leaving aside the truth or otherwise of the chaos premise, (noting merely that in the marine realm only a few phytoplankton time series have been examined so far and no evidence for chaotic dynamics has been found, Godfray & Blythe, 1990; Ascioti *et al.*, 1993), others argue that there is a flaw in Wilson's argument. Fogarty (1995), for example, suggests that chaotic variability doesn't matter because long-term predictions are generally not required for fisheries management. Catch or effort controls are usually based on pre-season surveys or other short-term forecasts of population size, so provided these are sound the strategy ought to work. To a degree this is true, but variability and uncertainty in fish populations does cause fundamental problems for the current management approaches, and obtaining sound data is a herculean and expensive task.

Why is it so difficult to achieve the biological objectives?

Fisheries biologists cannot be held solely responsible for the failure to manage fisheries successfully. Sadly, basically sound scientific advice is often not implemented because political and economic interests overturn it. The frustration of scientists serving on fisheries assessment panels is a testament to this situation. In the ICES community, for example, it was the failure of the management authorities to implement what scientists viewed as sufficiently low-catch quotas, that led them to change the form of their advice and make recommendations about effort reduction, rather than allowable take. This might seem to be a subtle shift, but it was felt that giving advice in terms of percentage effort reductions, without specifying the mechanism by which this should be achieved, would place more of an onus on political administrators and improve the likelihood that appropriate fishing regimes would be implemented. Similarly in the USA efforts have been made to make an explicit distinction between 'conservation' and 'allocation' decisions. By separating scientific and social matters the National Marine Fisheries Service hopes that it will avoid past failures.

1.5 Economic and social objectives

While biological objectives have been the focus of fisheries biologists, and sustainability of stocks is clearly a primary consideration, economic and social aspects of fisheries management can also have profound effects on the choice of management regime and the rigour with which it is imposed.

One model that is widely used to determine the economic status of a fishery is the Gordon-Schaefer model (FAO, 1993). In essence, this model defines an economic yield and an economic cost curve for the fishery, with respect to fishing effort (Fig. 1.3). Yield can be defined here as the value of the catch, and its curve is generally dome shaped with a peak at Maximum Sustainable Yield (MSY). As noted earlier, most fisheries scientists feel that natural variability and other factors make MSY a risky target to aim for owing to the likelihood of fishery collapse. In contrast to the yield curve, the cost curve generally increases monotonically with effort because more effort simply means more cost. (The cost curve is often assumed to increase linearly, but it need not necessarily take this form). With such yield and cost curves, the model defines a point of Maximum Economic Yield (MEY), where the distance between the two curves is greatest. At MEY, the net economic return to society ('economic rent') is maximised.

When a fishery is unregulated or when there are no opportunities for fishermen to undertake economic activities other than fishing, neither MSY nor MEY are stable

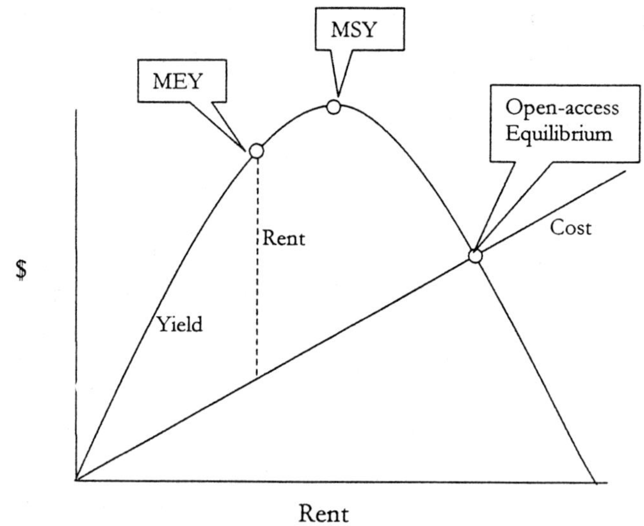

Fig. 1.3 The Gordon–Schaefer fixed price bioeconomic model of revenue versus fishing effort, including both a yield curve and cost line. With open access, lowering the cost of fishing leads to more intense exploitation.

equilibrium points. In other words, there are no forces that operate to keep the fishery at either of these two points and they will drift into another state. In fact, the most likely point the fishery will arrive at is the point where the cost line crosses the yield line – the bioeconomic or open-access equilibrium point. This point usually lies well beyond what is generally considered to be either economically or ecologically desirable. Unfortunately, many fisheries are operating at or close to this level, particularly if one factors in the huge financial subsidy to the global fishing fleet (p. 14).

If alternative non-fishing employment is available for fishermen, the income that is lost through not undertaking that nonfishing activity is added to the total cost of fishing. This addition is called the 'opportunity cost'. In effect, opportunity cost raises the cost curve and moves the equilibrium point in a way that imposes little burden on those who remain in the fishery. This is easy to see if one assumes that fishermen who are likely to make more money doing something other than fishing will switch activities, thereby reducing fishing effort. In contrast, when there are no opportunity costs (i.e. the only alternative to fishing is unemployment), the bioeconomic equilibrium is the point at which costs match income so that the average fisherman does not make a profit. Although at first sight it seems to make little sense for the average fishermen to operate without gain, the mechanism is easy to understand – when there is no restriction on who can enter the fishery, unemployed people will join the fishery until the bioeconomic equilibrium is reached.

In a discussion of coral reef fisheries, MacManus (1996) argues that there are abundant examples where MEY has been estimated and that it generally lies at about 40% of the effort level at which bionomic equilibrium obtains. I suspect that this figure is reasonable for other fisheries also and it implies that fishing effort would need to be reduced by about 60% to reach MEY. However, even if MEY could be achieved (and we know that this is difficult, particularly in open access fisheries), it is important to recognise that such a target is not necessarily the one that should be aimed for. There is growing recognition for example, that optimising standard of living may be more important than economic production *per se*. As MacManus (1996) points out, the yield per fisher virtually always declines as the number of fishers increases. A society may at some scale make more overall profit with more fishers, but the lifestyle of the individuals and the welfare of smaller-scale communities may well decline before MEY is reached. This to me is the essence of the problem – the difficulty of deciding how the profits from fishing should be distributed. To limit access – a decision that most would agree is essential for the sustainability of the fishery and perhaps the ecological well-being of the system – is effectively a decision to make a few people relatively wealthy and deny others the opportunity to earn a living from fishing. The fact that, if others were to enter, the system would collapse to bionomic equilibrium and everyone would suffer, is little comfort when potential fishers who are denied access look from the outside at the chosen few. It also seems counter-intuitive to those on the sidelines.

A thorough treatment of this subject within the context of the European Common Fisheries Policy is provided by Holden (1994). Holden argues that one of the main reasons why it has been so difficult to prevent over-exploitation is that, when stocks are over-exploited, improving the situation requires that fishermen accept a short-term financial loss in return for a longer-term gain. Such losses are usually unacceptable to fishermen, which tend to make the politicians charged with implementing fisheries policy reluctant to impose changes to the *status quo*. To rectify the problem, Holden (1994) argues for the incorporation of alternative objectives based on economic and social considerations. If and when social and economic factors are formally incorporated into management objectives, the political arguments should then seem strong enough for imposing measures which, while leading to short-term losses, will reap longer-term gains.

The problem remains, however, of defining what those management objectives should be. Most of us are familiar with the rather bland statements in official documents that state in general terms their objectives for 'doing good'. One might despair at how difficult it is to translate these general aims (with which no one would disagree) into particular management actions. Since every sector's societal, economic and biological optimum is likely to be different, perhaps the best we can hope for from legislation, is an exposition of general principles that, if acted upon, should take us in the right direction.

In this regard, an important point to stress is that management objectives must be stated by the authorities that seek advice from scientists. In a comprehensive analysis of the European Common Fisheries Policy, Holden (1994) argues that, even within the framework of a relatively sophisticated political structure, politicians have never explicitly stated what they are aiming for, other than in broad terms which mean all things to all men. As a consequence, responsibility for setting objectives is usually left to the scientists, most of whom are biologists. These objectives tend to be purely biological, with little or no account being taken of economic or social factors; an understandable outcome since these areas lie outside the competence of biologists. Holden argues that this is a major deficiency of the management process.

Economic and social considerations are undoubtedly important and the case for political issues leading to the discounting of basically sound biological advice can certainly be made. There are dangers, however, with simply blaming the politicians and adopting the view that the fisheries biologist is basically doing the right thing. First, the separation of science as an entirely objective activity from the social milieu in which it is undertaken seems to me to be untenable. The way data are collected, the weight that is given to various types of knowledge and the way data are interpreted are all affected by the social and political context in which the scientist operates. Finlayson (1994) gives an interesting account of the sociological perspective underlying the collapse of the Northern Cod Fishery, which includes a series of lively rejoinders from one of the assessment biologists involved at the time. Such social biases can certainly alter the outcome of scientific deliberations. Second, it may well be that while the paradigms adopted by fisheries biologists to manage single stocks are sound, greater attention needs to be paid to the system in which these stocks are embedded. Perhaps a case can be made for widening the focus of fisheries management.

1. 6 A changing emphasis

Perhaps the major impetus for the current wave of concern about fisheries management is two FAO documents, the first a review of global marine resources (FAO, 1994), the contents of which were summarised earlier. The second an analysis of the capital structure of fishing fleets (FAO, 1993) which show how massively over-capitalised the fishing industry is. To catch $US 70 billion worth of fish annually, it is estimated that the fishing industry incurred costs of $US 124 billion. The $54 billion deficit is met in large part by government subsidies such as fuel tax exemptions, price controls, low interest loans and development grants for fishing gear or associated infrastructure (Safina, 1995). These subsidies arise from the efforts of many governments to preserve employment, despite the effects on many fisheries. Alarmingly, between 1970 and 1990 the rate of growth in the world's industrial fishing fleet was twice that of the global catch with both the total tonnage and the number of vessels doubling in that period (Safina, 1995).

Given such figures and the status of stocks it is not surprising that there is widespread perception that it is time for a change in fisheries management, with many recent authors writing in despair about the failures of fisheries biology and fisheries management. Perhaps then the adoption of a wider perspective, with greater emphasis on the relationship between fisheries and the ecosystems and communities in which they are harvested, is timely. Indeed, agreed management objectives for marine systems are beginning to include this perspective. For example, the Convention for the Conservation of Antarctic Living Resources, the main management instrument in the Southern Ocean contains statements of principle such as:

> 'maintenance of the ecological relationships between harvested, dependent, and related populations of Antarctic marine living resources and the restoration of depleted populations'
>
> 'prevention of changes or minimization of the risk of changes in the marine ecosystem which are not potentially reversible over two or three decades, taking into account the state of available knowledge of the direct and indirect impact of harvesting'.

Similarly the Code of Conduct for Responsible Fisheries contains a number of references to protection of biodiversity and ecosystems (FAO, 1996). Fulfilling such objectives requires that (1) we understand which ecological relationships with systems are affected by fishing, and (2) that we devise management strategies to mitigate undesirable effects. I hope that the following chapters assist in that endeavour.

Chapter 2
By-catch and Discards

2.1 Overall levels of by-catch and discards

Definition of terms

One of the banes of ecology is that in some areas it suffers from a proliferation of terms and in others terms are used so loosely that they are of questionable value. In the case of by-catch and discards, it is the latter. Fortunately, however, a workshop was convened in 1992 to sort out these problems (McCaughran, 1992), and with a few minor modifications proposed by Alverson *et al.*, (1994), I propose to adopt the definitions recommended at that time (Table 2.1). (As a general rule I would rather avoid such hand-wringing, but in this case it seems necessary.)

Finfish and shellfish

One of the first detailed estimates of world by-catch and discards was by Saila (1983). Although there were no data available for many marine species and no information was presented on the by-catch of more charismatic taxa such as birds, marine mammals or turtles, Saila's estimated discard total of 6.72 million tonnes (approximately 12% of total landings) was startling. Among other things this surprisingly large figure stimulated efforts to gather better data. With better data the estimates of discard mortalities have risen. For example, Andrew & Pepperell (1992) estimated a global discard figure of 16.7 million tonnes for shrimp fisheries alone, exceeding Saila's earlier estimate by a factor of 2.5.

A report by Alverson *et al.* (1994), undertaken under the auspices of the FAO, is the most recent comprehensive assessment available. These authors assembled all the published information they could find (a total of 820 publications) which contained discard and by-catch information. From these papers they constructed a database containing records for a total of 55 target species/species groups and 63 by-catch species/species groups. Approximately 29% of these records are associated with target fisheries for shrimp or groundfish. This analysis resulted in a global mean estimate of 27 million tonnes discarded, based on a target catch of 77 million tonnes. Using maximum and minimum observations, the authors estimated a global discard range of 17.9–39.5 million tonnes. This figure did not include by-catch from

Table 2.1 Definition of terms for by-catch and discards (McCaughran, 1992).

Term	Definition
Target catch	The catch of a species or species assemblage which is primarily sought in a fishery.
Incidental catch	Retained catch of non-target species.
Discarded catch	The portion of the catch which is returned to the sea for economic legal or personal reasons. The discards may be either of target species (commercial discards) and/or non-target species (non-target discards).
By-catch	Discarded catch plus incidental catch.
Incidental catch rate	The proportion of the total catch which is incidental catch.
Discard rate	The proportion of the total catch which is discarded. Rates may be computed for individual species or combined groups of species.
Discard mortality rate	The proportion of the discarded catch that dies as a result of catching or handling processes.
Discard mortality	Discard mortality rate multiplied by the discarded catch.
Prohibited species	Any species which must, by law, be returned to the sea.
Unobserved fishing mortality	Mortality imposed on a species by the encounter with fishing gear that does not result in capture.
High grading	The discard of a marketable species in order to retain the same species at a larger size or price. The discard of a marketable species in order to retain another species of higher value. The retention of only those species or individuals within a species complex having the greatest market value; less valuable species or individuals are discarded.
Individual discard quota (IDQ)	A quantity of prohibited species which is initially apportioned among individuals or groups of individuals to allow effective harvest of non-prohibited species. IDQ may or may not be traded/sold within this group or between this group and other groups.
Discard quota	An amount of prohibited species allowed to be caught by a particular gear type before constraining a fishery. It may be the sum of all IDQs if such a system were in place.
Discard mortality quota	The discard quota multiplied by the discard mortality rate.
Discard ratio	The ratio of discard to actual retained catch (e.g. discard/retained catch frequency given as a percentage).
Unobserved fishery mortality	Death resulting from fishing that cannot be documented from observations of the on-board catch (e.g. deaths resulting from fish passing through webbing, freeing themselves from hooks, ghost fishing etc.).
Black fish	Non-reported catch.
Grey fish	Catch for which area or species caught is misreported.

recreational fisheries which could add substantially to the total removals. Mean shrimp by-catch was estimated at 11.2 million tonnes, but the maximum estimate for the shrimp fishery was very close to that obtained by Andrews & Pepperell (1992) (16.4 *cf.* 16.7). When adjusted for the by-catch fraction retained, the mean estimate fell from 11.2 to 9.5 million tonnes.

One important aspect of the data used by Alverson *et al.* (1994) is the biased coverage due to the interests of particular researchers. For example, a disproportionate number of publications deal with by-catch species such as salmon and halibut, which have high economic value. Marine mammals also draw considerable attention, owing to the high public interest in these taxa. Alverson *et al.* (1994) note that these three groups together account for almost twice as many records in their database as all other records for the top ten by-catch species; they also account for 30% of all records in the database. The authors are also quick to point out other gaps in their data (many regions of the world are under-represented) and the fact that discard rates for a particular fishery can be highly variable from year to year. This variation is often related to patterns of recruitment, changes in regulations and fishery practices. Inadequacies in these data meant that only a ratio-based assessment was possible (i.e. one where discards are considered as a proportion of retained catch) and no confidence intervals could be calculated. From these ratios estimates of absolute discards were obtained using available data on landings from each region.

Figure 2.1 shows how these discard figures break down on a regional basis. Just over one third of the total discards occur in the northwest Pacific, arising from fisheries for crabs, mackerels, Alaskan pollock, cod and shrimp, the latter accounting for about 45% of the total. The second ranked region is the northeast Atlantic where large whitefish fisheries for haddock, whiting, cod, pout, plaice and other flatfish are the primary sources. Somewhat surprisingly, capelin was also a rather important contributor to the total, primarily because capelin are discarded due to size, condition and other market related factors. The third place in world rankings is the west central Pacific, arising largely through the action of shrimp fisheries. These fisheries, prosecuted mainly off Thai, Indonesian and Phillipine coasts, accounted for 50% of the total by-catch for the region, although fisheries for scad, crab and tuna were also substantial contributors. Interestingly, the southeast Pacific ranks fourth, not because the fisheries in the area have high discard ratios (on the contrary, the ratios for the major anchoveta and pilchard fisheries are between 1 and 3%), but simply due to the enormous size of the total catch. For the remaining tropical regions, by-catch is again dominated by the actions of shrimp fisheries, although some crab fisheries are also significant.

One characteristic difference between temperate and tropical fishery discards is worthy of note. In the tropics, where shrimp fisheries dominate the statistics, discards mainly comprise small-bodied species which mature at under 20 cm and weigh less than 100 g. In contrast, for the temperate and sub-arctic regions discards are generally dominated by sub-legal and legal sizes of commercially important,

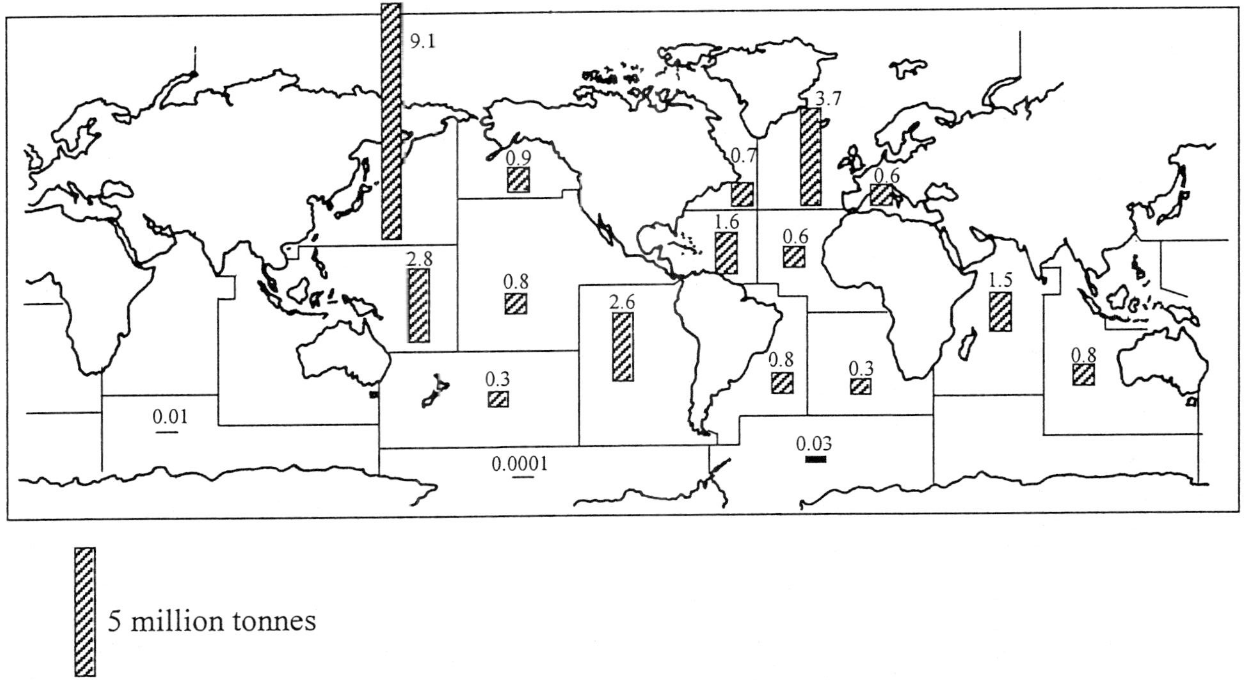

Fig. 2.1 Map showing regional distribution of discards. (Adapted from Fig. 2 and Table 5 of Alverson *et al.*, 1994. Reproduced with permission of the Food and Agriculture Organization of the United Nations.)

larger-bodied species. Thus in the temperate zone discarding is not only an ecological issue, it is also a fisheries management issue in the strictest sense. Fish are being discarded which, if left alone, would form part of the future commercial catch.

Of course, one can also break down the discard statistics on the basis of species groups or by fishing gear type. On both a species and gear type basis, the shrimp fisheries continue to dominate the statistics (Fig. 2.2). Crab fisheries have the second highest discard ratios, but because the total landings are relatively small the total level of discards is ranked fourth. For the north Pacific, the majority of discards in this fishery comprise individuals that are immature and/or below the legal landing size. The highest discard ratios for finfish species are associated with flatfish fisheries, although once again the total discards from the fishery rank lower owing to the

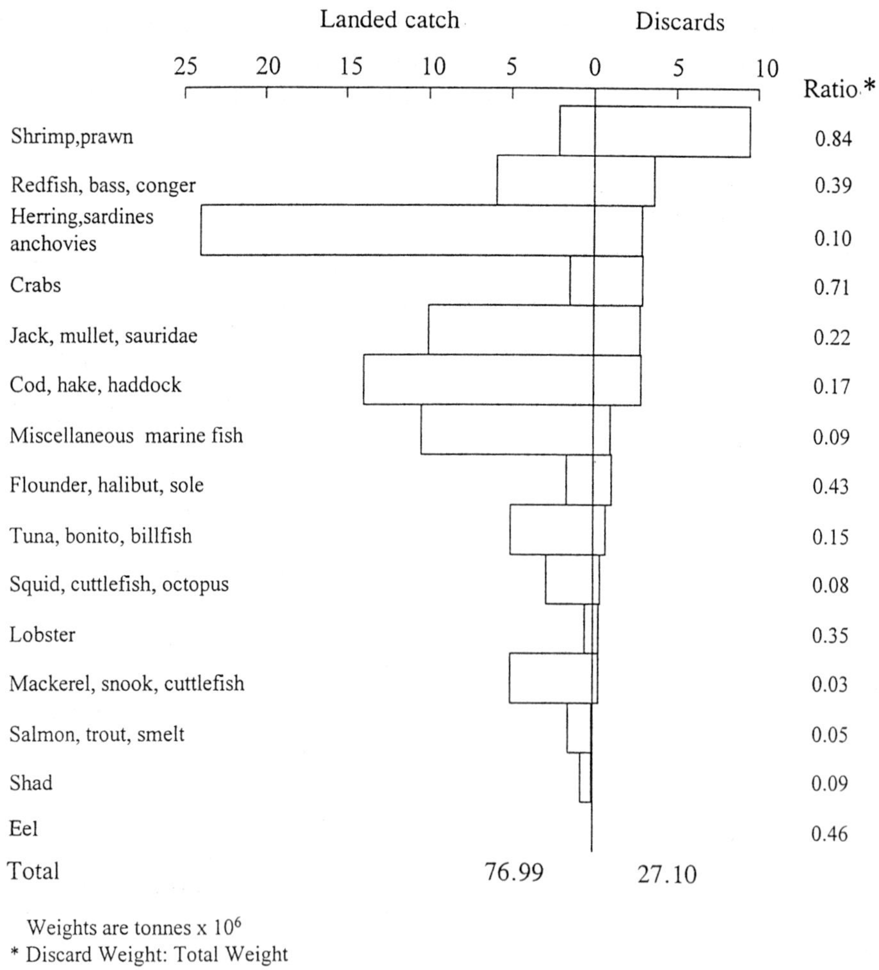

Fig. 2.2 Figure showing the global marine discards by species. (Adapted from Table 6 of Alverson *et al.*, 1994. Reproduced with permission from the Food and Agriculture Organization of the United Nations.) Weights are given in tonnes $\times$ 10^6. *Discard weight: total weight.

lower total catch. Discards associated with cods, hakes and haddocks represent approximately 9.5% of the global total, of which 52% are contributed by the whitefish fisheries of the northeast Atlantic. Whiting, cod (Atlantic and Pacific), pout and Alaska pollock fisheries account for the vast majority of the whitefish discards. The major pelagic species (redfish, herrings, sardines and anchovies) comprise the other group with discards in excess of 1 million tonnes, but this is largely a consequence of the very high landings, since the by-catch ratio associated with these schooling species is comparatively low. The other species of note are the tuna and billfishes, with a total by-catch of almost 750 000 tonnes. However, new practices employed by this fishery, such as setting nets around floating logs, or other deliberately set fish aggregation devices, mean that discards for this fishery may rise in future (see below).

The above assessments of by-catch and discard levels were based on weight or numbers discarded for finfish and some invertebrate species. However, one of the most serious public ethical concerns to be raised with respect to marine fisheries management is by-catch of non-fish taxa, i.e. sharks, marine mammals, turtles and birds.

Sharks

Migratory oceanic sharks are rarely the target of commercial fisheries, but several species form a large component of the by-catch of coastal and high seas longline, purse seine and drift-net fisheries for tuna and billfish. Indeed, these species can be a valuable component of the by-catch owing to the high prices paid for shark fins and the decreasing availability of some of the larger teleost species. As with the other non-teleost taxa for which by-catch effects are a concern, the life-history characteristics of sharks make them particularly vulnerable to fishing pressure. Slow growth, late age at maturity, low fecundity and natural mortality, and a close stock recruitment relationship all conspire against these taxa. Stevens (in press) reports that much of the shark by-catch of high seas fleets is comprised of blue shark (*Prionace glauca*), oceanic whitetip (*Carcharhinus longimanus*) and silky shark (*Carcharhinus falciformis*). In addition, shortfin Mako (*Isurus oxyrinchus*), thresher (*Alopias* spp.) and porbeagles (*Lamna* spp.) are also taken.

Available catch data for shark species suggests that large numbers of individuals are being taken globally. However, the reported discrepancy between catches and total landings is likely to be large, perhaps adding 50% to the reported total world landings (Bonfil, 1994). In terms of total shark catch, Bonfil (1994) arrived at a figure of some 84 000 tonnes for the central and south Pacific in 1989. With respect to catch composition blue shark comprised the greatest proportion with an estimated total annual catch by the high seas fleet of between 6.2 and 6.5 million individuals (Bonfil, 1994).

With the moratorium on high seas drift-netting (see below) by-catch of sharks ought to have declined, and a more recent estimate of the annual catch of oceanic sharks from the Pacific has been made by Stevens (in press) (Table 2.2). Stevens

Table 2.2 Estimated catch (× 1000 tonnes) of oceanic sharks by Pacific high seas fishing fleets in 1994. (from Stevens, in press).

Fishing method	Blue	Oceanic White Tip	Silky	Short-finned Mako	Thresher
Long line	137.8	45.1–232.4	72.4	4.1	2.7
Purse seine	2.3	7.0	11.7	?	?
Drift-net[1]	21.152	—	—	0.52	0.35
Total[2]	140.1	52.1–239.4	84.1	4.1	2.7

[1] 1989–1990 estimate.
[2] Excludes drift-net.

emphasises that his estimates should be considered as a rough guide only because he makes various assumptions about catch rates, the distribution and amount of fishing effort, average sizes and mortality. However, they are no more or less questionable than many other estimates in the literature. With respect to shark catches associated with purse seines, it is ironic that catches appear to have increased with the shift away from setting nets around dolphins (owing to conservation pressure) and towards sets around logs where sharks appear to be more abundant.

Marine mammals

Unfortunately, although we know that a wide variety of interactions occur between marine mammals and fisheries (Northridge, 1991), the quantities involved are less well known than for the commercially harvested species of fish or shellfish. Indeed, although considerable qualitative data are available, few reports include specific by-catch ratios, and for most fisheries only a small proportion of the total mammal by-catch is recorded (Northridge, 1991). Sufficient data are available, however, to give an indication of the potential magnitude of the problem.

There seem to be few mammal taxa that escape some degree of by-catch. In 1990, International Whaling Commission (IWC) scientists identified 54 species/populations/regions (SPRs) for which abundance estimates and data on incidental mortality in passive gears were available. Bear in mind that this list does not include species taken by actively towed gears or species from other taxonomic groups such as seals, sea lions, manatees and dugongs. IWC workshop data from 1991 suggest potential marine mammal discard mortalities ranging from 65 000 to 86 000 animals annually. However, these figures are probably conservative because: (1) no information was available for seven SPRs, (2) for a further 86 SPRs the IWC could only note that by-catch occurred, and (3) for many others only a very rough figure could be provided.

Catch rates of marine mammals are generally highest in gill-net fisheries and gill-net fishing effort (and consequently mammal by-catch) increased dramatically in the 1970s and 1980s. Increases in gill-net fishing for salmonids, squid and tuna in the north Pacific in the 1970s and 1980s were particularly notable, and came about for three basic reasons. First, because of the introduction of durable and inexpensive synthetic materials, which allowed cheap lightweight nets to be easily constructed and maintained. Second, through the establishment of exclusive economic zones by nation states which excluded foreign fleets from traditional fishing grounds. Third, due to over-exploitation of traditional fishery resources in more accessible waters. In the light of concerns about the huge by-catch of finfish, mammals, turtles, sharks and seabirds that has been documented, the United Nations adopted a resolution in December 1991, which called for a global moratorium on all large-scale drift-net fishing on the high seas. This moratorium became effective in December 1992.

Notwithstanding these legislative changes, problems still exist. In the Mediterranean, for example, large numbers of cetaceans are still taken in the drift-net fishery for swordfish and albacore. It would appear that despite legislation which states that European Union countries should not use nets >2.5 km long, nets of 10–12 km are still being used. Indeed, despite a ban on the fishery by the Italian Government in 1990, subsequent amendments and activities by other states have led to a situation which may now be worse than before; in 1994 over 800 vessels from Italy and 200 from Morocco were still operating. Sadly, no good estimates of by-catch exist for the region, although prior to 1990 the by-catch from the Italian fishery alone was thought to number several thousand. The striped dolphin, in particular, is considered to be at risk from these activities (Donovan, 1994).

In addition to passive gears, a considerable incidental by-catch is also reported for active fishing gears, in particular in purse seine fisheries (Fig. 2.3), although the situation is much improved. For example, in the eastern tropical Pacific tuna fisheries, it was often the practice to deliberately set purse seine nets around dolphin schools because several species are known to often be associated with high densities of yellowfin tuna. This practice developed in the early 1950s when purse seiners began replacing pole and line fishing and in the early years of the fishery it is estimated that nearly half a million dolphins were killed each year. However, with the introduction in the 1970s and 1980s of strategies to allow dolphins to escape the net, the numbers

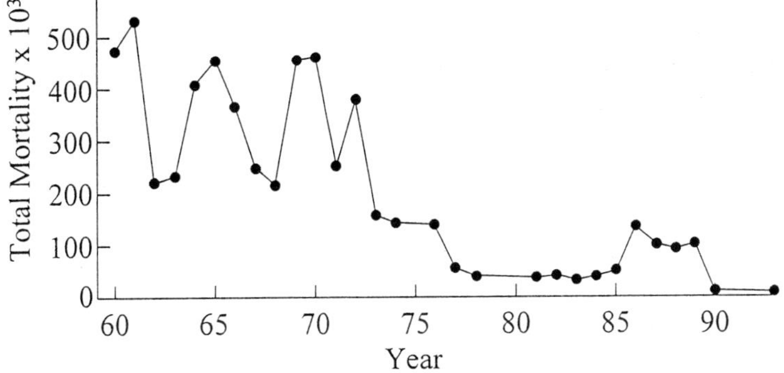

Fig. 2.3 Estimated discard mortalities for the eastern tropical Pacific tuna purse seine fishery from 1959–1993.

killed fell substantially, and in the last few years annual mortalities have been of the order of 5000 individuals. (With these data, as with others it is important to recognise that for almost all fishery/cetacean interactions we have only rough (and usually minimum) estimates of by-catch levels). Notwithstanding the gains made in reducing dolphin mortalities in purse seine fisheries, however, it is important to realise that the required changes in fishing practices leads to a concomitant increase in the discards of young tuna and other pelagic fish species (Table 2.3).

Table 2.3 Discards in the eastern tropical Pacific tuna fishery associated with two different net setting strategies; setting on tuna schools and on fish associated with dolphins. (From Table 11 of Alverson *et al.*, 1994. Reproduced with permission from the Food and Agriculture Organization of the United Nations.)

Species	School sets (n = 1060)	Dolphin sets (n = 764)
Dolphins	8	4500
Small tuna	5 230 000	833 000
Mahi mahi	1811	628
Sharks	17 632	5104
Wahoo	208	13
Rainbow runners	255	
Other small fish	1708	523
Billfish	2349	759
Yellowtail	47	39
Sea turtles	471	143
Other large fish	1151	
Trigger fish	283	

Information on other marine mammals such as seals and sea lions is even less well quantified than for cetaceans. Northridge (1991) notes that most marine mammals have been recorded as incidental victims in one type of fishing gear or another, but data for many species are very sparse (see Section 2.3). Alverson *et al.*, (1994), while reluctant to put a figure on it, suggest a conservative estimate of several hundreds of thousands of animals involved annually. They also note, however, that technological developments in the eastern tropical Pacific, the disappearance of several major high seas drift-net fisheries, national regulations to limit marine mammal by-catch, and fishermen's education programmes may have helped to reduce the total levels of mortalities in recent years. Despite these improvements, however, it is argued that several marine mammal species are in danger of extinction, at least in part due to by-catch mortality. Other species are apparently declining and may become threatened if by-catches are not reduced (see Section 2.2).

One area in which more comprehensive work on mammal by-catch levels has been undertaken is the Bering Sea. Trawl fishing in the Bering Sea and Gulf of Alaska has a long history and marine mammals have been caught since its inception. Exploratory fishing first began in the late 1920s and Japanese and Soviet vessels continued this activity until the late 1950s when commercial operations started.

Early activity focussed on Pacific Ocean perch and yellowfin sole, but as catches started to decline in the 1960s attention shifted to walleye pollock. Unfortunately, data on mammal capture rates for the early years of the fishery are sparse. A few records were collected from 1973, but it was not until the passage of the Magnusson Fishery Conservation and Management Act in 1976 that foreign vessels were required to carry observers. Observers are present, primarily to monitor fish catches, but from 1978 a standard protocol for reporting marine mammal captures has been in place. Unfortunately, although US vessels also fish in the area, no observers are required for these vessels so there is only an incomplete record of mammal catches for the region. We must, therefore, rely on the records for foreign vessels to give an indication of the scale of the problem.

Of the 17 mammal species caught by fisheries in the area, Northern sea lions (*Eumetopias jubatus*) are the most common. This species constitutes 87% of total marine mammal by-catch reported by US observers between 1973 and 1988. Fig. 2.4 shows the estimated by-catch of northern sea lions in foreign and joint venture groundfish trawl fisheries in the US EEZ of the Bering Sea, Aleutian Islands, and the Gulf of Alaska for the period between 1978 and 1988. Although data prior to this are somewhat sparse, Perez & Loughlin (1991) also attempted to extend the time series backward to 1966 (Fig 2.4).

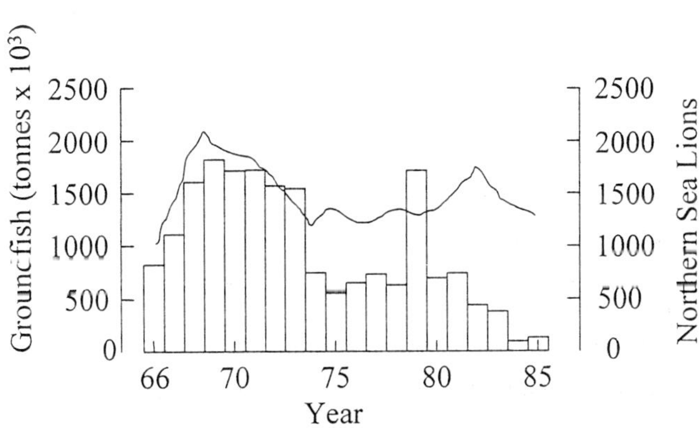

Fig. 2.4 Total groundfish catch (bars) and the estimated number of incidentally taken northern sea lions (line) which died during the fishing operations of all Alaskan joint venture fisheries between 1966 and 1985. (From Perez & Loughlin, 1991.)

Turtles

Another group for which concern has been raised are the turtles which are caught in both trawl and line fisheries. Populations of many species of turtle are in decline (Anon., 1990), with factors such as loss of nesting habitat, indigenous harvesting and mortalities from fishing all making a contribution. Turtle populations, in common with marine mammals, are likely to be more at risk from such factors owing to their life history characteristics; high age at first maturity and low fecundity both conspire to ensure that populations cannot be maintained without high adult survivorship. In the USA, the two species which are most affected by fisheries activity are Kemp's ridley and loggerhead (Anon., 1990).

Unfortunately, turtles often show a close association with the target species of fisheries, particularly in shrimp fisheries where turtle by-catch appears to be especially problematic. In a comprehensive review of turtle population status in USA atlantic waters and the Gulf of Mexico, Anon. (1990) examined the mortality rates associated with fishing (particularly shrimp trawling) and other activities. The USA shrimp fishery is a complex of fisheries operating from Cape Hatteras, North Carolina, to the Gulf of Mexico at the USA border. About one-third of the fishing effort occurs in bays, rivers and estuaries and 92% of the total effort is in the Gulf of Mexico. However, although fishery turtle interactions are complex, there are a number of lines of evidence which point strongly towards fishing as a major mortality agent. First, the proportion of dead or comatose turtles that occur in shrimp trawls increases with the time of the tow; very few are caught in poor condition in tows below 40 minutes, whereas 70% are moribund in 90-minute tows. Second, in some areas it has been shown that the numbers of stranded carcasses on beaches increased and decreased when shrimp fisheries opened and closed. Available data suggest that 70–80% of the turtles stranded at those times and places were caught and killed in shrimp trawls. Third, declining populations trends in some areas correlate with fishing activity.

Taking all the available evidence into account, Anon. (1990) estimated that mortality from shrimp trawling lies between 5000 and 50 000 loggerheads and 500–5000 Kemp's ridleys each year. Collectively, other fishing gears, both passive and active, accounted for an additional 500–5000 loggerhead deaths and 50–500 Kemp's ridleys. Note that the upper and lower bounds of these estimates vary by an order of magnitude – a level of uncertainty which highlights how difficult it is, despite a comprehensive assessment effort, to obtain reliable data on mortality rates. Drawing robust conclusions about effects at the population level then becomes doubly difficult. Nevertheless Anon. (1990) did conclude that for juveniles, sub-adults, and breeding adults in the coastal waters, shrimp trawling accounted for more deaths than all other sources of human-induced mortality combined.

Similarly large mortalities associated with shellfish trawls also occur in other areas; in the Australian Northern Prawn Trawl fishery, for example, Poiner *et al.* (1990) showed that several thousand turtles are caught each year. However, although trawl fisheries (particularly shrimp) are the most studied with respect to turtle captures, net and line fisheries are also implicated. Long-line sets are especially attractive to pelagic turtles, and although data on the numbers caught by this type of fishing method are extremely sparse, in the Hawaiian-based long-line fishery it is believed that up to several tens of thousands of sea turtle might be taken each year (Dayton *et al.*, 1996).

Seabirds

Seabirds are an abundant and highly visible component of many ecosystems and they are often subject to considerable popular interest. As such, they are particularly suited as indicator species for the status of coastal ecosystems. For most coastal breeding birds it is probably fair to say that the primary stress on populations is habitat loss (or modification) and disturbance and predation at colonies. In the more industrially developed regions of the world, however, bio-accumulation of fat soluble toxins is also an issue. Nevertheless, for some bird populations and breeding colonies, the effects of fisheries activities have been especially detrimental. Paradoxically, however, it should also be recognised that, perhaps more than for any other group, fishing activity has had marked positive effects on seabird populations through the provision of additional food by discarding.

Seabird mortalities

Seabirds can suffer mortalities from many types of fishing gear, but on the whole it is static gears of various kinds that cause most problems – in particular, gill-nets and long-lines. As with other groups of course, the situation on the high seas with respect to gill-nets has improved with the moratorium that was imposed in 1992 (see Section 2.2). Prior to this, however, high seas drift-netting was responsible for considerable seabird mortalities. In the north Pacific, for example, large numbers of seabirds drowned annually in gill-nets set for both fish and squid. Similar problems also occurred on the northwest Atlantic coast between Newfoundland and Greenland. For example, in an analysis of the conservation status of seabirds in Greenland in 1984, Evans (1984) concluded that, after direct hunting, incidental kills in fishing nets was probably the greatest threat to seabird colonies. In southwest Greenland mortality was estimated at between 215 and 500 thousand guillemots per year between 1969 and 1971, although with the introduction of salmon quotas, between 1976 and 1980 mortality rates fell to 10–20% of their earlier values. This is a good example of how improved management of the target resource can lead to concomitant reductions in the level of effect on other taxa. It is unfortunate that in this particular case, despite a promising reversal of trends with the introduction of quotas, subsequent changes in fishing practice (i.e. greater fishing offshore, a switch to fishing both day and night with the introduction of larger vessels, and a shift towards activity later in the year when more birds arrive in the area on breeding migrations) by-catch mortalities started to rise again.

By far the greatest by-catch mortalities for seabirds occur when fishing activities coincide with concentrations of foraging animals. For example, Piatt & Nettleship (1987) describe the situation for Newfoundland, where the inshore cod gill-net fishery is conducted between early June and mid-August. The peak in fishing effort for this fishery occurs during a 4–6 week period from mid-June to late-July, which corresponds to the period when the capelin occur inshore for spawning on coastal beaches. The cod move in to feed on the capelin, fishermen exploit the cod concentration and coincidentally catch seabirds which also concentrate in the area to feed on capelin. Thus, the majority of shearwaters, gannets, common guillemots and puffins (along with various marine mammals) are taken during this 4–6 week period in each year, with the exact time depending on the chronology of inshore capelin spawning. Piatt & Nettleship (1987) also suggest that the between-year variability in by-catch mortality could be explained by changes in the annual abundance of capelin. Greater numbers of birds occur in the area when capelin are abundant leading to greater entanglement in nets (Fig. 2.5). While, such data cannot prove that by-catch mortality and capelin abundance are causally related, the fact that catches of species with less or no dependence on capelin showed no relationship, and that fishing effort was similar between years strongly suggests that this interpretation is reasonable. Piatt & Nettleship (1987) also report that similarly marked concentrations of catches occur in other areas.

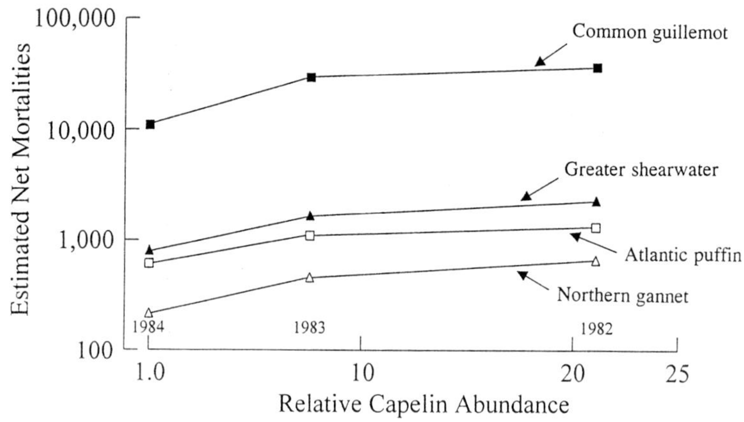

Fig. 2.5 The relationship between by-catch mortality of seabird species and capelin relative abundance in Eastern Newfoundland between 1982 and 1984. (Constructed from Table 3 of Piatt & Nettleship (1987) *Marine Pollution Bulletin* **186**, 344–9. Reproduced with permission from Elsevier Science.)

One area where fishing activity appears to have had rather widespread effects is in the north Pacific with the gill-net fishery for salmon and squid. Until the early 1990s the Japanese, Taiwanese and Korean fleets prosecuted these fisheries on the high seas. In 1991 the Japanese agreed to stop fishing, followed by the Taiwanese and Koreans in 1992, but prior to this time estimated mortalities were considerable. Table 2.4 summarises the most recent available data from the high seas drift-net fisheries of the north Pacific in the years just prior to the moratorium. Figure 2.6 illustrates how the mortality rate declined for one of these fisheries in the years preceding the cessation of fishing.

A key point with respect to the above data is that mortality rates in earlier years are likely to have been much higher. For example, both the size of the fleet and the

Table 2.4 Mortality rates for seabirds during the latter phase of driftnet fishing activity in the North Pacific, (from De Gange *et al.* (1993) Table 1).

Fishery	Fishing effort (km × 10^3 yr^{-1})	Catch rate (birds km^{-1})	Mortality rate (deaths × 10^3 yr^{-1})	Species affected[a]
Japanese mothership				
1981–1987[b]	58.30–137.00	0.60–2.30	32.00–250.00	STSH, TUPU, HOPU
1989[c]	15.82	0.60–1.10	9.50–17.40	CRAU, TMBU
Japanese land-based				
Large vessel 1987	52.25	0.80	56.00	STSH, TUPU, TMBU
Small vessel	Unknown	Unknown	18.00–60.00	Unknown
Squid drift gill-net	2850.00	0.01–0.20	875.00	SOSH, STSH, TUPU, HOPU, LAAL

[a] STSH = Short-tailed Shearwater; SOSH = Sooty Shearwater; TUPU = Tufted Puffin; HOPU = Horned Puffin; TMBU = Thick-billed Murre; CRAU = Crested Auklet; LAAL = Laysan Albatross.
[b] Four motherships and 172 catcherboats fished in 1981–1986; three motherships and 126 catcherboats fished in 1987.
[c] One mothership and 56 catcherboats fished in 1989.

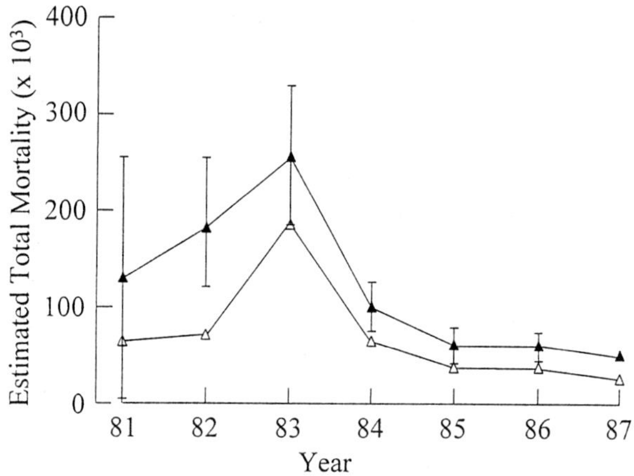

Fig. 2.6 Mortality of seabirds in the Japanese mothership salmon fishery, 1981–1987. By-catch in subsequent years reduced further and the fishery ceased in 1991. (Redrawn from Fig. 1 of De Gange *et al.*, 1993). Data to 1983 are for the Japanese EEZ only, thereafter they are for the EEZ and the region to the north. ▲ = All species; △ = shearwaters.

fishing effort of the large vessel component of the land-based fishery peaked in the mid-1960s at more than 295 000 km of net. This value is approximately six times the 1987 figure. De Gange & Day (1991) estimate that in the mid-1960s more than 260 000 seabirds were killed annually, compared to 21 250 in 1989.

Perhaps, more than any other, it is catches of albatross that have highlighted the effect of fisheries on seabirds. For the most part this is an issue for the Southern Ocean fisheries because, of the 14 species of albatross, 10 are confined to this region (3 to the north Pacific and 1 to the tropics). Albatrosses roam widely across the oceans of the world feeding mainly on squid and fish. Unfortunately, as with many other seabirds, they often follow ships such as long-line vessels. This behaviour along with the high likelihood of encounter with fishing vessels due to their wandering habit makes them particularly vulnerable.

The most startling albatross catches are in the Southern Ocean, where it is estimated that 44 000 albatrosses are killed annually by the Japanese long-line fishery (Brothers, 1991). This estimate was based on observations made on seven cruises off southern and eastern Tasmania in the winter of 1988 and the author argues that the figure may be in fact be double this. More recently, Klaer & Polacheck (1996) arrived at substantially lower figures for the Japanese long-line fishery in Australian waters (0.41 birds per 100 hooks cf 0.05–0.15, depending on area and year). The authors suggest that a likely explanation for this substantial reduction in catch rates from 1988–1992 is the increased implementation of seabird by-catch mitigation measures. A similar conclusion was drawn by Murray *et al.* (1993) who observed reductions in by-catch in New Zealand waters between 1988 and 1992. For both of these studies, of course it is possible that the reduction in catch rates occurred because overall population numbers have declined. At present, however, no direct information exists on abundance trends and no quantitative assessment of seabird abundance is available for these two areas.

For the most part studies of by-catch of birds have been conducted at relatively local scales, often with reference to individual colonies. This makes sense since a number of studies have shown that the greatest numbers of birds are caught around the breeding colonies or in closely adjacent feeding areas. Indeed, it appears that in many cases catch per unit effort declines markedly with distance from colonies (Piatt & Nettleship, 1987). Nevertheless, even at the smallest spatial scales, catches can be quite large. In the English Channel, for example, inshore gill-nets have proved to have particularly high catches in some times and places. In St Ives Bay there are reports of set nets having caught almost all of the diving birds (mostly razorbills *Alca torda* and divers *Gavia* sp) which foraged in the area. In eight days of fishing in January 1988, about 900 auks were killed in these nets. While these kinds of numbers are not significant in a North Sea context, at a local scale they are considerable (Anon., 1994).

In the UK as a whole, it would appear that most entanglement events are sporadic and localised (Anon., 1994). This is probably a consequence of the nature of the gill-net fishery which is prosecuted from small vessels. Unfortunately vessels below 10 m length are not required to keep log books so gathering sound data on entanglements is difficult. Estimates of mortality rates in other European fisheries include 25 000 diving seabirds that drowned in herring and cod gill-net fisheries in the Kattegat between 1982 and 1988. The majority (90–95%) of these mortalities was from the migratory guillemot (*Uria aalge*) population which breeds on the Scottish North Sea coast.

Another example of localised effects is in central California, particularly the area around Monterey Bay, where gill-netting for halibut and flounder has been undertaken since the 1930s. In the late-1970s, however, there was a 400–500% increase in fishing effort in some areas and a switch towards monofilament nets, which increased the numbers of birds caught as by-catch. Most of the increase in effort was focussed on Monterey Bay where a new fishery for white croaker developed. It soon became apparent that this fishery was catching large numbers of birds, with a rough estimate of 70 000–75 000 birds caught in central California between 1979 and 1987. In one year 90% of the 22 000 seabirds that were washed ashore had been drowned in nets. Most of these mortalities occurring among Sooty shearwaters (*Puffinus griseus*) and Common Murres (*Uria aalge*) (Jehl, 1984). A strong connection was also noted between areas and periods of gill-net fishing mortality and declines in individual bird colonies. In general, the degree of decline for a given colony was related to its proximity to Monterey Bay (Takekawa *et al.*, 1990). Largely in response to the large by-catches of birds (but also to mammal captures) legislation was passed in 1982 imposing a permanent ban on gill-net fishing in the waters of Monterey Bay less than about 18 m deep. In 1984 this legislation was extended to water less than about 27 m deep, but in the following two years bird by-catch remained high. Writing at around this time King (1984) concluded that central Californian populations of Murres and, perhaps, other species were unlikely to sustain themselves at the existing mortality rates. This and other assessments led finally to the introduction of even stronger legislation in 1987 with gill-net bans in water to depths of 73 m. These regulations resulted in the virtual elimination of the nearshore halibut, shark and white croaker fisheries north of Monterey Bay and will no doubt help to preserve local seabird colonies.

Consumption of discards

Fishing has provided increasing opportunities for some seabirds to take advantage of foods that would otherwise be unavailable to them. Furness *et al.* (1988) point out, for example, that in certain localities around the British Isles and at certain times of year, adult demersal fish such as cod, haddock and whiting can form the bulk of the diet (in both energy and mass terms) for some seabird species. This occurs despite the fact that the birds are incapable of diving to the depths at which the fish are usually found and that they do not normally occur in their diet. The only source of demersal fish for these species is from fishing vessels, either scavenged from behind the vessels as they are discarded, or in some cases, stolen from the catch as it is landed. Offal (fish livers and intestines) from gutting carried out on board is also taken by some species (particularly Northern fulmars).

Similar patterns of exploitation of waste are also observed in other parts of the world. In the northern Pacific, for example, species have been observed feeding on fish wastes around fish processing vessels, long-line vessels and trawlers. Such feeding flocks can be very large; Jones & De Gange (1988) report estimates of over 30 000 northern fulmars in the vicinity of a large groundfish processing vessel in the Bering Sea.

The subject of discard exploitation by seabirds has probably been most intensively studied in the North Sea, where there have been a number of attempts to quantify the scale of the phenomenon. Direct observations from fishing vessels in the North Sea show that species tend to specialise on particular sizes and types of discard, and that there are feeding hierarchies between different species and age classes. This can result in substantial proportions of the discards for some species being consumed by seabirds and relatively small proportions of others. For example, Hudson & Furness (1988) showed that between 60 and 70% of the discarded gadoid species were consumed during observations off Shetland, while flatfish and gurnard species were largely ignored. On the whole, however, a large fraction of total discards can be consumed, with estimates for the Shetland fishery of up to 75%. Admittedly the above estimate was for a fishery prosecuted very close to major seabird colonies, but other observations of consumption rates for experimental discarding in the southern North Sea also show rates of up to 88% for roundfish and 38% for flatfish (Anon., 1994). The evidence suggests that such high rates might be generally applicable (at least to regional seas).

Similar calculations can also be made with respect to the offal discarded from fishing vessels when fish are processed at sea, rather than landed whole. In some parts of the world, the amounts can be a substantial; in the North Sea, for example, Anon. (1994) report that approximately 12.5% of the mass of roundfish and about 6.5% of the mass of flatfish processed for market is discarded as offal at sea. In 1989 this corresponded to a total of approximately 69 000 tonnes. More recently, a comprehensive study of the consumption of discards and offal by seabirds in the North Sea (not including the Channel) was summarised by Garthe *et al.* (1996). This work included experiments to determine the fate of fish individually discarded during daylight from research vessels. Garthe *et al.* (1996) estimate that a maximum of 3.4×10^{12} kJ (800 000 tonnes) of discarded items was available to seabirds. However, discarding is often done in bulk, thereby swamping the ability of seabirds to eat the food before it sinks, or at night during which time some bird species do not feed.

In an effort to overcome some of these biases Garthe *et al.* (1996) scaled the estimated amounts of offal available, using data from observations of consumption rates for the different offal types in various regions and at various times. From these data he estimated that around 55 000 tonnes of offal, 206 000 tonnes of roundfish, 38 000 tonnes of flatfish, 2000 tonnes of elasmobranchs and 9000 tonnes of benthic invertebrates were consumed by seabirds in the North Sea. However, these figures are still likely to be overestimates since the effects of some aspects of discarding practice could not be accounted for. Interestingly, the behaviour of fishermen in different regions could also lead to substantial differences in the proportion of discards that are taken. Off eastern Canada, for example, fishermen clean whitefish by cutting off the head and discarding it with the guts attached. Because this sinks rapidly, substantially less of the total is likely to reach the bird population.

Benthic scavengers

In the relatively shallow waters of continental shelves where most fisheries are prosecuted, some discarded material will sink beyond the reach of seabirds, pass through the water column without being taken by pelagic scavengers such as fish, sharks and cetaceans and land on the seabed. These amounts can be considerable. Studies on seabird scavengers in the North Sea, for example, indicate that more than 7800 tonnes of offal, 56 200 tonnes of roundfish, 261 200 tonnes of flatfish, 13 000 tonnes of elasmobranchs and 140 800 tonnes of benthic invertebrates sink beyond the reach of seabirds (Garthe *et al.*, 1996). Estimating the amount of material taken in mid-water is very difficult and likely to be highly variable, so one cannot confidently estimate the proportion that reaches the seabed. In many circumstances, however, it is likely to be substantial. Some discarded organisms will survive the process of being caught and discarded, but the mortality rate of many species is very high. Dead or moribund animals are likely to be a rich food supply for bottom-dwelling fish and invertebrates.

Australia provides some of the rare examples where an effort has been made to quantify the amounts of discards taken by taxa other than birds. In Moreton Bay,

Queensland, for example, Wassenberg & Hill (1990) examined discarding by a prawn trawl fishery. As noted earlier, prawn trawls are among the highest discarding fisheries per kg of retained catch. It follows, therefore, that the scavenging fauna in areas where prawn fishing occurs could be expected to benefit considerably from the practice. On average approximately 8500 kg of material was discarded every night, of which only the fish and cephalopod component is taken at the surface. The 65% or so (by weight) that comprises crustaceans and echinoderms sinks to the seabed. In principle, this represents 5500 kg of material reaching the seabed, although many crustaceans are still alive and may survive trawling. Using a combination of baiting experiments, underwater TV and laboratory studies Wassenberg & Hill (1987) showed that the crab *Portunus pelagicus* was the dominant scavenger in this area. By comparing foregut contents from periods when trawling was under way and at weekends when there was no fishing, they also showed that the fresh fish and crustacean material was only present during periods of trawling. The authors concluded that about 30% of the diet of this species may be supplied by discards and that such food supplementation may allow greater population densities to be supported than would otherwise be possible.

The authors also showed that, in some trials, dolphins could obtain 3.4 kg of fish from a single trawl, between 30 and 50% of the daily food requirements of an individual. In theory then, about 10 h of trawling could support the daily food requirements of five individuals. In practice, however, because the Moreton Bay fishery mainly discards small fish that are not taken by dolphins, the proportion usually taken by the population was small. However, in contrast to Moreton Bay, in the Torres Straits, up to 30% of discarded material is intercepted in mid-water by various pelagic predators (Hill & Wassenberg, 1990).

2.2 Effects on populations

Of course, figures on mortality levels are not in themselves especially helpful if we are concerned with the impacts on populations. One can certainly take an animal welfare perspective and argue that no mortalities should occur, particularly with respect to the charismatic megafauna such as turtles or marine mammals, but for most people, even with these taxa, it is the conservation of populations that is the primary concern. Unfortunately, however, inferring population level consequences is fraught with difficulty, particularly when good estimates of population size are lacking for many species.

A number of pieces of information are required if a quantitative assessment of the impact of by-catch on populations is to be made. As a minimum we require:

(1) Reliable estimates of by-catch numbers
(2) Knowledge of stock identity and migration
(3) Reliable estimates of population size.

Knowing these three things for a species would allow us to estimate current by-catch levels as a proportion of current population size. Unfortunately, obtaining such information is often difficult and expensive and it is rare to find that these three requirements are met. In the case of cetaceans, for example, appropriate methods for estimating population size and guidelines for conducting surveys have been developed (e.g. Hammond, 1986), but the costs of comprehensive surveys can be huge. A survey of the North Sea and adjacent waters conducted in 1994 is estimated to have cost in excess of £1 million. For other taxa such as seabirds or turtles, which nest on land, estimating population size is less difficult, but still not without its problems.

However, even if we have such information, although it is a good start, is not really sufficient to determine the significance of the by-catch level the population is experiencing. To obtain the full picture, we need a clear understanding of the species' population dynamics. Only then can we fully determine the level of by-catch that the population can sustain. We also need such information if we are to relate current levels of abundance to the levels obtained before the fishery was operating. The question of stock identity is also one that bedevils efforts to assess population impacts for some taxa. For example, our knowledge of stock structure for most cetaceans is poor and there are no simple ways to address the problem, given the difficulty of adequately sampling individuals. However, despite the hurdles, there are some cases where there is sufficient data to at least make an educated guess at the population consequences of these activities.

Elasmobranchs

Of the elasmobranch species caught as by-catch on the high seas, it would appear that blue shark comprise the greatest proportion. In comparison with other sharks this species is fecund and fast growing, making it less susceptible than some to fishing pressure. Unfortunately, however, although little is known of blue shark population sizes or stock structure it seems difficult to believe that removing six million individuals annually will not have taken its toll (Stevens, in press). Nakano & Watanabe (1992), who used Cohort Analysis to examine effects on north Pacific populations of blue sharks, provide the only assessment of high seas fisheries on shark populations. The authors concluded that catch rates in the late 1980s did not have a significant impact on the population, although Wetherall & Seki (1992) suggest that appropriate data for the kind of analysis performed are lacking. Indeed, although high seas drift-netting was phased out, and total catches should now have declined, effects on some shark populations might still be being felt (Bonfil, 1994). Nevertheless, it should be recognised that other analyses of CPUE data have also failed to show declining catch rates (Stevens, in press).

Rays are also subject to by-catch mortality from trawl gear on continental shelves. From first principles, one might expect marked effects on their populations because of the large size at hatching for most species. In the North Sea, for example, Raja species hatch at between 9 and 24 cm in length, making all immature age classes vulnerable to the fishery. Walker (1996) examined the effects of North Sea fisheries on ray populations by examining catch rates of rays from fishery survey data from 1929–39 and 1991–95. The objective of the analysis was to determine whether observed shifts in species composition could be explained by the life-history characteristics of the species concerned. Specifically, he examined whether the species with the lowest age/length at maturity had come to dominate the community. Based on a standard life-table analysis, data on fecundity and age at maturity were used to calculate theoretical values for the instantaneous total mortalities that would lead to static populations for each species. (Similar calculations were not possible for a further two species which occurred in samples). From these calculations, Walker arrived at the following ranking with respect to the total mortality that species could withstand before the population would start to decline:

R. batis < *R. clavata* < *R. montagui* < *R. naevus* < *R. radiata*

Figure 2.7 shows the relative abundance data for four areas of the North Sea in the two time periods. A number of features of these data are notable. First, species

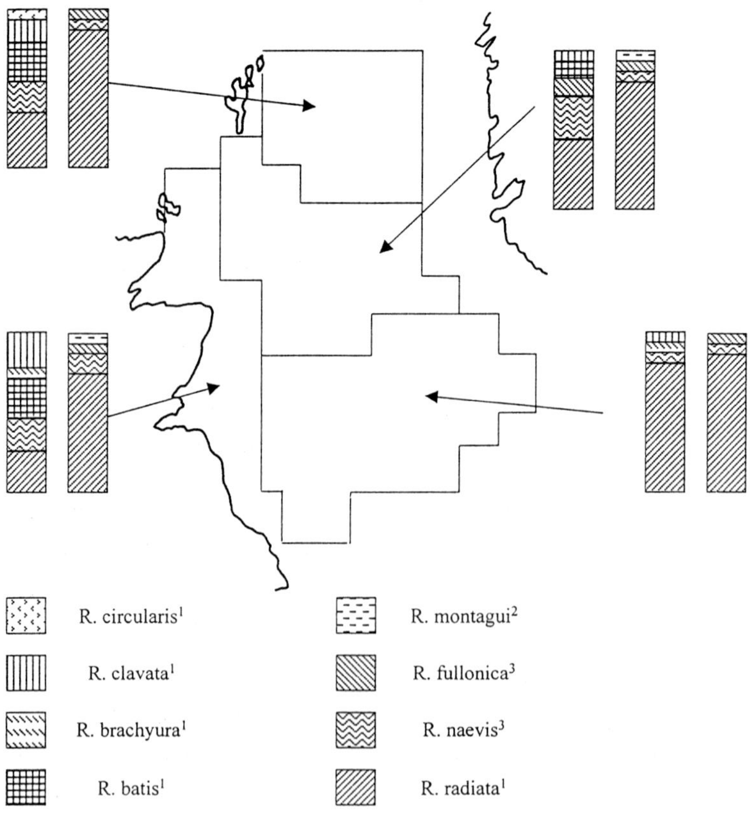

Fig. 2.7 The relative abundance of ray species in four sectors of the North Sea for the periods 1929–1939 and 1991–1996. (Adapted from Figs. 1 and 4 of Walker, 1996.) [1] Species found in 1929–39 only; [2] species found in 1991–95 only; [3] species found in 1929–39 and 1991–95.

richness was higher in the 1929–39 period with seven species of ray recorded, compared to four between 1991 and 1996. Second, the relative and absolute abundance of the starry ray *R. radiata* has increased in most areas, in line with the expectations from the life-table analysis. In contrast to a number of the other skates and rays, this species is rarely retained as by-catch and discard survivorship may be also be quite high. Analysis of length frequency data also showed that larger individuals are now less common in samples. The spotted ray (*R. montagui*), which was not caught between 1929 and 1939, also appears to have increased in abundance since it is regularly caught off the Scottish coast today. Of the species that have declined in abundance, perhaps the common skate (*R. batis*) is most notable. Walker concludes that the declines in trawl survey catches of this species probably coincide with a decrease in the total stock size, even though the North Sea represents the south-eastern limit of the distribution of the species. Brander (1981) has also observed declines in this species for the Irish Sea; it used to be a relatively common and commercially important component of the landed by-catch, but is now rarely observed. Taking all of the data into consideration, there seems little doubt that fishing activity has markedly altered the absolute and relative abundance of ray species in the North Sea. (see also Section 3.3).

Marine mammals

Cetaceans

There seems little doubt that by-catch has significantly depleted a number of marine mammal populations. Of these, perhaps the species that are most under threat are the small coastal porpoise species, which are especially vulnerable to gill-net fisheries. Indeed some of these species, such as the Vaquita (*Phocoena sinus*), a species of harbour porpoise which inhabits the Gulf of California, appear to be close to extinction. Although the gill-net fishery in the region has been banned, both legal and illegal fishing and the stock assessment fishery continue to remove some individuals – a pressure that the population can probably ill afford (Brownell *et al.*, 1989).

While the Vaquita is probably the most prominent of the marine mammals under threat from fisheries by-catch, other species of small coastal porpoise are also under pressure. For example, Hectors dolphin (*Cephalorhyncus hectori*) in New Zealand are experiencing mortality rates that are higher than the population can sustain (Dawson & Slooten, 1993). Similarly, harbour porpoise kills in the Gulf of Maine sink gill-net fishery for groundfish are thought to be at least equivalent and perhaps greater than the population growth rate (Murawski, 1994). Recent legislation requires that annual incidental catches of harbour porpoise must be significantly reduced to 1% or less of the estimated population size in the near future (Donovan, 1994).

The effects on cetaceans of passive gears such as drift-nets were examined by the IWC in 1991 when 7 of the 54 regionally distinct species population regions

(SPRs) were concluded to be experiencing mortality rates that were unsustainable. These populations included the Vaquita (see above), along with the Indo-Pacific humpbacked dolphin (*Sousa chinensis*), two populations of bottlenosed dolphins off the coast of South Africa, harbour porpoise in the northwest Atlantic, and striped dolphin in the Mediterranean. Although, not marine, the status of various river dolphin species is also perilous and worthy of note. The baiji (Chinese River dolphin, *Lipotes vexillifer*) is believed to be down to about 300 individuals, while the Indus River dolphin (*Platanista minor*) has a total worldwide population of only 500 individuals (albeit with 400 in a protected reserve). Other species of river dolphin are also under considerable pressure (Brownell *et al.*, 1989).

In Europe also, there has been considerable interest in the scale of cetacean by-catch in fisheries. The ICES study group on seals and small cetaceans in European seas recently reviewed knowledge on the by-catch of harbour porpoises *Phocoena phocoena* in bottom-set gill-nets in the North and Celtic Seas, and the relationship of the by-catch to estimates of stock size (Table 2.5). For a variety of reasons, these by-catches were almost certainly an underestimate of the total number drowned in these fisheries. Further studies of harbour porpoise by-catch are being undertaken in the North, Baltic and Celtic Seas.

Table 2.5 Estimates of harbour porpoise population size, and by-catch in gill-nets in north-west European waters (from Anon., 1996).

Fishery	Population estimate	By-catch
North Sea Danish cod & turbot gill-net	268 500	4449
Celtic shelf gill-net	36 280	1937

The situation on the high seas has certainly improved since the ban on extensive drift-netting. Nevertheless the legacy of these activities needs to be borne in mind. For example, by the mid-1980s more than 800 vessels from Japan, Taiwan and the Republic of China were active in the North Pacific. In total, these fleets could lay up to 40 000 km of net each night. A depressing characteristic of these fisheries is that they developed with little or no control, other than that of the market. Thus, there was no attempt to determine the level of fishing effort that would provide a sustainable yield from the resource, let alone any effort to estimate or minimise the by-catch of non-target species. Importantly, we have absolutely no data on the trends in abundance for the majority of species that are likely to be affected by this fishery and it is, therefore, impossible to determine what the system looked like before the fishery began, when it ended, or indeed today. The numbers of individuals killed was certainly large and for marine mammals, turtles, sharks and birds, all of which have characteristically low reproductive rates, it is difficult to believe that there were not marked effects on populations.

Northern sea lions

One group for which more formal analysis of population status has been undertaken is the northern sea lion. Using a stable population life table for adult female northern sea lions and estimates of the adult female population size, Perez & Loughlin (1991) attempted to calculate the effects of by-catch mortality on the sea lion population since 1966.

From counts of individuals at haul out sites on land in western Alaska, it would appear that the abundance of both adults and juveniles of this species have been declining by about 4.5% per year between 1956 and 1989. In other areas declines have been even more dramatic. In the eastern Aleutian Islands, for example, the count has declined by 93%, from approximately 41 000 in 1960 to less that 3000 in 1989 (Perez & Loughlin, 1991). The exact reasons for the decline are largely unknown, but interactions with fisheries is clearly likely to be a contributory factor. The minimum estimated decline due to by-catch mortalities was 6.7% between 1966 and 1989 for the Bering Sea/Aleutian Islands, with a mean estimate of 16.3%. A similar estimate for the Gulf of Alaska gave a minimum decline of 3.6% since 1966 and a mean estimate of 6.4%.

There is some considerable doubt, however, about the estimated catch levels in Alaska and it seems likely that the levels used to estimate these population declines were highly conservative. Perez & Loughlin (1991) conclude that with upwardly revised estimates, the contribution of by-catch to the decline in sea lion numbers since the 1960s could have been important, although it was unlikely to be the sole cause. As with many stories of this kind it is a sad fact that the necessary monitoring programmes to estimate mortalities were not put in place until well after the bulk of the population decline had occurred. The authors also conclude that the high estimated catch of northern sea lions during the 1970s by foreign vessels may partially account for the decline in population numbers in the Aleutian Islands region. However, in more recent years the incidental catch from joint venture fisheries has been generally low, yet the populations have continued to decline.

Turtles

The status of sea turtle species is perhaps best indicated by the long-term trends in individual populations. These are obtainable from counts of nesting females because individuals repeatedly return to the same location to spawn. Although other ways of estimating population sizes have been used, (e.g. counts from oceanic aerial surveys, strandings data, levels of by-catch per unit fishing effort, or tortoiseshell shipments to foreign markets) nest counts probably require the fewest assumptions about the biology of the species. Even so, considerable caution is required in interpreting any short-term trends in the numbers of nesting females because species differ in their nesting frequency; some species, such as Kemp's ridley turtles apparently nest each year, while others such as Green turtles only return every two or three years. Thus, between year variation can vary markedly between species and wide year to year fluctuations in the numbers of nesting turtles can make interpretation of short time series very difficult. Indeed, Anon. (1990) suggest that time series of less than a decade are of limited utility for the estimation of population trends, because it is only with longer data sets that the inter-annual fluctuations can be distinguished and the underlying population trend determined.

Notwithstanding the above difficulties there seems little doubt that for some species there used to be an awful lot more turtles around than there are now. At one site in Mexico, for example, estimates of trend have been obtained from a single datum in 1947 from film footage, a few years in the late 1960s and annual estimates from 1978 to 1988. These data suggest that the number of nests has declined to 1% of the 1947 level (from 92 000 to 800 nests). This is the most severe population decline documented for any turtle species and the numbers have continued to decrease (Fig. 2.8).

From the data sets for other species examined, it would appear that populations of nesting Loggerhead turtles have also declined over the last 20–30 years on US nesting beaches in Georgia and South Carolina. In contrast, for southern beaches in Florida the available data suggests that, if anything, the numbers may be increasing.

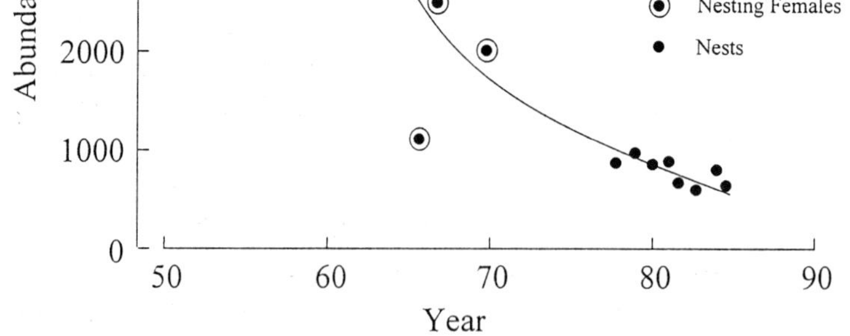

Fig. 2.8 Estimated decline in the Kemp's ridley turtle population at Rancho Nuevo, Mexico, based on number of nests per year or number of nesting females per year. (Adapted from Fig. 3.1 of Anon. (1990). *Decline of the sea turtles. Causes and prevention.* Copyright 1990 by the National Academy of Sciences. Reproduced with permission from the National Academy Press, Washington DC.)

Given that fishing is the single largest source of mortality – it has to be an important factor in these declines, at least in recent years. Earlier initial declines may well have occurred through coastal development etc., but it seems likely that the biggest current threat to population viability is fishing in many instances. For Kemp's ridley turtle in particular, the threat appears to be considerable. Dayton *et al.*, (1996), for example, state that the species is considered to be under threat of extinction with an estimated worldwide population of only 1000 breeding adults – a figure which may well be close to the minimum viable population size. There are no grounds for complacency regarding the threat to this species – a view that is clearly held by the US fisheries management authorities who now insist on the fitting of an authorised turtle exclusion device to all shrimp trawls.

In contrast to the above conclusions studies on some other turtle populations have suggested that prawn trawling was unlikely to be important. One such example is in the Australian Northern Prawn Trawl fishery where Poiner *et al.* (1990) showed that, although several thousand turtles are caught each year, only 6% of individuals actually drown. Although the authors pointed out that relatively little is known of the population dynamics of the species concerned and assessing the impact of the trawl by-catch was problematic, they concluded that it probably was not a cause for concern for this fishery. Of course, this may be true in terms of risks to populations, but the public's perception from an animal welfare perspective may well be rather different. These contrasting findings in different situations highlight both the difficulties of generalising about these matters and the need for fishery specific studies.

Seabirds

Seabirds are also relatively easy to count, compared to fish. This, combined with high levels of interest in birds by both the general public and the scientific community, has meant that some populations have been tracked for over a century. Long-term trends in marine bird populations are usually monitored at breeding colonies. This approach has the advantage that most marine birds nest at a relatively small number of isolated colonies so the reproductively active part of the population is confined to a limited area that can be censused relatively easily. Ringing schemes also make it possible to measure recruitment to reproductive status. This is not to suggest that seabird monitoring is without difficulty – nocturnal species which nest in burrows or boulder screes are very inconspicuous and the precision of counts is consequently low. Moreover, not all birds breed at coastal sites and those that nest inland (e.g. herring gulls) are often less well covered by survey programmes.

Although, as with most other taxa, ascribing unambiguous causes to trends in bird populations is difficult, for some bird species the role of fisheries seems quite clear. Below, I describe examples of both increases and decreases for which fishing effects are strongly implicated.

Population declines

One of the most serious concerns with respect to birds is for albatrosses. For these species (as with many other bird taxa) a number of features of their life history exacerbate the problem – all species lay only a single egg and have a rather high age at first maturity (between 9 and 12 years). In addition, several species of albatross breed only every second year. Unfortunately, neither an advance in the age at first breeding, nor an increase in fecundity (for example, by an increase in the proportion of the population breeding every year) would be able to compensate for the substantial declines in survival rate that fishing imposes. In wandering albatross, for example, a decline in survival rate of 1% would require a compensatory advance in the age of first breeding of 6.5 years (Weimerskirch *et al.*, 1987). One startling observation illustrates the seriousness of the problem – if one looks at the at-sea recoveries of South Georgia ringed wandering albatrosses since 1975, the tuna long-line fishery killed 75%. In the previous fifteen years no ringed birds were recovered in this way. This mortality is almost certainly the main factor accounting for the marked decline in wandering albatross breeding populations around the Antarctic.

Despite this gloomy conclusion, it is encouraging that the Japanese long-line fishery has introduced modifications to fishing practices which appear to significantly reduce albatross mortality and the number of fish lost. Efforts to increase the sinking rates of hooks so that they become unavailable more quickly, and to deter birds from the area immediately behind the ship by deploying streamer lines from a stern-mounted pole, both appear to be successful strategies for reducing albatross by-catch. One simple and effective strategy appears to be to set the lines at night (Brothers, 1991), although this may result in unacceptable reductions in tuna catch rate or an increase in shark by-catch. Conversely it appears that other long-line fisheries are developing which do not adopt such measures. Thus, albatross populations may continue to be under considerable pressure. For the southern ocean, Kock & Shimadzu (1994) report that albatross, southern giant petrel and Antarctic tern have shown consistent trends of decline over their entire breeding range (albeit at variable rates).

One area where there has been considerable effort to quantify by-catch mortalities is Newfoundland (e.g. Piatt & Nettleship, 1987). Despite what appear to be rather high mortality rates, Piatt & Nettleship (1987) found it difficult to draw firm conclusions about population trends, particularly for species that ranged widely outside the study area. Nevertheless, it did appear that mortality was probably insignificant compared to total population sizes for species such as shearwaters, petrel, cormorant, eider, scoters and gulls. In contrast, the situation for gannets, razorbills and guillemots seemed more serious. In the case of gannets it was estimated that approximately 2.1% of the population of about 20 000 birds (occupying three breeding colonies) was killed in nets annually during the 4-year study. At one site, however, about 9.3% of the population was killed in one year alone. Similarly, about 22 000 guillemots were estimated to be killed annually in nets, about 2% of the adult population. As with gannets, however, local impact appears to have been much greater. For example, for one colony it was estimated that 16.3% of the breeding birds were killed. Since mortality rates in stable populations appear to be about 12%, once again there is cause for concern. Worrying figures were also offered for razorbills. About 1600 individuals bred in Newfoundland at the time and (assuming that all birds killed were adults) about 12.4% of the population was killed annually. Piatt & Nettleship (1987) noted that this mortality rate exceeded most estimates of annual adult mortality in stable razorbill populations in the North Atlantic (usually about 10%). At the very least, estimates of the kind Piatt & Nettleship (1987) were able to make should serve as a warning. They indicate the potential of net mortality to affect population trajectories.

Population increases

The possibility that fishing activities might have caused observed increases in population size for some seabird species has been the subject of speculation for many years. Fisher (1952), for example, argued that additional food supplies from whaling and trawling caused the southern spread and population increase of the northern fulmar in the southern Atlantic. However, such an hypothesis was not without its challengers; others suggested that the spread was a natural result of genetic or behavioural evolution, or a consequence of either gradual warming of the northeastern Atlantic or some other hydrographic factor (Furness *et al.*, 1988). Similar hypotheses regarding the effects of extra food can also be advanced for other populations. For example, the large increase in the numbers of great skuas around the Shetlands during this century may have been sustained by the availability of small haddock and whiting discarded from the local fishery.

For the North Sea, where seabird studies are particularly comprehensive, one of the most striking features of the seabird population trends is the enormous growth in the numbers of scavenging species (Fig. 2.9). Interestingly, despite the considerable

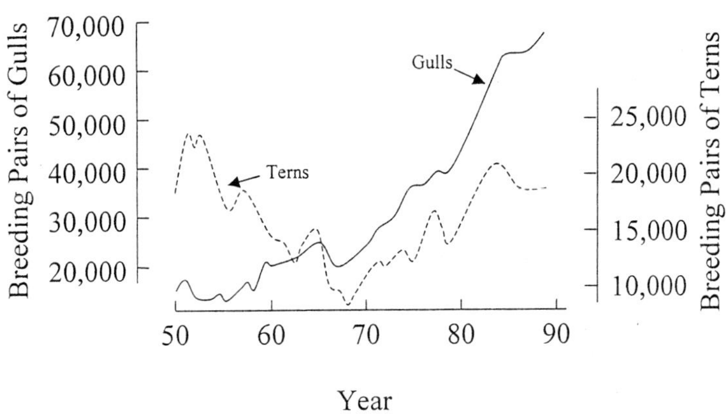

Fig. 2.9 Trends in the number of gulls (*Larus argentatus*, *L. fuscus*, *L. canus*, *L. ridibundus*) at 28 important breeding sites in the German Wadden Sea coast. (Redrawn from Fig. 4 of Dunnet *et al.*, 1990.)

speculation, no studies of the exploitation by seabirds of waste from fishing boats was undertaken until the 1980s.

In addition to an estimate of the amount discarded, determining the potential of fishery waste to support seabird populations requires an estimate of the energy demands of seabirds. The best measurements are from studies using isotopically labelled water, which suggest that expenditure is 2.5 and 5 times basal metabolic rate (BMR). Furness *et al.* (1988) settled on a value of about 3 and obtained estimates of BMR from allometric equations based on data for all seabirds, with some allowance made for the fact that seabirds from higher latitudes tend to have higher BMRs than those from the tropics. Using these estimates and estimates of the amounts and energy content of discarded material Furness *et al.* (1988) calculated the maximum number of 1 kg seabirds which could be supported by discards and offal in the North Sea (Fig. 2.10).

If one assumes that all the offal and discards were taken, some 2.8 million 1 kg seabirds could be supported around the UK in 1982. Of course, the birds cannot

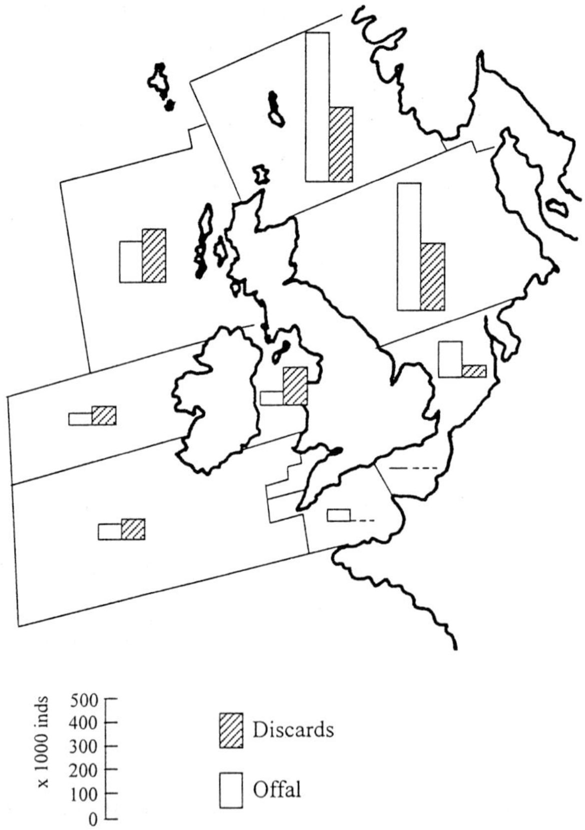

Fig. 2.10 Numbers of 1000 g seabirds that could be supported by offal and discards around the British Isles. Values for amount discarded were based on 1982 figures. (Adapted from Furness *et al.*, 1988.)

consume all of the available food; some must sink and be eaten by benthic scavengers, and some must be taken by other species such as marine mammals. Nevertheless, Furness *et al.* (1988) estimate that probably about 90% of offal and 75% of discards are taken by birds around the Shetlands (admittedly an area with substantial seabird colonies). On a wider North Sea scale Garthe *et al.* (1996) estimated some 800 000 tonnes of discards were potentially available to seabirds. Again this amount is sufficient to supply the energy requirements of the entire scavenging seabird population. However, this potential consumption could never be realised because most of the available discards are dumped in the southern North Sea, whereas most seabirds are present in the northwestern North Sea. Moreover, even if overlap were perfect, several other assumptions would also need to be met if fishery waste were to support all scavenging seabirds.

Although we have only imprecise estimates of both total seabird population sizes (including non-breeding birds) and of their movement patterns, the available data suggests that seabird numbers exceed by a considerable margin the 2.5 million or so individuals that fishery waste could support. The question, however, is whether the populations of these birds is larger than they would be in the absence of such additional food. Furness *et al.* (1988) cite additional evidence in support of such an hypothesis, suggesting that those species which compete best for fishery waste, such as Northern fulmar and gannet seem to have continued to increase around the Shetlands, while others have stabilised or declined. Firm conclusions on this subject will require much more information about the importance of discards and offal for the survival rates and breeding success of scavenging seagulls. On balance, however, it seems difficult to argue against the idea that offal and discards have had substantial positive effects on some seabird populations.

Benthic scavengers

Despite the reasonable expectation that benthic scavengers have increased in response to the dumping of discards, there are almost no data available to critically assess the hypothesis. One of the few examples of a marked population increase that has been partly attributed to fishing, however, is in the sub-tidal communities of Hong Kong where Nassariid gastropods have increased markedly in some areas (Britton & Moreton, 1994). Because both discards and fauna damaged in situ by trawls are available to benthic species, this topic is covered more fully in Chapter 3.

2.3 Conclusions

Discarding and by-catch are major problems for world fisheries. In many respects they could be considered to be the major problem. In the first place they are a very obvious testament to the unselective nature of many fisheries. Illegal undersized fish, or low value fish that would take up space in the vessels hold, are caught and discarded instead of being allowed to remain in the sea to grow to a harvestable size and non-target species are caught and discarded because they are of no commercial value. The practice is wasteful in terms of the potential useable protein that we fail to harvest from the sea, but also because of the levels of mortality imposed on taxa that we would never wish to harvest for commercial gain.

It would be quite wrong to imagine that fishers and fisheries managers are not also very aware of the problem and have a strong desire to do something about it. Many of the conflicts between fishing sectors arise because the incidental catch to one fishery is the target for another, and if fishers could avoid cluttering their nets with unwanted species they would. Indeed, efforts to devise more selective gears that retain the desired species and allow others to escape have been welcomed by many fishing sectors. Although concerns about the practicality of the solution for a working vessel and the potential for loss of revenue always loom large, there have been a number of highly successful research and extension programmes which have developed by-catch reduction devices and encouraged uptake by the commercial fleet.

From a fisheries management perspective also, the problem of discarding is a major impediment to sound practice, primarily because the amounts discarded are not usually recorded. Since most fisheries assessment methods rely on catch data for their operation, the uncertainty associated with unknown levels of discarding can be enormous. Indeed, the problems are so great that some assessment biologists feel that without observer programmes to quantify the levels of discarding, the data used to conduct assessments is of questionable utility. At present, there is no requirement for the reporting of by-catch in the oceanic zone or within the EEZs of most countries so there is little monitoring or management.

Others also rightly point to the undesirably high levels of mortality on species of high conservation interest. For some species levels do indeed seem to be so great that populations are under threat (e.g. some species of albatross and turtle species); but even if the mortality rates are not this great (or the data are inadequate) there is a legitimate animal welfare perspective which argues for strenuous efforts to limit the mortalities regardless of population effects. Few people like the idea of turtles or dolphins being needlessly drowned in fishing nets, regardless of whether they will go locally or globally extinct if we continue to do so.

On many grounds, therefore, finding solutions to the discard and by-catch problem is highly desirable for many of the sectors that have aspirations for the marine environment. Yet paradoxically, in at least one case when fishing practices have been modified, a potential conservation dilemma has emerged. Audoins's gull (*Larus audouinii*) is one of the few endemic seabirds of the Mediterranean region and in the early 1970s its population was estimated to be very small and threatened (Cramp & Simmons, 1985). In 1981, however, a new colony was established in the Ebro Delta, which by 1994 had grown to become the largest colony in the world with 70% of the world population (10 000 breeding pairs). It has been argued that this increase occurred because of the high levels of food that were available as a result of discarding by the local trawler fleet (e.g. Ruiz *et al.*, 1996). Since 1991, however, the trawling fleet has upheld a two month fishing moratorium which, because it coincides with the breeding season for gulls has affected food supply to birds and in consequence their breeding performance (Oro *et al.*, 1996). These effects varied between years, depending on the precise timing of the moratorium relative to the breeding cycle of the birds – in 1992 egg production was affected while in 1993 it was chick rearing. Oro *et al.* (1996) suggest that both adult survival and recruitment rates have been affected and that if the moratorium were to continue over many years, as is being proposed, the gull colony could be under threat!

With respect to cetaceans it appears to be impossible at present to determine reliably the impact of by-catch on populations (Donovan, 1994). Yet waiting until it can be shown with certainty that by-catch levels are unsustainable seems indefensible. Although, there has been some effort to examine the behaviour and sensory capabilities of cetaceans to develop means for reducing by-catch (e.g. Nachtigall *et al.*, 1995), technical fixes of this sort will take some time to be developed, and even then they may be of limited utility.

By far the most effective action for cetaceans would be to reduce the use of gillnets still further. Using an alternative fishing method, such as long-lining, would certainly have the desired effect for cetaceans, but the trade-off for such a decision is the increased mortalities that would be imposed on other taxa such as birds and turtles. Moreover, there are reports that long-line fishermen in some areas use dolphin meat as bait (Donovan, 1994).

Legislative measures to reduce by-catch problems are certainly one means for addressing the problem, but clearly they are only of value if they are enforced. On the high seas, for example, there has been a notable degree of success in reducing the dolphin by-catch from tuna fishing. However, unless enforcement is adequate there is a danger that, not only will undesirable by-catch levels be sustained, but that information on catch levels will become more difficult to obtain. This appears to be the situation that occurred in Peru when a law banning the capture and trade in small cetaceans was passed in 1990. A more recent law passed in 1994, which elaborated enforcement measures appears to be having more effect (Donovan, 1994). It is important not to underestimate the difficulties of enforcement, however, especially in fisheries with large numbers of artisanal vessels in developing countries. What seems certain is that compliance with the law is more likely if the reasons for the legislation are clearly understood and accepted by fishermen. Further discussion of possible by-catch and discard mitigation strategies is provided in Chapter 9.

Chapter 3
Effects of Trawling and Dredging on the Seabed

It is difficult to ignore the fact that trawls designed to catch demersal fish and shellfish scrape the seabed. Indeed, in reviews of the effects of trawling, it is now almost mandatory to cite a Commons Petition to the King of England from 1376 which expressed concern about a newly introduced fishing gear, the 'wondyrchoun', which ploughed the seabed so heavily that 'it destroys the flowers of the sea' (e.g. De Groot, 1984; Auster *et al.*, 1996). In earlier times concern was usually expressed by fishermen who did not possess the new (and often more effective) gear. More recently, however, many other groups have identified this form of benthic disturbance as a major cause of concern with respect to non-target benthic species and also for the longer-term viability of the fisheries themselves. Part of the reason for the increase in concern that has taken place in recent years can be attributed to an increase in the power of fishing vessels, and in consequence the size and weight of the gears that can be used. This factor applies particularly to beam trawls which weighed up to 3.5 tonnes in the late 1970s, but had increased to about 10 tonnes by the early 1980s (van Beek *et al.* 1990). In addition, hitherto untouched areas are now becoming accessible to fishermen with the advent of new technologies such as chain mats which protect the belly of the net. In Australia, for example, new fisheries are developing in deeper water down to depths of 1200 m (Judd, 1989).

As with other kinds of fishery effects, worries about trawling can be classified into those which concern interactions with other fisheries and those dealing with effects on non-target taxa. Until recently, studies of the former have probably been most common (Auster *et al.*, 1996), although they seem to be being rapidly overtaken by studies of the wider impacts. Investigations of fishery interactions have largely focussed on valuable fisheries such as lobster or scallops and various studies have shown the potential for interaction between, for example, pot and trawl fisheries (e.g. Jamieson & Campbell, 1985). These issues are somewhat peripheral within the context of this book. More relevant are the wider effects of trawling on non-target species and communities, a subject to which conservation bodies and the scientific community are paying increasing attention in many parts of the world. Such concerns are clearly legitimate – it would be quite foolish to deny that benthic communities can and are substantially altered by towing a fishing gear over them.

The problem, however, is to put the issue into perspective both in terms of the magnitude of the effect for different combinations of habitat type and fishing gears and in terms of the spatial extent of the disturbance. Only when this has been achieved can we rationally assess the scale of the problems and how the various aspirations we have for benthic environments can be reconciled.

3.1 Which gears cause the problem?

Any fishing gear that is towed over the seabed will disturb the sediment and the resident community to some degree. The intensity of this disturbance, however, is very much dependent on the details of the gear and the sediment type. Considerable effort has been expended to quantify these relationships for some types of gear (e.g. BEON, 1990, 1991) and, although the details for many other gear types remain unexamined, a qualitative ranking is generally possible. As an example, Table 3.1 provides a summary of the physical effects for towed gears used in the North Sea. The table shows that, at the scale of an individual fishing track, the various types of shellfish dredge and the heavy flatfish beam trawl disturb the seabed most intensely. For lighter gears such as the otter trawl, disturbance is largely restricted to the trawl boards. There have been many studies by fishing gear technologists which characterise and measure the physical pressure imposed by particular parts of a gear, and more recently acoustic methods for evaluating changes to sediment structure have been developed (e.g. Schwinghamer *et al.*, 1996). On their own, however, such data are of fairly limited utility because the type of sediment over which gears are towed is also important. Not surprisingly, penetration into soft mud will be considerably greater than into hard-packed sands and effects on infaunal benthos may alter accordingly. For example, Churchill (1989) estimated that coarse sand was typically penetrated to a depth of 1 cm by otter boards and resuspended approximately 39 kg sec^{-1}, whereas the figures for fine sand and muddy sand were 2 cm (78 kg sec^{-1}) and 4 cm (112 kg sec^{-1}), respectively. Others suggest much greater penetration depths than these under some circumstances. One advantage of using a parameter such as sediment resuspension rate is that it is relatively easy to remotely monitor suspended sediment load (or a proxy such as water turbidity) so that disturbance data can be correlated with fishing activity over the longer term. Moreover, since other natural disturbance events also suspend sediments one potentially has a basis for scaling an aspect of fishing disturbance against the natural disturbances that the community experiences (see Section 3.6).

3.2 Measuring mortality rates

The physical effects of the passage of a gear on sediments is clearly one way in which impacts can be assessed, but what we are usually after is an assessment of likely

Table 3.1 Summary of the effects and areas fished for different gear types used in the North Sea. (Adapted from Anon, 1995.)

Type of gear	Gear parts in contact with the ground	Typical width of major disturbing parts (m ship^{-1})	Towing speed (knots)	Penetration depth* (cm)	Estimated area fished (km^2 100 h^{-1})
Beam trawl (flatfish)	Trawl shoes, tickler chains or chain mats				
(a) Offshore (> 12 miles)		12.0 × 2	6	> 6 (6)	31.1
(b) Inshore (< 12 miles)		4.0 × 2	5	?	8.9
Shrimp beam trawl	Trawl shoes, ground rope with rollers	0.2 × 4	4–5	?	0.7
Otter trawl	Otter doors	1.5 × 2	3–4	8	2.0
	Ground rope	30	3–4	8–10	22.2
Industrial trawl					
(a) Single	Otter doors	1.5 × 2	3.5	8–10	1.9
	Ground rope	25	3.5	?	16.2
(b) pair	Ground rope	25	3.5	?	16.2
Demersal pair trawling	Ground rope	40	3	?	26.0
Mussel dredge	Blade and belly	1.75 × 4	2	5–25	2.6
Cockle dredge	Suction head	1.0 × 2	2	≥ 5 (5)	0.7
Scallop dredge	Tooth bar and belly				
(a) English		0.75 × 16	3	3–4 (2–3)	6.6
(b) French		2 × 5	3–4	< 10 (NA)	7.2
French clam dredge	Blade and belly	0.7 × 2	3–4	< 15 (NA)	0.8

* Parentheses denote penetration on hard ground

effects on the fauna. For some classes of taxa and types of trawling one can predict that there will be high mortality rates from first principles. In particular, erect foliose fauna or fauna which build reef-like structures of any kind are going to be destroyed by a towed gear. In temperate waters, species such as the tube worm *Sabellaria* or the corraline algae (Maerl) fall clearly into this category, while in other areas coral-like bryozoa are vulnerable (Bradstock & Gordon, 1983). One does not need much in the way of research to appreciate the threat that trawling poses to such species, or to the species which exploit the physical architecture such taxa provide: if they are trawled, these animals will be destroyed. The more difficult problem is identifying whether (probably small and patchily distributed) reef or foliose structures remain in fished areas so that, if one chooses, they can be protected. Good maps showing the distribution of such patches are also required for unfished areas. With such maps we can gauge what we may have lost from fishing grounds and identify areas of conservation interest.

Efforts to make direct measurements of impacts on benthos have been widespread, but until recently relatively little of this information appeared in refereed journals. A fairly comprehensive bibliography of the earlier literature is provided by Redant (1991) and for readers interested in the early work on this subject, this is a good starting point. Measurements of mortality rates are usually made from short-term experiments. These studies can take a variety of forms, but most have been restricted to looking at the benthic species retained in the net and the survivorship of those that are thrown back (e.g. De Groot & Apeldoorn, 1971; De Groot, 1984; Creutzberg *et al.*, 1987; Kaiser & Spencer, 1995). Data from this kind of study can provide a lower bound on mortality estimates for species that are actually caught. However, the most useful type of mortality estimate is obtained when the densities of species are first measured *in situ*, a fishing gear is then towed over the site and a second estimate is made shortly afterwards.

One of the more comprehensive exercises to measure mortality rates is that of the BEON group in the Netherlands (BEON, 1990, 1991). This study adopted a range of approaches to examine the effects of beam trawling for flatfish, and showed that survival rates for the larger infauna and epifauna caught in the net varied markedly. Whelks and hermit crabs were largely unaffected, starfish experienced 10–30% mortality, while up to 90% of the bivalve *Arctica islandica* died. With regard to the fauna that are not retained in the net, samples taken inside and outside trawl tracks showed significant reductions in the abundance of more fragile taxa, such as the heart urchin *Echinocardium* sp. and tube-building polychaetes. Overall, the results suggested that a relatively high proportion of the fauna in the path of a beam trawl is killed.

Another example of the approach is that taken by Eleftheriou & Robertson (1992) who examined the effects of scallop dredging in a high energy sandy sub-tidal environment. By taking samples at various time points during a sequence of controlled hauls, the authors were able to obtain a reasonable assessment of the consequences of scallop dredging in this kind of habitat. Their results reinforce the idea that effects depend markedly on the kind of sediments and the hydrodynamic environment in which trawling is taking place. For example, in contrast to other studies of the physical effects of scallop dredging that have been conducted in deeper water (e.g. Caddy, 1973), no effects on the vertical distribution of sediment grades were observed. With respect to the fauna, the results indicated that most groups of infaunal taxa are also unaffected, owing to the behavioural and morphological adaptations which make them well suited to the natural physical disturbances that occur in such a high energy environment. In contrast, and as one would expect, more sessile and larger infaunal and epifaunal taxa did have reduced abundances post-trawling.

With a few exceptions (e.g. Bergman & Hup, 1992; Eleftheriou & Robertson, 1992), mortality estimates from experiments are scattered through the grey literature. Recently, however, there have been some efforts to synthesise available information for areas such as the North Sea (Table 3.2). As one might expect, such syntheses suggest that it is the shallow dwelling and epibenthic species that suffer most. But as we will see later, generalities are difficult to make owing to the wide range of mechanisms by which the passage of a gear can lead to the death of organisms. For example, small-bodied shallow-dwelling bivalves seem to be relatively immune to trawling, perhaps because the pressure wave in front of the gear blows them out of the way (Gilkinson *et al.*, 1998). Species vulnerability may also

Table 3.2 A compilation of mortality estimates (expressed as a percentage of the initial density) for benthic species in the path of beam trawls. (Adapted from Table 3.3.2 of Anon, 1994.)

Group	Species	Type/habit	Mortality (%)
Polychaetes			
	Aphrodita aculeata	Shallow burrowing sea mouse	<9
	Polychaetes	Infaunal burrowing, varying depth	15–25
	Pectinaria sp.	Shallow tube living	60
Crustacea			
	Callianasids	Deep burrowing shrimps	4
	Corystes	Small bodied epifauna	30
	Eupagarus sp.	Hermit crabs	50–75
Molluscs			
	Abra alba	Small shallow living bivalve	85
	Donax vittatus	Small shallow living bivalve	40
	Spisula sp.	Small shallow living bivalve	35–50
	Placopecten	Epibenthic scallop	8–25
	Arctica islandica	Large deeper living bivalve	35
Echnioderms			
	Asterias sp.	Starfish	4–5
	Ophiura sp.	Brittlestars	7–30
	Echinocardium	Heart Urchin	0–50
	Psammechinus miliaris	Urchin	50

change due to its physiological state (or age), which affects depth distribution within sediments. Sediment type itself is also important, in part because this also determines how individuals are vertically distributed, but also because gears penetrate more deeply into softer sediments. With respect to the effect of sediment type itself, for example, work on non-capture mortality of the target species by scallop dredges showed that mortality rates can range from 8% on sandy substrates to 25% on gravel (Shepard & Auster, 1991).

Much more work needs to be done on mortality rates to provide the basic data for inferring the large-scale effects of trawling. However, despite continuing efforts to measure mortality rates, it is important to note that most field studies are carried out in places that have been intensively trawled in recent decades, so rates are generally only determined for those species that remain reasonably abundant.

3.3 Population consequences

The studies described in the previous section show that, although technically demanding, it is perfectly possible to make direct estimates of mortality rates for benthic taxa. On their own, however, these estimates are of limited utility and, as with the by-catch issue discussed in Chapter 2, we must turn to the effects on populations. Nowhere does this question become more problematic than with the benthos, where even basic understanding of recruitment processes and population dynamics is poor for most species. Of course, as mentioned earlier, the effect of trawling on some species populations, particularly those that build fragile reef-like structures, is self-evident but for others we are very much in the dark. The conventional wisdom is that the larger, slower growing species will suffer at the expense of smaller, faster growing ones. Whether such a general model for how fishing disturbance might operate is valid remains to be seen, but what would certainly be useful is a method for quantifying the vulnerability of particular taxa. If such a method were available, it would provide a better basis for deciding which species one might focus on in a given habitat.

Estimating vulnerability

Recently, fisheries scientists have started to consider how their assessment techniques can be extended to evaluate the vulnerability of both non-target by-catch species, and non-target benthic species. To do so, one problem that has to be overcome is that traditional fisheries assessment tools that rely on catch data are of little use for smaller epibenthic or infaunal animals that are not retained in the net. Thus, the concept of fishing mortality has to be adapted to include the mortality imposed by fishing on that part of the population that is damaged by the gear and is left dead or dying on the seabed. This is one reason why the mortality estimates discussed in Section 3.2 are so important.

Whatever the difficulties, the first step in the process of determining population vulnerability is to decide on a quantitative measure that allows species to be compared on a common scale. MacDonald *et al.* (1994) suggest that one such measure of relative vulnerability can be obtained by comparing the current level of fishing mortality to the one under which the spawning population per recruit would be reduced below a certain proportion of the virgin level. One might, for example, define a critical value of fishing mortality (F_{crit}) as the value that would reduce the population to 5% of its unfished fish level. Although, for species that are not actually caught, using catch data to reconstruct the population at size by catch data is impossible, the principle used in providing an index of relative vulnerability can still be applied (Anon., 1994). The underlying idea remains the estimation of a critical F value (F_{crit}) which, if exceeded, would put the population under the risk of collapse (potential jeopardy) given unfavourable recruitment conditions. F_{crit} can be estimated based on modest life history information for the species and Table 3.3 shows some very tentative first estimates for molluscan taxa made by the ICES Working Group on the Ecosystem Effects of Fishing Activities, (Anon., 1994). Although, very much a crude attempt, these values show a reasonably logical progression of relative robustness to fishing pressure with slower growing species such as the Quahog (*Arctica islandica*) being put at potential jeopardy at low levels of sustained mortality while smaller, faster growing species such as *Donax trunculus* are much more resistant.

Table 3.3 Tentative estimates of F_{crit} for molluscan taxa. F_{crit} is the fishing mortality that would reduce the population to 5% of its unfished fish level. (Adapted from Table 3.3.1 of Anon., 1994.)

Species	F_{crit}	% Mortality
Arctica islandica	0.05	5
Modiolus modiolus	0.16	15
Mytilus edulis	0.45	36
Venus striatula	0.47	37
Mya arenaria	0.62	46
Macoma balthica	0.74	52
Pecten maximus	1.05	65
Nucula minuta	1.08	66
Ceratoderma glaucum	1.27	72
Chlamys varia	1.27	72
Chlamys opercularis	1.43	76
Donax trunculus	2.34	90

Unfortunately, for many species there is a lack of any life-history data that would allow one to draw firm inferences about relative vulnerability. If such approaches are to be pursued in future, more reliable estimates for a number of life-history parameters for benthic species are needed. In particular we need data on natural mortality, length at first maturity, length at first capture and growth parameters.

Certainly, some information already exists, but it is spread through the literature and is rarely expressed in a form that is useful for the present purpose. We need to find a way to use this information; one possibility would be to use existing Production:Biomass ratios from populations in unfished areas to estimate the level of natural mortality for various species.

Assuming we can obtain an estimate of F_{crit}, if some estimate of the current F is also available then we can compare the two values to provide a tentative assessment of vulnerability. In normal fisheries we could determine the levels of fishing mortality directly from catch information. However, because we do not retain benthos in the net, an indication of current F can be obtained by considering fishing mortality to be proportional to the product of two more tractable components: the probability of being killed as a result of the encounter, and the probability of encounter of an individual of a species with the gear. The probability of being killed can be estimated using the approaches discussed earlier with mortality experiments that measure the abundance of species prior to and after the passage of the gear. The probability of encounter can be estimated if data on the time spent fishing, the spatial distribution of fishing effort and the area covered are available. At present, however, the spatial and temporal resolution of available effort data is probably insufficient for this purpose (see Section 3.6).

Ideas on how to determine vulnerability for benthic species are still in their infancy. Nevertheless, the approach developed within the ICES community is one of the few attempts to go beyond chanting well-worn mantras about older, slower growing species being most at risk and smaller opportunistic species less so. Of course, for some taxa quantitative estimates are unnecessary – we can deduce from first principles that effects on reef-building tube worm populations will be profound, for example, (provided we know how the population is distributed in space). We should also recognise that species of this kind may have been largely lost from large areas of some continental shelves. For most of the benthic taxa that remain, however, we are far from establishing how vulnerable their current populations are, although a trend of declining abundance for a population on a fishing ground might provide a strong indication. Thus, identifying long-term changes in benthic communities and linking these to trends in fishing effort is another potentially informative approach to determining fisheries effects.

Long-term trends

The focus in this section will be exclusively European, and for good reason – analyses of benthic time-series data from other areas are very hard to find, and for any that do exist, the prospects for dissecting the effects of fishing from other effects are very slim indeed. In fact, I know of no non-European sources that shed light on this issue.

One widely quoted analysis suggesting fishery-induced change is from the Wadden Sea (Reise, 1982; Reisen & Reise, 1982) where faunal survey data from the

1920s was compared with that obtained from samples taken in 1980. Some of the changes detected seem to be directly attributable to the effects of fishing, although probably not through physical disturbance effects *per se*. For example, all of the species that showed marked declines were originally associated with oyster beds that disappeared in the early part of the century as a result of over-exploitation. Since oyster beds are likely to provide important secondary habitat for other benthic taxa, their removal may have had important knock-on effects. Increases in abundance were recorded for 30 species out of the 101 common species that were originally found in the area and these increases were dominated by polychaetes, of which the majority was adapted to life in disturbed habitats.

It is often suggested that these papers indicate how sediment disturbance by fishing gears results in slower-growing, longer-lived species (particularly molluscs) being replaced by shorter-lived polychaete species, an observation that is consistent with the generally held model of how disturbance might be expected to operate at the community level. However, while such inferences are tempting, careful reading of the source papers shows that although the changes seem to be real, the evidence for ascribing changes to physical disturbance of sediments by fishing gears in preference to any other explanation is rather weak. Of the non-fishing explanations for the observed changes, perhaps the most important was the change in habitat structure associated with the loss of seagrass beds during the 1930s. In fact, most of the seagrass beds in the North Atlantic were decimated at this time, suffering from what has been termed the seagrass 'wasting disease'. Although there are a number of theories to account for the triggers and the true cause of the seagrass decline, fishing does not count among them (Rasmussen, 1977). The only clear case for an effect of physical disturbance by fishing in the Wadden Sea, other than that directly associated with oyster bed removal, is the loss of reefs built by the polychaete *Sabellaria* which were removed by the activities of trawl fishermen. Other fishery-related changes may have arisen, however, from the large-scale seeding of mussel populations into sub-tidal habitats.

Evidence for similar shifts to those reported by Reise (1982) i.e. towards smaller-bodied polychaete species, are provided by Kröncke (1990) for the Dogger Bank in the North Sea, who compared the results of surveys conducted in the 1950s with those conducted in the 1980s. Fishing activity was common in the area during the 1950s, but effort had declined markedly in the region by the late 1980s. Paradoxically, however, changes in the macrofauna over this period were dominated by an increase in the abundances of small, opportunistic, polychaetes – the opposite of what one might expect with a decline in fishing effort. Neither data for the Wadden Sea nor the Dogger Bank provide strong evidence for a long-term fisheries effect in sediment communities (*Sabellaria* and oysters excepted); trends on the Dogger bank are counter-intuitive, and the Wadden Sea data can be explained in other ways.

Another source which is often quoted as providing evidence for the effects of trawling is Holme (1983) who very briefly described changes in the benthic community of the western English Channel, based on data collected at various times between 1899 and 1982. The frequency of citations for this paper is very puzzling and I urge anyone interested in this subject to actually read it. In the three pages of text only the final paragraph mentions fishing as a potentially important additional factor that 'may have an increasing effect' on the benthos. Indeed, the majority of the paper deals with the importance of hydrographic and other factors as determinants of change. For the issue at hand, this paper offers almost no information and it appears that chain quotation has operated, where authors have simply repeated a citation they have seen elsewhere. Most of us have probably 'quoted unseen' at one time or another, but Holmes' paper seems to have been singled out for such treatment.

More convincing analyses of long-term trends are based on one of the most comprehensive benthic time-series for the North Sea – a site off Northumberland, UK (Buchanan & Moore 1986). The Northumberland monitoring programme involves biannual sampling at two near-shore stations, one of which lies within the Northumberland Nephrops (otter) trawl fishing grounds. This fishery has shown considerable development since the 1970s, when the herring fishery in the area started to decline and effort switched to this inshore trawl fishery. There is, therefore, a time-series of benthic data over a period when fishing effort has increased from moderate to high levels – an almost uniquely valuable data set, that offers prospects for exploring relationships between trends in benthic time-series and trends in fishing effort. There are, however, a number of signals to disentangle, some of which appear to be stronger than fishing.

Figure 3.1 summarises some of these data that are available. During the first 10 years of observations there was a clear biennial cycle of species abundances (note the data for numbers of individuals fall alternately above and below the smoothed trend). Buchanan & Moore (1986) concluded that this result occurred through density-dependent regulation of winter survival. In 1981, however, this biennial cycle broke down, and macrofaunal abundances became much more variable. Interestingly, this change in dynamics correlated with a change in phytoplankton dynamics (Buchanan, 1993), with the strongest correlation between trends in the benthic community and phytoplankton occurring when a 2-year lag in the phytoplankton data were used (Frid *et al.*, 1996). The explanation for this correlation probably lies in the fact that phytoplankton standing stock gives a crude indication of organic flux to the benthos. Overall this correlation explained over 55% of the inter-annual variation, and for individual species abundances the relationship with phytoplankton two years previously often explained >80%. However, fishing effort also positively correlated with the phytoplankton index, so differentiating between the effects of the two is difficult.

The fact that *both* fishing activity and changes in phytoplankton correlated with the change in the benthos two years later illustrates the difficulty of examining time-series for a fisheries effect. While it is more difficult to explain why there should be a lag in the benthic response if fishing disturbance is causing the effect, formally distinguishing the relative contribution of each to the observed changes is probably

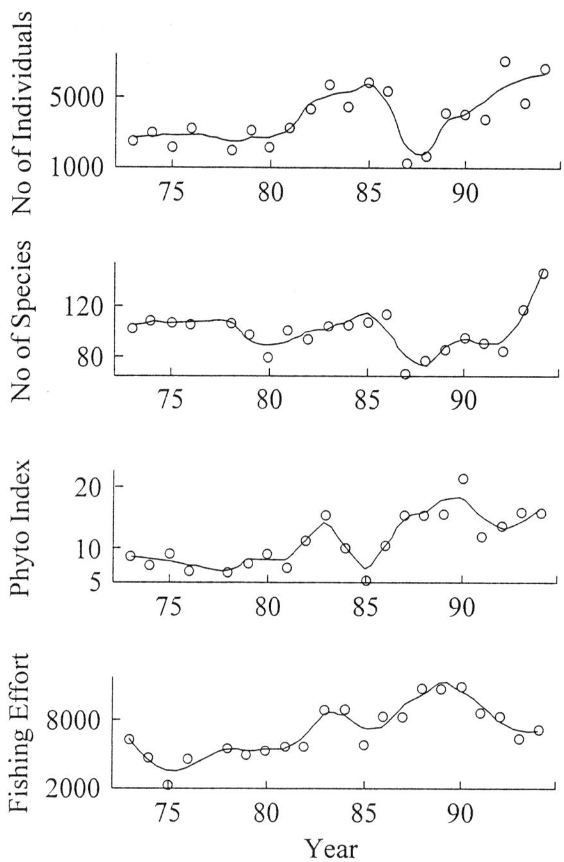

Fig. 3.1 Trends in numbers of species, and numbers of individuals from samples taken at a single site off Northumberland, UK, in March each year. Data are shown along with trends in phytoplankton standing stock (greenness index) and trends in fishing effort for the 30 mile × 30 mile statistical rectangle in which the benthic station is located.

impossible without additional data. Moreover, because the level of spatial resolution on the fishing effort data is low, one is relating the benthos at a single station sampled by a 0.1 m^2 grab to changes in effort over an entire 30 mile × 30 mile ICES statistical rectangle. This too makes relating the two time-series problematic.

Another time-series which gives us a perspective on changes in benthic populations comes from the Netherlands. In 1930, the Zoological Station at Den Helder (now the Netherlands Institute for Sea Research, Texel) started a scheme to collect information on North Sea species caught as by-catch by the demersal trawl fleet. Specimens were delivered by fishermen to the laboratory on request, or when a rare species was found, and were paid for the material. Using the records from this programme (which continued until 1990) and data on fishing effort by the Dutch fleet, Philippart (1997) has been able to piece together trends in some of the benthic species over the period 1945–1983. First, Philippart assumed that immigration and emigration of individuals was negligible for the area and the populations were in equilibrium, i.e. in any given year birth rate was equal to death rate. Then a model was constructed to predict the number of individuals in the population each year. The exact details of the approach need not concern us, but the model only required three parameters: the efficiency of the two gears used in the fishery (beam and otter trawl) and the initial population at the start of the time series (1945). With these three parameters and the fishing effort data, the population each year and expected by-catch could be calculated. Then, by adjusting the parameter values in the model to get the best fit between the predicted by-catch time-series and the actual by-catch observed, a best estimate of population trajectory was obtained.

Figure 3.2 shows the predicted population trends for larger epibenthic taxa. These graphs indicate quite clearly that many of the taxa have declined substantially on fishing grounds. Within 15 years after the Second World War, the otter trawl fishery appears to have caused marked declines in elasmobranchs, some fish, and larger-bodied invertebrate species. The introduction of beam trawlers in the 1960s appears also to have increased fisheries mortality to such a level that most of the remaining by-catch species fell to very low levels within the next 20 years.

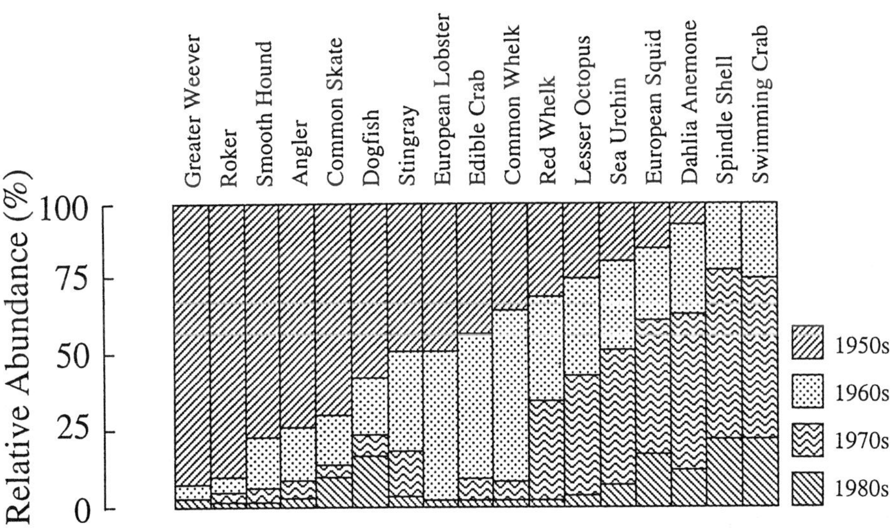

Fig. 3.2 Estimates of long-term trends in relative abundance of demersal fish and benthic invertebrates in the southeastern North Sea between 1947 and 1981. (Adapted from Philippart, 1997.)

Another study that has suggested fishing as a potential explanatory factor is that of Lindley *et al.* (1995) who noted that echinoderm larvae have dominated the North Sea plankton since the early 1980s. This period of domination coincides with an increase in beam trawling effort that started in the early 1970s and has continued to the present day. The authors argue that such a shift is consistent with a fisheries effect for the following reasons. First, ophiuroid and asteroid echinoderms are noted for their regenerative abilities and so they suffer less mortality during trawling than many other benthic taxa. Second, fishing may have reduced predation pressure on some species by reducing the abundance of fish predators, and third, it may have provided additional food for scavenging species (see Section 3.4). Certainly the trends are consistent with what one might expect of fishing, but it is difficult to judge this explanation against any other non-fishery related mechanism. For example, since we have also seen other shifts in phytoplankton communities at around the same time, one wonders whether hydrographic changes leading to increased survivorship of echinoderm larvae might have occurred.

Taken as a whole, there is evidence for long-term trends in benthic communities of the North Sea. However, while correlations between effects and potential causes will always be open to criticism and doubt, the case for invoking fisheries as a primary cause for the *recorded* changes is not very strong. There have simply been too many changes in other parts of the system that have occurred in parallel and which fishing would be unlikely to have affected (e.g. Aebischer *et al.*, 1990). This is certainly not to say that fisheries are causing no effect, simply that fishing impacts are confounded with other effects. As a primary explanation for the changes we have, data for a fishing related explanation is no more compelling than any other.

3.4 A boom for scavengers?

For towed gears two food sources are generated for benthic scavenging species – dead discarded material and exposed and damaged fauna. Studies on the former were dealt with in Chapter 2, but with respect to the exploitation of exposed and damaged fauna, there is surprisingly little published information. Yet the common practice of fishermen to re-trawl an area shortly after it was fished the first time tells its own story – for fish at least, scavenging in trawl tracks is a worthwhile pursuit (notwithstanding the risks of being caught second time around). One of the few studies to examine the magnitude of this response is that of Kaiser & Spencer (1994). These authors described the behaviour of scavenging species that move into an area following beam-trawling. The approach was simple, trawl an area and retain fish species (which presumably were not feeding in trawl tracks produced by someone else) for stomach analysis. Then after a couple of hours fish the same area and compare the stomach contents of the fish caught with those caught the first time. These studies showed that, for some species such as gurnards (*Aspitrigla cuculus* and *Eutrigla gurnardus*) and whiting (*Merlangius merlangus*) the total number of prey items in stomachs was significantly greater after trawling (Fig. 3.3). For other species, such as dogfish, while there was no increase in stomach contents, some prey species were only observed in stomachs after trawling had occurred, suggesting perhaps that they were previously unavailable. Similarly, body parts such as bivalve feet, that are rarely found in fish stomachs were also present in post-trawling samples, presumably because fish could feed on damaged prey.

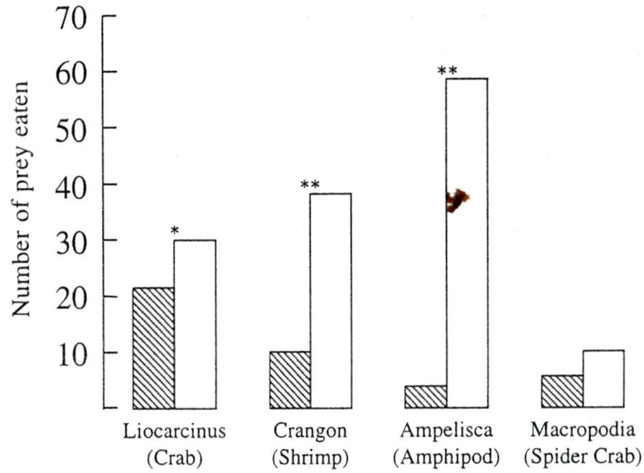

Fig. 3.3 The number of prey in the guts of Gurnards on fishing grounds before and after bottom trawling. Adapted from Fig. 2 of Kaiser & Spencer (1994). Statistically significant differences between before and after samples denoted by * $p < 0.05$, ** $p < 0.01$. ▨ = Before; □ = after.

One prey item that was especially prominent in the diets after trawling was a tube-building ampeliscid amphipod. Kaiser & Spencer (1994) hypothesise that the tubes are damaged by the passage of the trawl and individuals are, therefore, exposed. At an average homogeneous density of 140 ind. m^{-2}, at their study site, and assuming that all tubes were damaged, this implies that an hour of fishing with a beam trawl could expose 4.2×10^6 individual amphipods – a difficult figure to put into any kind of context, but interesting nevertheless.

It is impossible at this stage to assess the effects of food made available by trawl damage to benthos on populations of benthic feeding fish or invertebrates and some studies reveal that one does not always see a large scavenger response following trawl disturbance (e.g. Ramsay *et al.*, 1996). However, it is perhaps noteworthy that one group of species which seems to have increased in the North Sea is the gurnards (Greenstreet & Hall, 1996, see Section 7.2.3), which Kaiser & Spencer's study showed to be the most obvious beneficiary of the food made available. Another group that largely depends on benthic invertebrates for food is the flatfish and there are suggestions from both the Gulf of Thailand and the North Sea, that these too have increased in relative abundance in trawled areas (see Chapter 6).

3.5 Comparative and experimental studies

Perhaps the ideal approach to determining the effects of trawling on seabed communities is to combine observational studies that contrast similar habitats in fished and unfished areas with experiments to document community responses to repeated trawling impact. There are two impediments to this approach. First it has proved rather difficult to unequivocally identify locations that have been unfished so that comparisons can be made. Second, the spatial and temporal scales upon which experiments can be conducted are usually small relative to the scale at which fishing occurs. Despite these problems and the technical difficulties of conducting trawl impact experiments, there is a growing catalogue of studies that have adopted comparative and experimental approaches, some of which have shed valuable light on the question of trawling effects on benthic communities.

Comparative approaches

The idea of comparing fished and unfished areas to understand fishing impacts is an old one. Unfortunately, the simplicity of the idea belies the difficulty of implementing it. In a nutshell, it seems one can rarely be certain that an area is truly unfished. Moreover, even if one is certain, there is always the suspicion that it remains unexploited because it differs from productive fishing grounds in some way. On either count the comparison is confounded. An alternative, and at first sight more convincing, approach is to contrast areas within fishing grounds that have been protected from fishing in some way. Such might be present through regulations of some kind (e.g. fishery closures), but more often it is from structures, such as wrecks on the seabed, which are a hazard to fishermen and are, therefore, avoided.

One of the earliest to adopt such an approach was Graham (1955) who postulated that circular areas (0.75 mile radius) around light vessels would rarely be trawled and could be used as sites for comparison with adjacent fishing grounds. Graham failed to detect differences between protected control sites and adjacent fished areas, but this failure could have been due to the inferior sampling technology and design that could be adopted at the time. In 1987, I and co-workers re-employed Graham's basic idea was tried again but this time with much more sophisticated technology (Hall *et al.*, 1993). The dynamic positioning capabilities of the research vessel and acoustic positioning systems for the grab sampler allowed us to examine patterns in benthic community structure around a wreck with much greater resolution and accuracy (± 1 m) than had previously been possible. The wreck we chose was in 83 m of water about 80 km off the Scottish east coast in the heart of a demersal trawl ground and three transects running outwards from the wreck were sampled (Fig. 3.4a).

In essence, we were looking for spatial patterns in the benthic community that were consistent with a hypothesised trend of increased fishing disturbance with increasing distance from the wreck; but, like Graham before none was found. Anyone with even a passing interest in experimental design will recognise that there are many things wrong with this study – not least the fact that only one wreck could be studied owing to bad weather, and that the sampling scheme left the spatial pattern in the data unresolved (Fig. 3.4b). More important, however, is the fact that one could only assume that the sediments near the wreck were disturbed less often than areas further away. In fact, I know of no studies which have adopted the wreck

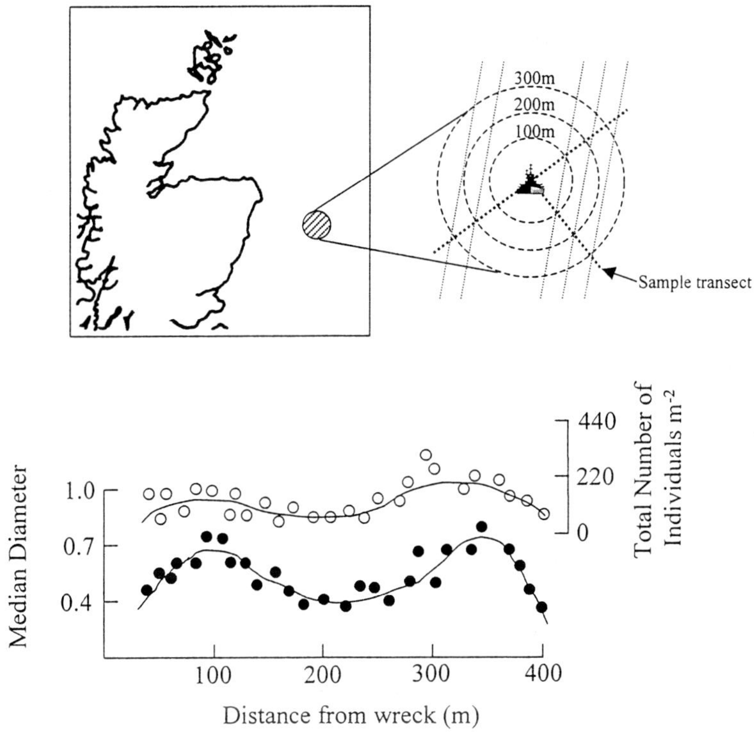

Fig. 3.4 Figure showing the location of the study site, the distribution of samples around the wreck, and the generalised trend in mean sediment diameter and the number of individuals along sample transects. Because transect samples radiated out from the wreck the study was unable to distinguish whether the observed bimodal pattern with increasing distance from the wreck was a result of patterns of concentric rings centred on the wreck or linear waves running through the study site. (Adapted from Hall *et al.*, 1993.)

approach (and they are still being done) where the distribution of fishing effort is actually measured. Moreover, even if we do see a change in the benthic community that is consistent with a hypothesised effort distribution, fishing activity is only one of the factors which could provide the explanation. One might expect, for example, confounding effects due the presence of the wreck itself.

Although a weak test by any standards, it is worth pointing out that our study did reveal some very interesting spatial patterns in the benthic community, none of which could be attributed to any plausible pattern of effort distribution (Fig. 3.4c). Indeed, one would, perhaps, have expected such patterns to be disrupted if fishing had been a major disturbance process in the sampled area. However, such conclusions are tentative at best, and given the doubts about how to interpret any data one obtains, there seems little to commend this kind of approach.

In contrast to most European fishing grounds, many of the areas fished by trawls and dredges on the continental shelf off the east coast of North America include substantial areas of gravel, and one might expect the fauna in these habitats to be more vulnerable to trawling impacts. Such habitats tend to support richer benthic communities than sandy areas, with the larger epibenthic organisms associated with a biogenic cover of attached epifauna, including colonial worm tubes, bushy hydroids, bryozoa etc., see Thouzeau *et al.* (1991). Thus, the baseline from which effects might be judged is higher. In particular, the encrusting epifaunal community associated with the hard substrate is likely to be damaged as rocks abrade one another when they are disturbed by the passing gear, or when the hard parts of the gear contact them. Importantly, these attached epifauna provide food and shelter for many animals, including the prey of commercially important fish species, so there is the potential for indirect effects on other groups.

This gravel pavement habitat has been examined in a recent study on Georges Bank by Collie *et al.* (1997) who surveyed five (5 × 10 km) sites by side-scan sonar to determine the densities of trawl tracks at each site. Fortunately (or unfortunately, depending on your perspective) trawl tracks in this kind of habitat are likely to persist for a considerable time, so analysis of side-scan pictures is likely to give a good index of trawling intensity. Sites were clustered at depths of 40 and 80 m and at each depth level sites were chosen with contrasting levels of trawling. Video transects and still photographs were taken at each site. On the basis of the side-scan records three sites had either none or a few trawl tracks and were classified as unfished, while two sites with high densities of tracks were classified as fished. Taking depth into account, this gave two 'unfished' and one fished site in deep water (approx. 80 m) and one of each type in shallow water (approx. 40 m). Within each site between two and six stations were sampled, with still photographs chosen at random from each station. No one would argue that this is an ideal survey design, but as a contrast between areas of differing trawling intensity, it is better than most that have been possible to date.

Differences between sites were readily apparent in photographs, with the deep-water unfished site in particular being characterised by a high percentage cover of a colonial fan worm. This subjective impression was confirmed by statistical analysis which showed that there was significantly greater cover of fan worms and plant-like animals (e.g. erect hydroids) at unfished sites and that there was also a depth effect for fan worms, with more present in deep water (Fig. 3.5). Another important result was that there was an interaction between fishing and water depth indicating that fishing appeared to have greater impact in deep water. Indeed, for plant-like animals the effect of fishing was actually reversed in shallower water, with slightly greater densities at fished sites. Non-colonial taxa were also affected by fishing, with most of the responses being negative. Non-burrowing anemones, shrimps, sponges, nudibranchs and tube worms were all less frequent at the fished sites. These organisms are all relatively fragile and likely to be killed by a trawl. In contrast, scavenger species such as hermit crabs were more abundant at fished locations.

Taken as a whole Collie *et al.*'s results are, in their own words, 'consistent with the

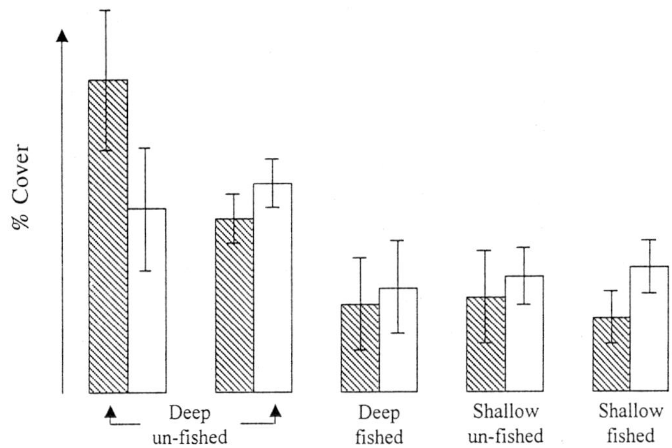

Fig. 3.5 Percentage cover for erect sessile fauna at deep and shallow water sites classified according to fishing intensity. (Adapted from Collie *et al.*, 1997.) ▨ = Colonial fan worms; □ = plant-like animals.

hypothesis that gravel habitats are very sensitive to physical disturbance by bottom fishing and that the primary impact is the removal of emergent epifaunal taxa' – one can only agree. Although other factors such as water depth and tidal currents are certainly important, fishing activity can and almost certainly has substantially changed the nature of the communities present on gravel fishing grounds.

Experimental approaches for trawl fisheries

Notwithstanding the logistical difficulties of doing them, and with certain caveats about the scale at which they can be conducted, experiments which manipulate the intensity of fishing disturbance, and which contrast changes in benthic communities over time, are usually the least equivocal means for determining effects. There is now a reasonable catalogue of such experiments for trawl fisheries and for this section I review a number of case studies and discuss their findings and limitations. This is certainly not a complete review of all that has been done and other work is in progress, but it is representative of the current position.

Northern Europe

The opportunity to conduct experiments at locations that one knows have been protected from fishing for extended periods is rare indeed. Fortunately, however, some opportunities still exist. One such fell to myself and co-workers who were able to study Gareloch, a sea loch on the west coast of Scotland, which has been closed to fishing owing to the presence of a naval submarine base. Anecdotal evidence suggests that although good catches have been taken from the loch in the past, very little trawling took place prior to restrictions on fishing that were imposed in 1967.

Thus, it is almost certain that the area has remained unfished for 30 years. The continued absence of fishing in the loch offered a unique opportunity to experimentally examine the long-term impact and recovery of the benthic community from repeated fishing disturbance. Another important distinguishing feature of this study is that the sediments of the area were poorly sorted fine silts. This is in contrast to the majority of studies on fishing impacts which have been conducted on sands or gravels.

In studies of environmental impact the type of design that is most often advocated is a before/after, control/impact, or BACI design (Bernstein & Zalinski, 1983; Stewart-Oaten *et al.*, 1986), which usually involves replicated sampling over time of a putatively affected area and one or more reference areas. Having multiple reference sites is greatly preferred because one limits the possibility that any observed differences between control and impacted areas are simply the result of chance differences between sites (Underwood, 1992). However, as with many large-scale studies, compromises in experimental design are necessary and in this case multiple reference sites were not available. This does not invalidate the results of the study, but it does mean one has to be circumspect when trying to draw general conclusions about the likely responses at other sites.

The first part of the study involved a preliminary survey to select a treatment and reference area for the experiment. Underwater video, side-scan sonar, acoustic mapping of sediments, sediment grain-size analysis from core samples and an infaunal survey based on grab samples were all undertaken at this time. Then, for the area selected as a treatment site, experimental trawling was undertaken from a locally chartered fishing vessel. For one day each month 10 tows of approximately 45 mins duration were made over the treatment area, using a standard trawl gear, but with no trawl attached. It may, at first sight, appear rather odd to remove the net from the fishing gear to undertake such experiments, but removing the net allowed us to achieve the high levels of disturbance we required in the time available. The study site was also one in which continuous sustained trawling could have markedly reduced the densities of larger predators and scavengers. Not using a net ensured that only the effects of physical disturbance were studied and that the densities of species which feed on damaged or exposed prey were not depleted. The trawling treatment was repeated each month for 18 months with a full faunal survey of both treatment and reference sites every 6 months. The two sites were then followed over a further 18-month period to examine the recovery process.

The main physical effects of the otter trawling were the trenches left in the sediment by the trawl doors. This differs from beam trawls and scallop dredges, which flatten seabed features. About 18 months were required following cessation of fishing before the physical characteristics of the site returned to their pre-dredged state, but such features are unlikely to be as large or long-lasting in less sheltered areas with coarser sediments (see, for example, Currie & Parry, 1996). The experimental disturbance did not appear to have any effect on sediment particle size or organic carbon levels.

Trawling disturbance led to an increase in the number of species and individuals relative to the reference area while reducing diversity and evenness (Fig. 3.6). The fact that the measures of diversity decreased while the number of individuals and species increased suggests that there was a disproportionate increase in the abundance of a few dominant species in the treatment area – a response that one would have predicted for a disturbed area. The time-scales over which the different effects became obvious differed, suggesting varying levels of sensitivity to trawling disturbance. Measures of diversity which use the relative abundances of species in samples to derive an index were the most sensitive to the initial changes in the community. Each index revealed significant differences between the sites after only 5 months of disturbance, but indicated that the sites were no longer different after 12

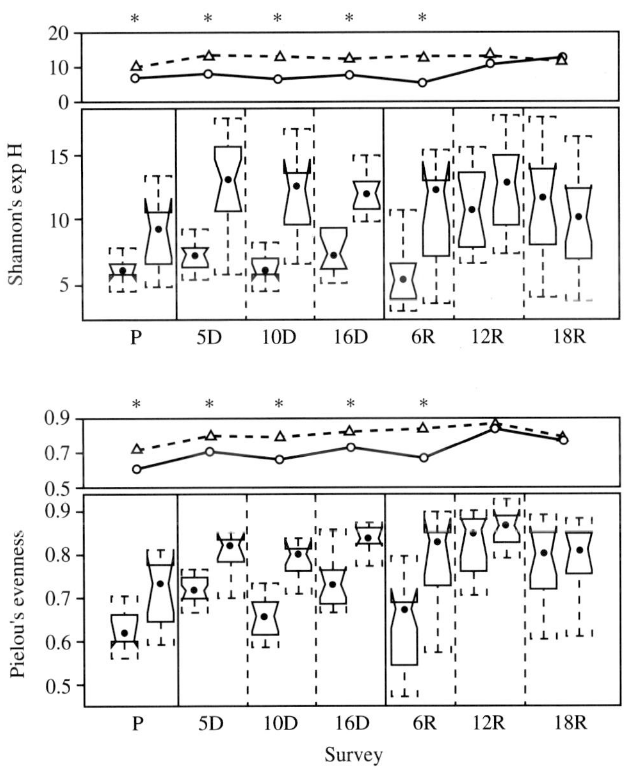

Fig. 3.6 Box plots of diversity and evenness (lower panels), along with the time-series for the median values for each survey (upper panels). Box plots are arranged in pairs in time (survey) order, with the reference plot on the right for each pair. Surveys in which medians of two sites were significantly different are marked by an asterisk. Surveys labelled as follows: P – preliminary survey; 5D – 5 months disturbance; 10D – 10 months disturbance; 16D – 16 months disturbance; 6R – 6 months recovery; 12R – 12 months recovery; 18R – 18 months recovery. The notches in the boxes indicate 95% confidence intervals of the median. (Adapted from Fig. 4 of Tuck *et al.*, 1998.) (Note: Although diversity and evenness were significantly different between sites in the preliminary survey it is the statistical interaction between trawled and control sites over time on which conclusions must be based.)

months recovery. In contrast, species richness (simply the total number of species in samples) only showed a significant difference between sites after the 16-month disturbance period, remaining different throughout the monitored recovery period.

Analysis of the individual species abundances provided information on which species may be considered indicators of disturbance (those that increase in abundance) and which are sensitive to disturbance (those that decrease in abundance). The main species that showed a consistent increase in abundance in association with disturbance belonged to the cirratulid and capitellid worm families, and are generally considered to be opportunistic in nature. These are small-bodied species, which are able to reproduce rapidly to increase population size when resources become available (for example, if environmental conditions change and species less able to survive are killed). The time-scales over which the species showed significant differences varied, with some remaining more abundant in the treatment area beyond the end of the experiment, while others showed no differences in density between sites after 18 months recovery. Interestingly, spionid polychaetes increased in density immediately after the disturbance finished, and remained more abundant in the treatment area for up to 18 months after trawling ceased. This result suggests that although some species are unable to take advantage while trawling continues (possibly due to a sensitivity to physical disturbance), they are opportunistic and can rapidly increase in numbers when trawling ceases.

Species which declined in abundance were less obvious than those which increased, but the data suggest that a number of species could be identified as being sensitive to physical disturbance in this habitat. The bivalves *Nucula nitidosa* and *Corbula gibba* both declined in abundance relative to the reference area following disturbance as did a few polychaete species. One common feature for many of the species that declined is that they have previously been reported to be sensitive to high rates of sedimentation, something that may well be an issue in the fine sediment habitat of Loch Gareloch. From a broader perspective, multivariate analysis of the infaunal community data also showed that the two sites became significantly different after only five months disturbance, and remained so throughout the experiment.

For epifaunal species, no long-term effects on the total number of species or individuals were detected, but some individual species did show responses. In particular there was an increase in the density of the brittlestar *Ophiura* sp. and a decrease in the density of the plumose anemone *Metridium senile* and the dog whelk *Buccinum undatum*. As noted earlier, with respect to ophiuroids, it has recently been suggested that fisheries may have contributed to an increase in echinoderm populations in the North Sea (Lindley *et al.*, 1995). However, while the observed increase in the density of *Ophiura* sp. in the treatment area is consistent with this hypothesis, one would need to know more about the population biology, immigration and emigration of *Ophiura* sp. before firm conclusions could be drawn. Anenomes were probably damaged or killed by physical impact with the fishing gear, but a scavenging species such as *Buccinum undatum* might be expected to benefit from fishing disturbance, through increased food availability. Recent studies in the North Sea, however, suggest that trawls also kill this species and from our results it would appear that the benefits of increased food availability may be outweighed by higher mortality rates. One might also have expected marked effects on the seapen *Virgularia mirabilis* another common epifaunal species. Although *Virgularia* stands erect above the sediment surface, it was not affected by the disturbance, perhaps because it is able to withdraw into the mud very rapidly, and may, therefore, avoid being damaged as a trawl passes.

In summary, we found that effects on epifauna were evident during the period of fishing, but recovery was relatively rapid. For the most part, densities were indistinguishable from the reference area six months after fishing ceased. It should be borne in mind, however, that the epifaunal species assemblage in Loch Gareloch was relatively poor in the first place, with few of the erect sessile species such as sponges or corals that are likely to be especially vulnerable. Presumably these fragile taxa have never been present at this site. As noted earlier, the experimental design (i.e. a single treatment and reference area) has been criticised as being suitable only to demonstrate differences between locations (Hurlbert, 1984). Thus no formal statements about the effects of fishing outside Loch Gareloch can be made. Nevertheless, by sampling at repeated points through time, during a period of impact and recovery, the conclusions drawn for this site are more likely to be of wider relevance. Experimental trawling disturbance had clear long-term effects on the topography of the seabed and the infaunal community at this site. While physical effects were almost indistinguishable after 18 months recovery, community effects (measured by both univariate and multivariate techniques) extended beyond the life of the experiment. The reader must decide how generally relevant these results are likely to be, but I suspect that similar effects would probably be observed at other sheltered sites with fine silt sediments. Such long recovery times also suggest that even fishing during a restricted period of the year may be sufficient to maintain a community in an altered state, assuming that the intensity of disturbance during such a period was comparable to the monthly fishing disturbance imposed at Gareloch.

Australasia

The Great Barrier Reef: In tropical environments, perhaps one of the areas of greatest conservation interest is the Australian Great Barrier Reef (GBR). Here, the most important management issue with regard to seabed habitats is probably the effect of prawn trawling. The area to the landward side of the reef and the inter-reefal areas on the offshore side of the GBR support a valuable otter trawl fishery, with about 900 boats landing approximately 7000 tonnes of prawns and 1000 tonnes of scallops per annum. The effects of these activities have been documented most closely by a recent 5-year study conducted both in and adjacent to an area closed to fishing which extended across the shelf in the far northern sector of the reef (Pitcher *et al.*, 1996). The first part of this study adopted a comparative approach with a description of the seabed and associated communities in the area and sampling to compare areas open and closed to fishing. Unfortunately, as with most comparative studies, there were considerable doubts about the validity of the approach from the outset. First, the inshore part of the closed area was trawled before the closure was established. Second, because trawling activity is highly aggregated in the area so much of the open area may have remained untrawled. When few differences between closed and open areas were found these qualifying observations seemed particularly important.

In view of these inconclusive results an experiment was initiated. This experiment used the kind of before/after; control/impact (BACI) design described earlier and involved 24 plots of seabed. These plots were identified and surveyed before the experiment, half of them were then trawled and all were subsequently resurveyed both immediately afterward and six months later. The twelve 2.7×1.2 km treatment plots were trawled entirely on a single occasion. Importantly, plots were pre-selected to ensure that they included sessile epibenthic fauna that acted as structural components so that the experiment resembled as closely as possible trawling to establish new prawn grounds. Sponges are major components of this structural epibenthos, and are also of particular interest, owing to their probable role as providers of secondary habitat for other organisms. Interestingly, despite the scale of the experiment and the fact that trawling removed between 1 and 7 tonnes of benthic organisms from each treatment plot, very few statistically significant effects were observed! This result was somewhat surprising, but it suggests that a single pass of a prawn trawl should probably not be cause for concern. Of course, the question then arises of how realistic a single trawling episode is.

The logical next step, was to examine repeated trawling events at the same location and this was achieved with a trawl depletion experiment. This study showed that each pass of a trawl removed between 5 and 20% of the initial biomass of sessile benthos and that 13 trawls removed 70–90% from a plot. The conclusion from the study is clear – the impacts of prawn trawling on the Great Barrier Reef may be undetectable in areas that are sparsely trawled, but the cumulative effect of trawling in the same area may be substantial.

The Northwest Shelf: One analysis, which demonstrates the importance of habitat structure provided by benthic organisms, is that of Sainsbury (1988) who studied the fishery off the northwest shelf of Australia. This region is fished mostly by foreign fleets and it was first exploited by the Japanese trawl fishery (1959–1963) who fished mainly for emperors (*Lethrinidae*). From 1972 a Taiwanese trawl fishery developed, taking a range of species mainly breams (*Nemipterids*, 21%), lizardfish (*Sauridae*, 12%), snappers (*Lutjanidae*, 9%) and *Lethrinidae* (8%). A domestic Australian trap fishery began in 1984 targeting *Lethrinidae*, *Lutjanids* and *Epinephelus*. Research survey data available from 1960 onwards has shown that, while the total biomass of fish has not changed as these fisheries have developed, the composition of the fish community has altered, with *Lethrinids* and *Lutjanids* declining and *Saurids* and *Nemipterids* increasing (Fig. 3.7).

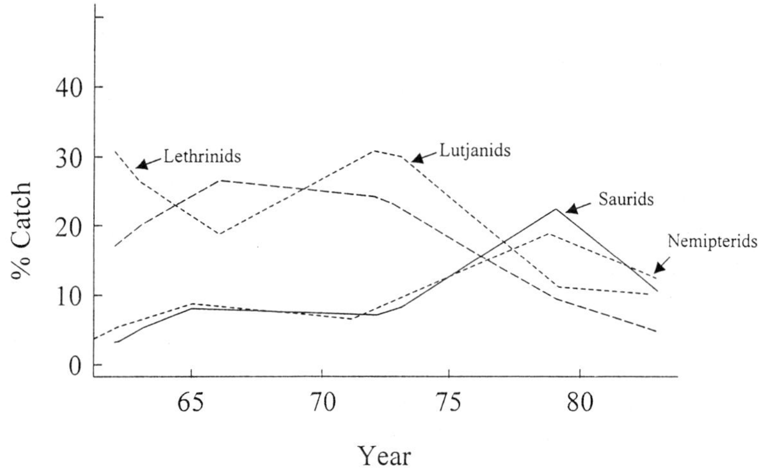

Fig. 3.7 Trends in abundance of the four major exploited fish taxa on the Australian northwest shelf. (Adapted from Fig. 14.2 in Sainsbury, 1988.)

The available data also indicate that the benthic environment has altered over the same period. In particular, the quantity of epibenthic fauna caught in trawls (mainly sponges alcyonarians and gorgonians) is now considerably lower than it was prior to the development of the trawl fishery (Sainsbury, 1987). Using video data from cameras attached to a survey trawl's headline, Sainsbury (1988) identified four habitat types on the basis of dominant benthic fauna. Fish catch data was then examined in relation to habitat type, and the results indicated that *Lethrinids* and *Lutjanids* were almost exclusively associated with habitats supporting large epibenthos. In contrast, the *Saurids* and *Nemipterids* were only found on open sand.

This situation gave rise to an important management issue for the region. First, there was the question of whether the change in community composition could be reversed. Then, if changes were reversible, was it worth attempting to do so given the uncertainties of the outcome and the time frame over which the change would occur? From a pure fisheries management perspective the monetary returns from increased value of the catch also needed to be included. Finally, if an attempt was to be made, what management measures were most appropriate to achieve the goal?

The key to resolving these issues lies with understanding the mechanisms responsible for the changes in the first place; four alternative hypotheses were formulated:

(1) **Intra-specific dynamics:** the observed changes result from independent responses of each species.
(2) **Competitive release due to fishing:** there is a negative influence of *Lethrinus* and *Lutjanus* on the population growth rate of *Saurida* and *Nemipterus* so that when the *Lethrinids* and *Lutjanids* were removed by fishing the latter experienced a release from competition and increased in abundance.
(3) **Competitive depression:** *Saurida* and *Nemipterus* have a negative influence on the population growth rate of *Lethrinids* and *Lutjanids* and the abundance of these species declined because the former increased for reasons independent of the fishery.
(4) **Habitat modification:** habitat characteristics determine the carrying capacity of each genus separately so that trawl-induced modification of the abundance of the habitat type alters the carrying capacity of the different genera.

All four hypotheses are ecologically reasonable and were consistent with the available data. It is important to recognise, however, that each has markedly different management implications. Hypotheses (1) and (2) imply a relatively low productivity of *Lethrinids* and *Lutjanids* with reductions in the biomass of these taxa being viewed as a consequence of fishing. Accordingly, even if stocks could be rebuilt, the sustainable yield from the fishery would need to be low to prevent the same decline happening again. In contrast, hypotheses (3) and (4) imply a relatively high productivity for *Lethrinids* and *Lutjanids* under some circumstances. Selective harvesting of these taxa under hypothesis (3) and harvesting without damage to benthic habitat structure under hypothesis (4) would result in high sustainable catches. These differing implications make determining which mechanism operates much more than an academic exercise.

While specifying hypotheses is a relatively simple matter, distinguishing between them using available data certainly is not. What is especially commendable about this study, however, is that the authors adopted a formal evaluation procedure by stating the above hypotheses as explicit mathematical models. Establishing such models is very worthwhile because they permit one to formally evaluate which hypothesis is most likely to be true given the available data. This evaluation could be partially achieved from the existing data using a branch of statistics known as Bayesian statistics. Bayesian methods seem especially suited to the kinds of problems we are dealing with here and more details are given in Chapter 9. Here, it is sufficient to say that initially all four hypotheses were considered to have equal probability and that these probabilities were then updated using the research data. This analysis suggested that there was a relatively low expected present value from continuation of the existing licensed trawl fishery, and the additional information that could be gained from monitoring the outcome of continued trawling would not help make future decisions about what to do for the best. In fact, even though the probabilities that could be assigned to the various models at the time were relatively low, there appeared to be clear benefits from an immediate switch to a domestic trap fishery. However, it was also shown that some experimental management regimes, involving cessation of trawling in some areas and the introduction of trap fishing in some of the areas closed to trawling could provide a higher expected present return from the resource.

It is encouraging that, partly on the basis of this work, the management agencies for the northwest Australian shelf agreed to conduct such an experiment by subdividing the area into three zones. One part of the area was left open to trawlers, a second part was closed to trawlers in 1985 and a third was closed in 1987. Trap fishing was permitted throughout. It was hoped that closing part of the area to trawls would allow this fishery to develop to exploit species which are found in less disturbed habitats.

Unfortunately, the resulting contrast in trawl fishing effort were not exactly as planned owing in part to the unexpected development of a domestic trawl fishery for which no legal instruments to prevent access to one of the closed areas were in place. Changes in the management agency responsible for the area and other factors also played a part in upsetting the original design. Despite these difficulties, however, an adapted experiment is still continuing and sufficient data was collected from two of the areas to allow the four hypotheses to be evaluated. The data from these two areas provided a contrast between a closed area and one open to trawling between 1985 and 1990. Figure 3.8 shows how the area closed to fishing experienced an increase in the density of *Lethrinus* and *Lutjanus* and in the abundance of small benthos. The abundance of larger epibenthos stayed the same or perhaps increased slightly. In the area open to trawling, the abundance of fish declined along with the small and large epibenthos.

These results in themselves provide a valuable perspective on fishery effects, but it is in the further formal evaluation of the above four mechanistic hypotheses that the real strength of this study lies. This is because the data from the experimental period allowed the probabilities assigned to each of the four hypotheses to be further updated. Once this update was done, the probability placed on the habitat modification mechanism increased substantially. The value for this hypothesis rose to about twice that of the closest contender. This result indicates that a high value *Lethrinus* and *Lutjanus* fishery could be established on the northwest shelf if the habitat could be protected and that changes in fish community structure can probably be attributed in large part to habitat modification by trawling.

The northwest shelf is a good example of where an interaction between fisheries and the structure of benthic communities may lead to both an enhanced fishery and a less disturbed benthic community. Such mechanisms may not happen everywhere, indeed, the habitats in which they operate might be quite restricted, but we should be alert to the possibility. One can but hope that rational analysis of other systems might lead to similar conclusions. Unfortunately, however, in the case of the northwest shelf, it is apparent that the time-scales for recovery for epifaunal benthos are slower than previously thought. Rather than taking 6–10 years for sponges to grow to 25 cm it now appears that at least 15 years are required. Moreover, video

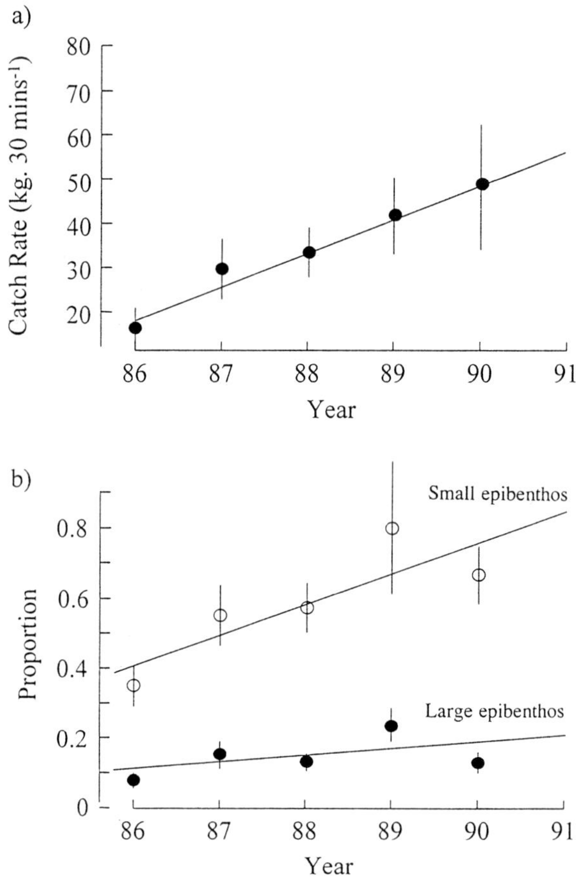

Fig. 3.8 (a) Changes in the abundance of fish (*Lethrinus* and *Lutjanus*). (b) Changes in the proportion of large and small epibenthos in areas closed and open to fishing. (Adapted from Fig. 2 of Sainsbury *et al.* (1997) In *Global Trends: Fisheries Management* (K. Pikitch *et al.*, eds), pp. 107–12. American Fisheries Society, Maryland.)

analysis of the effects of the trawl ground rope indicate that about 89% of encounters lead to dislodgement of sponges and almost certainly subsequent death. This slow recovery dynamic and the apparently high probability that large benthos will be removed by a trawl mean that measures to protect the habitat would need to be very effective to maintain the habitat structure required to support this high value fishery. In many ways, the formal evaluation procedure adopted by Sainsbury and co-workers provides a model for the way research on fisheries effects issues should proceed.

Experimental approaches for shellfish fisheries

Fisheries for shellfish are usually more localised than traditional towed gears and they are often restricted to shallower coastal environments. Dredges to catch shellfish fall into two basic classes: hydraulic dredgers which fluidise the sand with water jets located immediately in front of the dredge, and conventional dredges, which are towed in the normal way. In the case of hydraulic systems, fluidising the sand either makes it easy for the gear to penetrate deeply enough to catch the shellfish or the fluid power is used to suck sediment and prey on board the vessel. Conventional dredges usually have a toothed bar which digs the sediment and a steel mesh or ring bag to retain the catch.

Of the shellfish gears that have been studied, the scallop dredge has received most attention and in recent years the quality and comprehensiveness of the studies that have been undertaken has increased considerably. In particular, a study by Currie & Parry (1996) which examined changes in benthic infauna caused by scallop dredging at a site in Port Phillip Bay, southeastern Australia, is worth describing in some detail. The Port Phillip Bay scallop fishery started in 1963 and after a rapid expansion of fishing effort collapsed in 1968. Today fishing effort is much lower than in the mid 1960s , but 84 vessels are still licensed to dredge. Zoning regulations put in place since 1985 have prevented fishing in waters less than 10 m deep, thereby concentrating effort into areas 10–20 m deep. Commercial dredgers in the area typically work closely together in groups of 5–50 vessels and within each group dredges are dragged in the same direction for 500–2000 m before being emptied and dragged in the opposite direction. Using data on fishing effort for various sectors of the bay, Currie & Parry (1996) were able to estimate the proportion of the area within each sector that had been dredged 1, 2, . . . , n times each year from 1980–1990. Using this analysis and the information on trawling practices a study was designed to closely duplicate normal fishing activity.

The basic design of the experiment was as follows. Two adjacent 600 × 600 m experimental plots were located in 12–15 m of water. Then, at 3 months, 2 months and 2 weeks prior to any experimental intervention, each plot was sampled with replicate grabs to determine infaunal community composition. A fleet of six commercial vessels then experimentally dredged one plot over a period of three days, while the other was left undredged. The dredging regime ensured that the majority of the plot was covered twice. This intensity corresponded to a moderately high fishing activity, which at its maximum in the 1980–1990 period was experienced by about 127 km^2 (3%) of the 1920 km^2 area of Port Phillip Bay. Before the dredging commenced, a series of colour-coded rings were inserted into the experimental plot at different depths. Penetration depth of the gear was then determined from the colours of the rings retained in the net. These data showed that the dredges penetrated to about 20 mm, although depths of up to 60 mm were recorded. Following the dredge treatment, each plot was sampled on a further 6 occasions (immediately post-dredging, 3 weeks, and 3.5, 5, 8 and 14 months afterwards). In addition to the composition of the infauna, estimates were made of the abundance and sizes of sediment mounds built by burrowing shrimps and of the nature of sediment bedforms. During the dredging operations the fauna that were dislodged were also assessed by fitting a 1 mm mesh net to the upper rear section of the dredge.

With respect to the physical characteristics of the seabed, there were some clear visible changes following dredging which persisted for about a year (Table 3.4). For the most part, these changes were associated with the destruction of mounds and depressions caused by the burrowing activities of Callianasid shrimps. In turn, this flattening led to the removal of unattached weed and seagrass which tends to accumulate around such features. The spatial heterogeneity that mounds and patches of organic matter provide is often considered to be an important factor controlling the diversity and species composition of benthic infaunal communities (e.g. Hall *et al.*, 1994), so their destruction would be expected to have effects. Importantly, however, dredging activity did not lower the densities of Callianasids so burrow reformation could occur without the need for the recruitment and growth of juvenile shrimps.

Table 3.4 Changes in the topography of a site following scallop dredging and the time course of recovery. (Adapted from Currie & Parry, 1996.)

Time	Features
Pre-dredging	• Bedforms dominated by low relief mounds and depressions caused by burrowing Callianasid shrimps. Average density of features 1.2 m^{-2}. • Adjacent pits and depressions often trapped detached seagrass and algae.
8 days after dredging	• Most of the seafloor within the dredge plot very flat, mounds and depressions filled by the 'grader like' action of the dredge. • Dredge tracks apparent.
1 month after dredging	• Seafloor still flat. • Dredge tracks still distinguishable.
6 months after dredging	• Mounds and depressions now present over most of the area and similar to the pre-dredge condition. • Dredge tracks no longer visible. • Detached weed and seagrass present. • Some flattened areas remaining.
11 months after dredging	• No distinguishable differences between dredge and control plots.

Figure 3.9 summarises the basic findings for effects on the infaunal community. Before dredging, the number of species found on control and dredge plots was very similar. Following dredging, however, there was a statistically significant decline in the number of species on the dredge plot that persisted for 14 months. Although there was a strong seasonal signal in the trends in total numbers of individuals in plots, due largely to the recruitment of juveniles of a few species, no significant effects of dredging were detected. The third panel in the graph shows changes in the dissimilarity of the community between plots. Dissimilarity measures take into account the abundance of all the species found to come up with a composite

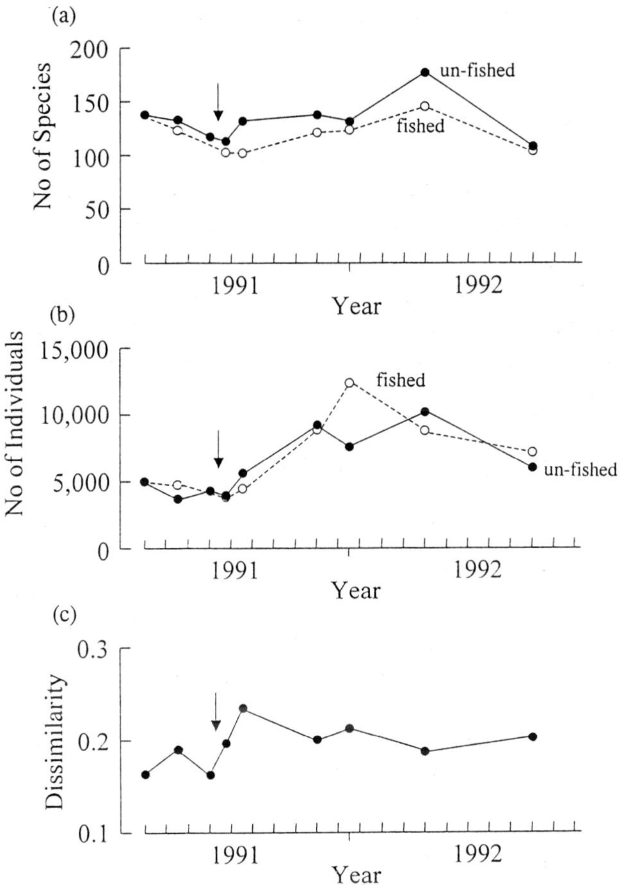

Fig. 3.9 (a) Number of species in fished and un-fished plots before and after experimental fishing. (b) Number of individuals in fished and un-fished plots before and after experimental fishing. (c) Dissimilarity between fished and un-fished plots. (Adapted from Fig. 4 of Currie & Parry, 1996.)

univariate measure of how closely the communities in two samples resemble one another. An increase in this statistic means that the communities on the two plots were less similar to one another and it is clear that the two plots diverged after the dredging treatment was applied. The statistic also remained higher than the pre-dredge levels for the rest of the experiment. This way of looking at the data appeared to be the most sensitive measure of dredging effect, indicating that effects can persist for up to 14 months after dredging. It is perhaps notable that this period of persistence is similar to that we found in Loch Gareloch (see page 67). Further analysis indicated that observed effects were largely due to a reduction in the number of rarer species that were common to both dredge and control sites following dredging. Other studies of scallop dredging have also shown persistent effects. For example, Thrush *et al.* (1996) showed that even in sites characterised by small-bodied short-lived species – the kind that one might expect in heavily fished areas, reductions in animal densities due to experimental fishing persisted for at least three months.

Currie & Parry (1996) also examined the effects of dredging on particular species, but of more general interest is their analysis of effects on broader taxonomic categories or functional types. At the coarsest taxonomic level (annelids, crustacea, molluscs, nemerteans) all groups experienced declines with dredging, but at the level of taxonomic class within each group there was no consistent effect. For functional groups, it appeared that deposit feeders were more susceptible to dredging than predators and scavengers, and suspension feeders became more abundant after dredging. However, these conclusions were strongly influenced by the responses of a single numerically dominant species in both categories. If these two species were excluded from the analysis, no clear pattern with feeding type was apparent. One might expect that species living near the sediment surface, would also be most susceptible to dredging effects. Interestingly, however, with the exception of a numerically dominant amphipod which builds tubes at the surface and experienced marked declines, the trend was the opposite – surface-dwelling animals appeared to increase in abundance following dredging, compared to those living deeper.

What can we conclude from Currie and Parry's study? First, dredging has clearly detectable physical effects, transforming an area characterised by sediment mounds created by shrimp burrowing activity into a flat plain. Since the burrows themselves have marked effects on the infaunal sediment community, their removal is also likely to have consequences. Importantly, however, dredging did not kill the shrimps so mounds were re-built – albeit over an extended period of 6–11 months. Other processes such as cyclone events can also flatten mounds yet leave the shrimps unharmed (Riddle, 1988). Second, effects *do persist*; in Currie and Parry's case for at least 14 months, a comparable period to other studies conducted with a similar design. Third, almost no general patterns emerged with respect to the susceptibility of fauna as a function of living depth or body plan. The authors suggest that this is because there are a variety of mechanisms that can lead to the death of an individual (e.g. direct damage by the dredge, burial by sediment, exposure to predators) and we do not yet understand these sufficiently well.

Although most shellfish fisheries are in relatively shallow waters, one example of a deep-water fishery is the dredging for Ocean quahog (*Arctica islandica*) on the continental shelf off the northeast USA. This fishery has only operated since 1974 and only a small proportion of the area over which the species is distributed has been fished. (*Arctica* is distributed over much of the shelf, but they are most abundant in the depth range from 37–60 m). In contrast, the other major fishery in this area is for surf clams (*Spisula solidissima*) which are fished from near shore down to approximately 37 m (MacKenzie, 1982). MacKenzie (1982) tried to determine the effects of fishing for *Arctica islandica* by sampling three sites on the northeast continental shelf of the USA. One of these sites had been fished for approximately one year, but had been abandoned for approximately three months before sampling. A second site had been fished for two years and was actively being fished at the time of sampling. A third site was unfished. Analysis of the invertebrate community from samples taken at these sites failed to reveal any differences between them and the authors concluded that this result may in part reflect the limited efficiency of dredging in deep water. However, the power of their study to detect effects may also have been low. MacKenzie (1982) concludes that these undisturbed sections can still represent a substantial portion of a fishing ground when the economic return from an area becomes marginal. The validity of this statement or the generality of it for other dredge fisheries has yet to be determined.

3.6 Disturbance effects in context

Fishing gear is only one of the many agents responsible for physical disturbance in the marine environment and it is informative to try and put its effects in context and to evaluate how various disturbance processes interact. Other forms of natural disturbance range in scale from a few centimetres in the case of bioturbation by infauna, through metre scales for animals such as whales feeding on the seabed, up to areas in excess of hectares when storms or icebergs scour the seabed (see Hall *et al.*, 1994 for review). All of these scales of disturbance might affect the regional scale composition of the community. It has been argued, for example, that a high frequency of small-scale disturbance (e.g. by predator feeding) can lead to a mosaic of patches on the seabed at different stages of recolonisation. Habitats with such a mosaic might therefore support communities that differ substantially from areas where no such small-scale disturbance processes operated. If they are frequent enough, however, large-scale disturbances at a site, might remove any mosaic effects by 'resetting' large areas to a uniformly disturbed condition. It is the relative frequency and intensity of disturbance at these different scales that determines the structure of the benthic community one observes. Clearly, the role that fishing plays must be set in context if we are to fully understand its importance.

The interaction between fishing disturbance and natural disturbance events is illustrated by the results of a study on the effects of suction dredging for the razor clam *Ensis* sp. on the associated macrofaunal community at a sandy site in 7 m of water in a Scottish sea loch (Hall *et al.*, 1990). Samples were taken from replicate fished and unfished 50 m × 100 m plots at 1 and 40 days after fishing. Disturbance by the dredger was certainly dramatic with 3.5 m × 0.6 m pits dug into the sediment in some locations. Immediately after fishing, correspondingly large effects on the infaunal community were observed with significant reductions in the abundance of a large proportion of the species at the site (Fig. 3.10). After 40 days, however, no differences between treatment and control plots could be detected – a result that was a considerable surprise. This rapid recovery was most probably a result of intense wave and storm activity in the period between samplings, which moved sediment and animals as bedload and in suspension. These movements led to a

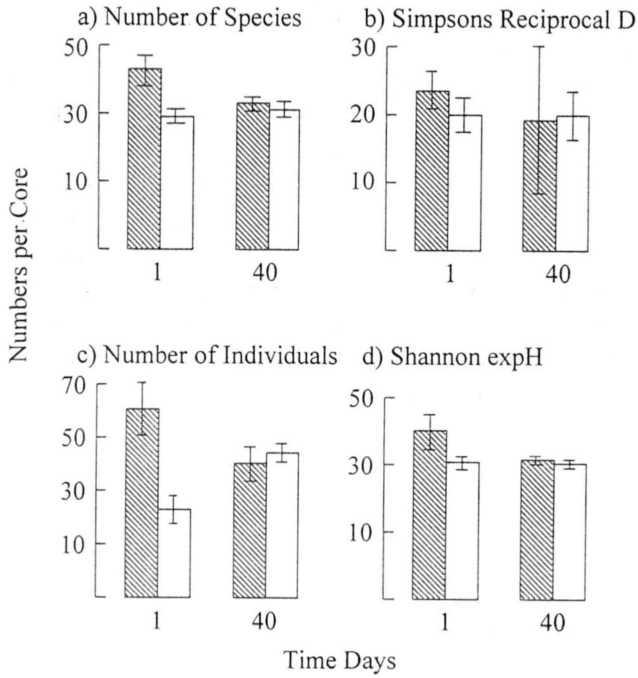

Fig. 3.10 Differences in the number of species, number of individuals and in two measures of species diversity for fished and un-fished plots, 1 and 40 days after fishing. (Adapted from Hall *et al.*, 1990.). ▧ = Un-fished; □ = fished.

'dilution of effects' within the fished plots as the sediments at the site were mixed by the effects of the storm. With all patches of this kind, local reduction in the infaunal population are only likely to persist if: (a) the sediments and fauna are not exposed to hydrodynamic conditions which move them, or (b) the affected area is very large relative to the remainder of the habitat such that the effect dilution process we observed could not occur.

Data presented by Churchill (1989) offer one of the few examples of a study that examined the relative intensities of two different disturbance processes, fishing and natural storm events. In this study, instruments were set up to remotely monitor salinity and suspended sediment load over a 3-month period at a 125 m deep site in the Middle Atlantic Bight off Long Island. Churchill concluded that most of the suspended sediment load did not result from local suspension at the site, but represented sediment that was advected from inshore. Storm effects operating in shallower water account for most of the pulses in sediment load that were observed, but the most dramatic suspension event, which occurred in mid-October and could not be accounted for by storms, coincided with intense fishing activity. It would appear then that trawling can re-suspend sediments in magnitudes and on spatial scales comparable to those caused by storms.

Churchill then went on to compare the relative contribution to the suspended load from trawling and waves and currents. This comparison was achieved using models that required data on wave climatology, currents and the distribution of fishing effort. The results of this analysis indicated that currents dominate sediment re-suspension in the 40–80 m depth range over an area of mud on the middle shelf. However, from 100–130 m the estimated suspended loads from trawls in the mud area exceeded that from currents and waves by a very considerable degree. In the most intensely trawled areas on the Nantucket Shoals, trawling contributes anything between 20 and 99% of the suspended load between May and December, but this is greatly overshadowed during January to April, the period of most intense storm activity (Fig. 3.11). This latter period is when the largest contribution to the year-averaged suspended mass is made.

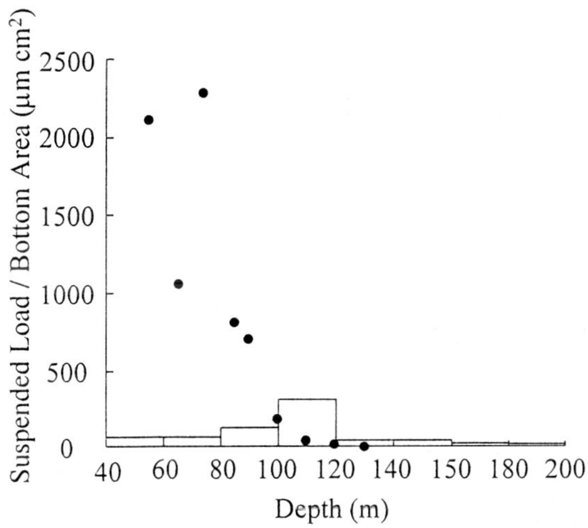

Fig. 3.11 Estimates of sediment load resuspended by currents from January to March 1984 (circles) and estimates of the time averaged sediment load resuspended by trawling (bars). (Redrawn from Churchill (1989) *Continental Shelf Research* **9**, 841–64. Reproduced with permission from Elsevier Science.)

Another study which illustrates how the natural disturbance processes an area might experience can alter the level of fishing effect one can detect is that of Kaiser & Spencer (1996). These authors compared experimentally trawled sites with unfished plots at two sites in the Irish Sea, one that was characterised by mobile sediments and which supported a spatially heterogeneous and impoverished fauna, and a second area with more stable sediments and a richer faunal assemblage. The authors found that in the more stable area trawling led to a 58% decrease in mean abundance of some taxa and a 50% reduction in the mean number of species per sample. Most of these effects were due to reduction or removal of the less common species at the site. In contrast, effects at the more disturbed site were much less apparent (Fig. 3.12)

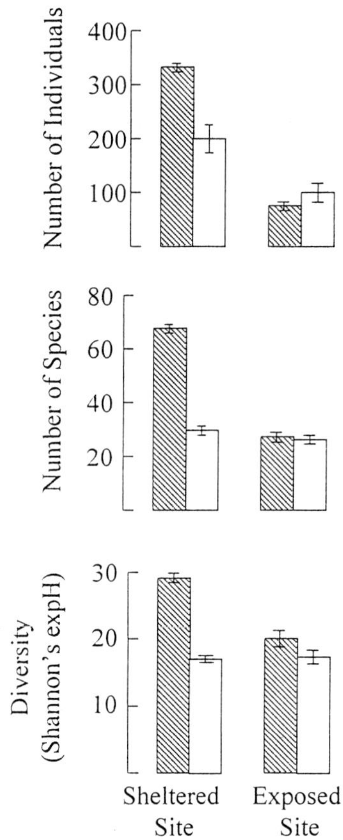

Fig. 3.12 Numbers of species, numbers of individuals and species diversity in fished and un-fished areas at sheltered and exposed sites in the Irish Sea. (Adapted from Table 1 of Kaiser & Spencer, 1996). [hatched box] = Un-fished; [open box] = fished.

The fallacy of natural variation

An often used argument in studies of environmental impacts is that, if the effects of man's activities (in our case physical disturbance by fishing gears) is small compared to the range of variation in natural disturbance processes, then the effect is of no ecological or environmental consequence. The life histories of organisms are closely linked to the regime of natural disturbance that prevails at a site, so one might expect areas of that are naturally highly disturbed to support faunas which are less susceptible to fishing effects. The results of Kaiser & Spencer (1996), for example, would seem to support such a conclusion. That a naturally disturbed area should be of less concern makes intuitive sense, but it is one that a number of people have argued is logically flawed. Peterson (1993) cites two excellent examples that illustrate the point. First, consider fish catches that vary naturally over years by a factor of two. This level of variation does not imply that a reduction of 10% in the average catch is inconsequential, despite the fact that such a reduction would fall well within the range of natural variation. Similarly, reductions in, for example, average light intensity can affect the species of plant that persist at a given location, yet light ranges from high to near zero on a daily basis. Thus, while it is important to appreciate the range of natural variation in disturbance from currents, wind and waves so that fishing can be put into context, the fact that the natural range is large *in itself*, gives no basis for arguing that the additional perturbation imposed by fishing is inconsequential. One might imagine that habitats which experience frequent natural disturbance events are more likely to support communities which are adapted to such stresses, but to assume that this removes the need to evaluate the role that fishing might play is unwarranted. The question is whether fishing impacts are additive sources of mortality.

3.7 Appropriate measures of fishing effort: a perennial problem

When attempting to assess the impact of trawling, two key pieces of information are required – the type of gear used and the frequency of disturbance. As discussed in Section 3.1, not all gears are equally destructive, so knowing the type of gear used and the kind of habitat it is used in is clearly important. Equally important, however, is the frequency with which an area is trawled and here we have much less data. At the crudest level we can use standard fishing effort data, which is usually expressed in terms of hours or days at sea. Although, there are few estimates of disturbance frequency of any kind, this is probably the most common and easily obtained. After making some assumptions about the proportions of the total time spent fishing, towing speed and trawl dimensions etc., one can easily calculate the total area of the seabed that is swept by the gear. Standard practice is then to determine the total area of seabed that is available and calculate the total swept area as a percentage of this value.

Adopting the above approach, Churchill (1989) estimated the cumulative area fished annually for the US east coast. The data were spatially resolved into 30′ latitude × 30′ longitude boxes, and in some of them more than three times the actual area was trawled on an annual basis, notably off Long Island and Narragansett Bay (Fig. 3.13). More recently, Auster *et al.* (1996) calculated that between 1982 and 1996 the fleet trawled an area equivalent to the entire US portion of the Gulf of Maine. An area 2–4 times the area of the US portion of Georges Bank was trawled between 1976 and 1996. In the North Sea, one of the few other areas for which such calculations are available, Rauck (1985) estimated that several areas are trawled three to five times per year (assuming of course that effort is evenly distributed throughout the area over which the calculation was performed). Similarly, Nielsen (1985), quoted in Reimann & Hoffmann (1991) gives values of twice per year for the Limfjord, while Rijnsdorp *et al.* (1991) estimated that some of the more

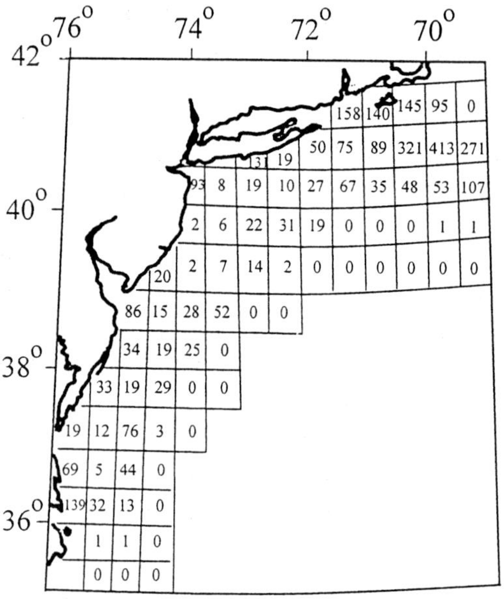

Fig. 3.13 Fishing effort data for Long Island and Narragansett Bay (adapted from Churchill (1989) *Continental Shelf Research* **9**, 841–64. Reproduced with permission from Elsevier Science.)

heavily fished ICES North Sea rectangles (30 × 30 km) are trawled seven times a year.

Although it is often the best that can be done with the data available, the above approach to estimating disturbance frequency is worrying because the scale at which the calculations are conducted is largely inappropriate for the question at hand. Moreover, our perception of what such statistics mean may be somewhat distorted. A simple example will illustrate the point. The UK Annual Abstract of Statistics tells us that 421.9 billion vehicle km were travelled in the UK in 1996. Assuming each vehicle sweeps a path 2 m wide, this represents a total swept area annually of approximately 350 times the area of the UK! In other words, if cars could drive anywhere, the entire area of the UK would be covered almost every day. In practice of course, cars drive on roads and most of the land area is untouched so the calculation makes no sense. In the case of trawling, if the total disturbed area was equivalent to the area we give over to roads on land, concern over its effects would be negligible. Of course this is a ridiculous example, but it illustrates an important point – knowing the fine-scale distribution of trawling activity could radically alter our perceptions of how potentially damaging the practice is.

There are a number of approaches for obtaining fine-scale effort data, each with its own suite of difficulties. One approach is to quantify visible signs of trawl activity on the seabed. For example, using a manned submersible, Caddy (1973) recorded visible tracks in the Gulf of St. Lawrence and estimated that 3–7% of the area showed evidence of disturbance. Using side-scan sonar records from the Baltic,

Krost *et al.* (1990) analysed a much larger area and estimated that the most disturbed regions had up to 35% of the area as visible tracks (mean value 25%). One of the highest estimates of this kind comes from the southern North Sea where the BEON group recorded 89 trawl tracks crossing a 3 km long side-scan sonar transect. Most of these tracks were double 12 m beams, which gives an estimate of 71% of the area disturbed. Of course, interpretation of such estimates is difficult because the persistence of visible tracks is uncertain and depends on sediment type and current regime. In this latter case, however, the water depth was only 34 m and natural erosion of such features was likely to be quite marked, suggesting that the proportion of the habitat and the frequency of disturbance by fishing may have been very high at their study site. Another problem with such approaches is that available ship time and technical limitations necessarily restrict the spatial coverage in such surveys so that a complete picture of how fishing is distributed over a region cannot usually be obtained. In the final analysis, without direct measurements of the spatial distribution and frequency of fishing disturbance the importance and scale of impact will be difficult to resolve.

One potentially valuable approach to obtaining a direct index of disturbance is to examine shell damage in long-lived bivalve species such as the ocean quahog *Arctica islandica. Arctica* is widely distributed over areas of the north Atlantic continental shelves and lives buried in the sediment with short siphons protruding above the surface. Internal growth marks on the shell reveal that some specimens reach ages of up to 100 years. Because individuals live close to the surface heavy gears such as beam trawls often catch *Arctica* or damage them *in situ*. For those that do not die from the damage, a scar is left on the shell, providing a record of fishing disturbance. Witbaard & Klein (1994) have undertaken a preliminary evaluation of the potential for using such marks. For empty shells sampled from a wide geographic area, the authors showed that in the northern North Sea about 40% were undamaged, while in the southern North Sea where beam trawling is much more common this value fell to about 10%, a strong indication that the approach may be valid. The authors also collected a sample of 52 individuals from the catch of a beam trawler in the southern North Sea and examined the history of damage by examining the frequency of scars present in each yearly ring relative to the total number of shells studied for that year (Fig. 3.14). These data show a marked increase in the frequency of damage from the mid-1970s, an effect that may be explained by the increase in the size and power of beam trawlers and the weight and towing speed of the gear employed. Although preliminary, these data look promising. Examining the frequency of damage may give us indices and a historical picture of trends in fishing intensity at finer spatial scales than is possible with traditional fishing effort statistics.

Another possible solution to the problem is to study the behaviour of fishermen – after all it is their activities we are trying to quantify so why not measure them directly? Such an approach has been adopted by Rijnsdorp *et al.* (1991, 1996) who examined the activity of a representative 24 members (about 10%) of the Dutch

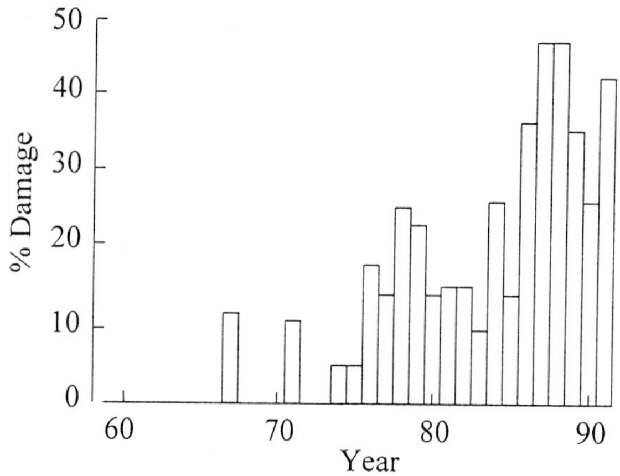

Fig. 3.14 Occurrence of damage in *Arctica islandica* shells from 1958–1991. Each bar represents the relative occurrence of scars on yearly shell growth increments. The number of shells with a scar in that yearly increment is given as a proportion of the shells studied for that year. (Adapted from Witbaard & Klein, 1994.)

Beam Trawl fleet using an automated position recording system on each vessel. By carefully calibrating the data on vessel speeds with logbook records of vessel activity (fishing, steaming, hauling etc.) it was possible to infer when fishing was occurring for any given vessel track.

The first thing this study showed is that over a 3-year period, even at relatively fine spatial scales effort was patchily distributed. For example, at the scale of 3 × 3 miles, 70% of the fishing occurred in only 20% of the area that was fished. Patchiness in fishing activity is observable then when data are aggregated at various spatial scales; the question is at what scale does it become random? We require this information because we have then defined the level of resolution required to make reasonable estimates of the area fished. To answer this question, Rijnsdorp *et al.* (1996) looked at the micro-distribution of effort in the 11 most heavily fished ICES rectangles (30 × 30 miles boxes) and analysed their data with increasingly finer subdivisions. They found that effort was patchy until they got down to the 1 × 1 miles scale at which point it became randomly distributed. Using this scale, and extrapolating from the sampled portion to the whole fleet, they calculated that, in all of the eight most heavily fished 30 × 30 miles ICES statistical rectangles, about 50% of the area was fished 1–5 times per year, and a further 15% was trawled between 5 and 50 times per year. It follows that 35% of the area in these eight rectangles remained largely undisturbed by beam trawls. Of course, because the study did not examine the entire fleet, there are concerns about scaling up effort from a sub-sample of the population. However, for sub-samples of their data, the authors examined the trawling frequency–surface area relationship and found that the area fished stabilised when sample sizes were greater than about 10 vessels. In other words, their extrapolations are defensible.

These are important results because they show that, even in the eight areas that are most heavily fished, over a 3-year period fishermen tended to go to the same places. Much of the explanation for the pattern in effort lies, of course, with the distribution of the target species, but a number of other factors are also important. Notably, proximity to the home port, the suitability of the seabed for trawling (physical obstacles and shipping lanes constrain activity) and the presence of the Plaice Box, an area closed to fishing by beam trawlers for the protection of juvenile plaice. Interestingly, the borders around the Plaice Box are especially heavily fished. Whatever the reasons, and despite remaining uncertainties about other fleets, Rijnsdorp *et al.*'s (1996) study suggests that we must look more closely at where fishermen go and we should be cautious about blindly accepting simplistic calculations of the area fished. Studies of this kind offer an essential perspective on the distribution of disturbance by fishing gears. With the advent of remote sensing technology and the possible need for closer policing of fishing effort for fish stock management purposes, we can perhaps expect new and better means for obtaining detailed maps of the distribution of fishing disturbance.

3.8 Conclusions

The issues of impact on benthic communities is one that has received considerable attention and public interest. Scientists have responded with many recent efforts to document changes and relate them to fishing activity, and to undertake experimental studies to look at effects. How much closer are we to understanding what has been or is being done? Probably quite a bit, but we are severely hampered by poor data on rates, distributions and intensities of fishing disturbance (see Section 3.6). Without these data, our necessarily small-scale experimental studies are difficult to put into a proper context. What we have achieved, I think, is a clearer picture of how communities respond to fishing disturbance. For the most part this response is consistent with the generalised model of how communities should respond, with losses of erect and sessile epifauna, increased dominance by smaller faster-growing species and general reductions in species diversity and evenness.

This agreement with the general model is comforting, but we have also learnt that not all communities are equally affected. For example, it is much more difficult to detect effects in areas where sediments are highly mobile, while boulder or pebble habitats such as those on the northeast coast of the USA seem particularly vulnerable. This vulnerability presumably arises because the erect sessile epifauna that pebble pavements can support can be easily killed when pebbles are turned over or abrade one another as a trawl passes. Such differences in the susceptibility of habitats are important and need more attention. In addition to identifying areas where trawling has marked effects, scientists have a responsibility to also be saying where they do not, or at least where the effects are much less marked. Unless one takes the view that trawling is an activity that should not be undertaken under any circumstances it is encumbent upon us to identify, not only sensitive areas, but also those for which concern about effects on benthic communities are fewer – the area of seabed occupied by such communities may well be considerable. Before identifying such areas, however, there is a longer-term historical perspective that needs to be thought about.

What did benthic systems use to look like?

Being realistic it is unlikely that we will ever unearth data sets that will give us a really good historical picture of how benthic communities have changed over the period when fishing developed. This gap leaves one question hanging continually in the air. If we don't see major trends in time-series or marked effects from experimental trawling, might it be that fishing removed or degraded the community to a considerable degree before we even started looking? If the bulk of the change occurred during the initial period of trawling development, we may now be unable to detect trends or impacts from fishing, either because the community is resistant to further effects, or the effects are relatively trivial compared to those that have gone before. There are many scientists who firmly hold this view, and one must recognise that it is perfectly possible for areas that have a long history of fishing activity. Of course, if we really have no data from unfished areas or from historic records, one's belief or disbelief in the hypothesis of marked initial change becomes an article of faith. There are, however, precedents for the kinds of rapid change that people are probably referring to when they talk about effects already having happened.

To me it is the issue of epifauna that is the key to these debates, because I imagine that the kinds of marked changes in benthic communities most people have in mind are shifts from a rich and luxuriant mobile and sessile epifaunal community to a largely infaunal community with few animals on the sediment surface. A community that the naked eye would appreciate as being markedly different from what we see in many places today. The argument goes that the first few passes of a trawl gear over an area remove this epifauna which, particularly in the case of slow growing sessile taxa with poor dispersal abilities, can never re-establish with repeated trawling. Perhaps the best example of such a response can be found on the Australian northwest shelf (see page 74) and similar effects may well have happened in other areas. Although I confess to being rather sceptical that all heavily fished areas used to support rich epifaunal communities, it is certainly plausible that many did. Moreover, my scepticism may reflect the fact that I am most familiar with northern European shelf seas that have one of the longest histories of fishing and may therefore have been most affected in earlier times. I find it difficult to believe, however, that we cannot find, nor have any records of areas in the North Sea which support(ed) the remnants of a rich epifaunal community on soft sediments. In Loch Gareloch, for example (see Section 3.5), the epifaunal community is rather poor, yet the site has been closed to fishing for 30 years. It has probably never supported a rich epifaunal community and many other parts of continental shelves may be the same way. Nevertheless, as in other areas, such communities may occur as a mosaic of isolated small patches in a sea of sand and therefore be difficult to find and map. Such a situation certainly obtains in the inter-reefal areas on the Great Barrier Reef and in the temperate waters of the Great Australian Bight. Of course, this is not to suggest that larger epifaunal taxa have not declined in soft sediments of the North Sea; the work of Philippart (1997) and others discussed earlier suggest that this is certainly not the case. However, one should probably make a distinction between declines in some mobile epifaunal populations which probably always occupied a largely featureless sand expanse, and the loss of luxuriant epifaunal gardens.

Perhaps ignorance makes it easier to accept that shallow tropical or sub-tropical systems, or temperate boulder and cobble systems are more likely to experience rapid degradation in the first few years of trawling than the sandy and muddy sediments of an area like the North Sea. Alternatively, there may really be such differences in the propensity of communities to be affected by trawling. In the absence of data, one can only speculate. All we could do to answer this question is close areas of soft-sediment substrate to bottom fishing, and wait for a potential response. This is an experiment that is well worth trying and is one course of action that deserves active promotion by the scientific community. The topic of Marine Protected Areas is re-visited in Chapter 9.

Habitat modification and fishery yields

Another feature of trawling effects which needs to be considered is its potential to effect changes in fisheries yields through habitat modification. It is well recognised that species which increase the three-dimensional architecture of the epibenthos provide secondary habitat in which many other species can be supported. This habitat may be especially important for the juveniles of some species since the additional cover can provide refuges from predation. In many areas this secondary habitat is patchily distributed, and the question is whether such localised refugia are sufficiently widespread and important to make a difference for the fishery as a whole. This will often be a difficult question to answer. We know, for example, that juvenile cod that occupy the shallow sub-tidal region around the coasts of Europe and North America benefit from the shelter that algal cover can provide. Yet the proportion of the total stock that survives and recruits into the fishery, which were afforded such protection is probably quite trivial compared to the numbers that survive from more open areas. On a per capita basis then, the margins may be better, but in terms of total contribution they are probably insignificant. When this is so, we might protect the habitat for other reasons, but enhancing fisheries would not be one of them.

Unfortunately, we have the additional question of whether habitats which we now consider to be devoid of structure, used to afford protection to juveniles (the issue discussed in the previous section returns!). On the northwest shelf of Australia, it seems clear that habitats were rapidly affected by trawling and, as noted above, there is no real basis for supposing that this has not happened elsewhere. It should be recognised, however, that if benefits to a fishery can be shown if habitats are protected or restored, it is likely that those benefits will only be obtainable if alternatives to trawling can be developed. Trawling and benthic habitat protection is largely incompatible. The one exception to this, of course, is if benefits accruing in an area closed to fishing spill over into trawled areas – an issue that is dealt with in Chapter 9.

The research agenda

Studying the effects of trawling on benthic communities is difficult, but not impossible and the number and quality of the experimental studies that are now appearing in the literature is encouraging. But experimental studies can never approach the temporal and spatial scales of real fishing grounds and putting experimental results into the appropriate context will always be difficult. How does a decision to trawl five times per year over an area of 1 km^2 in an experiment compare with the actual fishing intensity on fishing grounds? At present, we have much less of an idea than we should because the scale at which effort data are available is too coarse (see Section 3.6). This absence of fine-grained effort data and good maps of benthic communities for fished and unfished areas are the single biggest impediments to progress in assessing benthic effects. Without them we have a poor appreciation of the true scale of the problem and the appropriate locations to focus upon. This is important on a number of counts. First, I think our views on trawling should be similar to our views on ploughing a field – we don't want to do it everywhere but we are happy to have substantial areas turned over. Unfortunately, because the activity of fishermen is largely unconstrained at present we have no guarantee that we do have unploughed areas, or that they are representative of the kinds of habitats we would like to protect. In many areas, studies of fine-scale distribution may show that a relatively large fraction of the region does remain unaffected, or only very lightly trawled, but without the appropriate data it would be incautious to assume that this is the case. Second, we may wish to close certain areas to fishing for conservation, fishery management or scientific purposes; choosing appropriate sites, deciding how big they should be and policing them would be greatly facilitated by better effort data, and better maps of the benthic community. Satellite tracking of fishing vessels as part of a fisheries enforcement and monitoring programme could provide the effort data. It is an approach that is already being implemented in some areas. Obtaining benthic maps, however, is a much more difficult and expensive proposition. It may be that side-scan sonar or related approaches can provide an idea of the distribution of erect epifauna, but large-scale mapping will be expensive. With additional justifications, perhaps based on the potential importance of such habitat for fisheries recruitment, it is possible that research vessels will be deployed to undertake such work.

Chapter 4
Fishing at the Coastal Margins

In this chapter the direct effects of fishing on benthic communities that are not associated with sub-tidal trawling activity are considered. Included in this category are all methods of fishing in intertidal and very shallow (1–5 m deep) sub-tidal systems, such as hand collecting, digging and mechanised dredging, along with activities in deeper water, such as trap or line fishing, seining or dynamite fishing. Although some of these methods are prosecuted far from shore (e.g. on fringing reefs), for the most part they are nearshore activities.

As with other chapters in this Part 1, I have avoided dealing with the indirect effects of fishing that result from the ecological interactions consequent with the removal of the target species. These effects are dealt with in Part 2, but the reader should be particularly aware of this when reading the parts of this chapter which deal with rocky shores and coral reefs. Indirect effects often seem to be especially important in these habitats and reading this section alone would give a particularly incomplete account of the issues associated with these systems.

4.1 Rocky shores

Rocky shorelines are a ready source of animal protein for humans. Yet the potential impact of exploiting the fishery resources of rocky shores is seldom acknowledged, perhaps partly because the other threats to coastal environments (e.g. pollution, industrial development, population pressure) are so palpable. Moreover, as Castilla & Paine (1987) point out, the effects of exploitation may be difficult to recognise where human populations have been exploiting marine resources for thousands of years at unknown intensities. Evidence for pre-historic activity comes from shell middens, which can be found in many parts of the world, indicating that mankind has exploited a wide range of species for an awfully long time.

Human trampling

The most obvious direct effect of fishing on rocky shores (other than the removal of the target species) is trampling activity. The very process of examining or visiting a rocky shore is a disturbance and one can well imagine that large numbers of individuals searching for invertebrates and prising them from rocks could have marked effects on the remaining fauna. In a sense, this is directly akin to the inadvertent damage caused by towed fishing gears discussed in Chapter 3. Except of course that, unlike trawling, disturbance by trampling is by no means restricted to fishermen, particularly in more developed areas where visitors and locals alike derive considerable pleasure from exploring the seashore. Bait collection for recreational fishing, for example, can lead to intense trampling, particularly near large urban centres. Nevertheless, in many parts of the world it is fishing activity of one sort or another that brings most people to the shoreline. On Chilean rocky shores, for example, the muricid gastropod (*Concholepas concholepas*), which probably played an important role in the subsistence of pre-hispanic native populations is today a species of enormous economic importance to the country. Not surprisingly fishing pressure is intense. Duran & Castilla (1989) have shown, for example, that reducing fishing effort by protecting areas can lead to very high densities of *Concholepas*. As noted above, the ecological role that *Concholepas* plays in this system and the changes that follow as a consequence of its removal are discussed in Part 2. It is sufficient here to simply note that many people trample the shore to exploit the resource.

Although many studies of the effects of trampling have not been motivated by fisheries-related activity, it seem sensible to include such information to examine the general effects of trampling. One of the more recent efforts to do so was by Brosnan & Crumrine (1994) who studied two communities, an upper-shore algal-barnacle and a mid-shore mussel bed community. The authors trampled replicate plots at these two sites 250 times per month for a year and then allowed plots to recover for a further year. In the upper-shore community, the authors found that foliose algae were very susceptible and declined markedly shortly after trampling had begun, whereas turf-like species were resistant and increased in relative abundance in trampled plots. Barnacles were also crushed by trampling and declined in numbers. These results are in general agreement with a number of others (e.g. Beauchamp & Gowing, 1982; Povey & Keough, 1991), which show that trampling leads to a reduction in cover by foliose canopy-forming species which tend to be attached to the rock surface at a single point. While these species are highly resistant to wave action by possessing flexible stipes and fronds, which absorb wave energy, they are ill-adapted to trampling which easily dislodges them. In contrast, turf-forming algae are low profile, profusely branched and attached to the rock surface in many places, thereby making them resistant to trampling effects. The other important feature of Brosnan & Crumrine's study, however, is that the algal community recovered within a year after trampling ceased (Fig. 4.1a). In contrast, the experiments in the mussel community indicated that mussels were removed, leaving bare space and did not recover in the year after trampling stopped. Rather, bare patches created by trampling actually increased during this period (Fig. 4.1b).

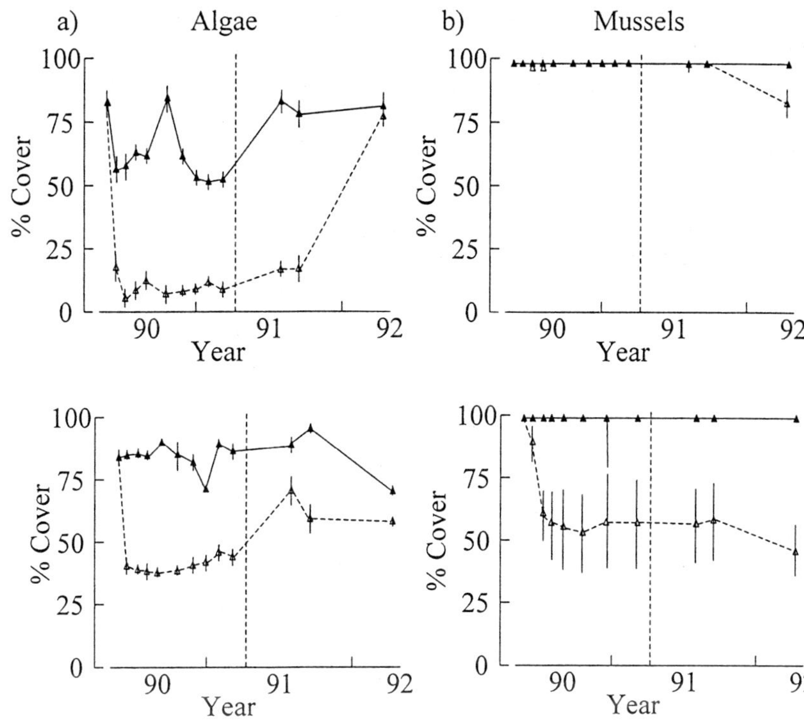

Fig. 4.1 Percentage cover of: (a) algae in upper shore algal communities, and (b) of primary space by mussels within mid-shore mussel beds. Panels in each column are for two sites and data cover a 1-year period of human trampling followed by a year where trampling was prevented. (Adapted from Figs 2 and 4 of Brosnan & Crumrine (1994) *Journal of Experimental Biology* **177**, 79–97. Reproduced with permission from Elsevier Science.) (—) Trampled; (---) control.

Divers in the rocky sub-tidal areas

The effects of fishing in rocky sub-tidal areas are less well documented than in the inter-tidal, but one rather depressing account is provided by Fanelli *et al.* (1994). These authors examined the effects of fishing by divers for the date mussel (*Lithophaga lithophaga*), a rock-boring mollusc that is widespread along the Mediterranean coast. This species lives in calcareous rocks and in southern Italy where divers exploit it by breaking open the calcareous rock with sledgehammers to expose the animals. The fishery has been illegal in Italy since 1988, but it still continues both here and in other parts of the Mediterranean. Clearly, such destructive methods are likely to have profound effects on the other fauna that live on the rock surfaces.

Fanelli *et al.* (1994) conducted repeat surveys in 1990 and 1992 along approximately 200 km of the Italian coast in an effort to quantify the scale of the effect on the sub-tidal communities. Diver transects were surveyed perpendicular to the shore, every 2 km along the coastline and the abundance of mussels and the extent of rock damage was evaluated. Surveys conducted in 1990 showed that 159 km of the surveyed coast was rocky and about half of this was affected by date mussel collection. Moreover 28% of it was classified as heavily damaged, exceeding 50% of the maximum damage that could be sustained. Maximum damage was taken as complete desertification (i.e. akin to a sandy bottom). By 1992 the area affected by mussel collection had risen by about 15% and the heavily damaged region comprised 37% of the total. These increases were associated with large increases in damage indices for areas that were previously less exploited, rather than increased effort on previously damaged areas. Figure 4.2 summarises the results of these surveys. One worrying prospect is that even where some rock remains, the effects of this destructive fishery may not be reversible since it is postulated that urchin grazing may prevent recolonisation of the bare rock.

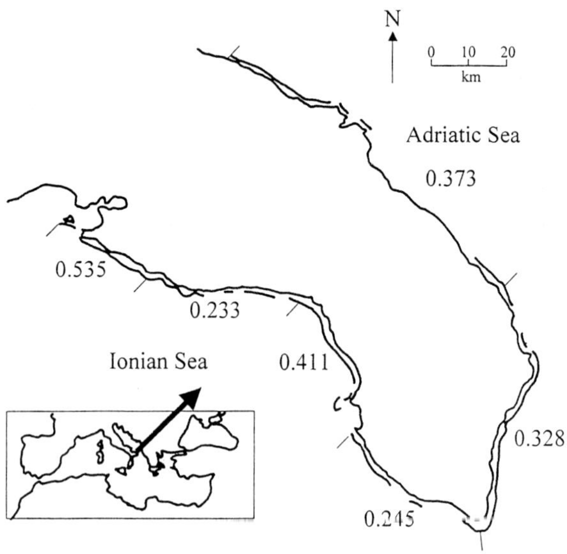

Fig. 4.2 Sectors of the Salento peninsula surveyed for damage by fishermen (dark lines denote rocky substrate) and the mean damage index for surveys in 1990 and 1991 for each sector. The damage index ranges from zero for no damage to 1 for complete desertification. Inset: the location of the Salento peninsula (Apulia, Italy). Adapted from Figure 2 of Fanelli *et al.* (1994).

4.2 Tidal flats

The extensive tidal flats that can be found in many parts of the world support substantial populations of shellfish and other invertebrates and are an important focus for coastal fisheries. Fishing in these areas is, of course, a very public activity and it can often raise the hackles of those living nearby. When the fishing method leaves highly visible signs of disturbance over large tracts of sand flat, concern by people living in the area is understandable, especially if they perceive the fishermen to be 'outsiders'. As with most other types of fishery there are three aspects to this concern: the sustainability of the fishery itself, the interaction between the fishery and other predators which depend on tidal flats for their food resources, and the direct effect of the physical disturbance caused by the fishing process on the non-target fauna. One notable feature of many inter-tidal habitats, of course, is that they are subject to substantial disturbance from natural storm events. As discussed in Chapter 3, one might suppose that the fauna in such areas are adapted to high levels of disturbance and the additional effect of fishing will be trivial. While as a general principle this assumption may have some merit, there are a number of studies which show that this may not hold under all circumstances.

Shellfish harvesting

Of the inter-tidal species harvested in Europe by mechanical means, the cockle (*Cerastoderma edule*) is perhaps the most widely exploited. In Britain almost all cockle beds have probably experienced at least some intermittent exploitation and notable cockle fisheries have been prosecuted in a number of areas. The European market for cockles was dominated by the Dutch fishery, which exploited stocks in the Wadden Sea until their collapse in the late 1980s. Thereafter, exploitation rates in the UK rose dramatically.

Cockle harvesting techniques vary widely, ranging from hand raking through to the use of various mechanical devices such as hydraulic or towed dredges. It is the more recent hydraulic and mechanical dredgers that cause particular concern. Hydraulic suction dredgers operate by fluidising the sand using water jets and then lifting the sediment and cockles. This sediment is then sifted through a grid and the cockles drawn up a suction pipe to a revolving drum on the deck of the boat. Small cockles and other debris fall through the screens of the drum, returning to the water, while the remaining cockles are sent down to the hold. In contrast, tractor dredgers operate at low tide and there are a number of variants on the basic design. Of these, the trailer-mounted devices are perhaps the most impressive. These devices possess a blade between 70 and 100 cm wide that penetrates the sediment to a depth of between 20 and 40 cm. Sediment is then carried on a conveyor into a large rotating cage drum which retains the cockles but allows the debris to pass through. The catch is removed through a hatch at the rear of the drum.

In 1993 I was involved in two experiments, one to evaluate the effects of a suction dredger, and the other a tractor dredger (Hall & Harding, 1997). In both cases, the design of the experiment was relatively simple with the recovery of replicated fished and reference plots followed over time. One additional feature of the tractor dredge experiment, however, was that we replicated plots at three different sizes to examine how dependent the results were on the size of the disturbed plot. Figure 4.3 summarises the results of this study. Comparisons between the two experiments was complicated by the fact that one experiment started in the spring (May), while the

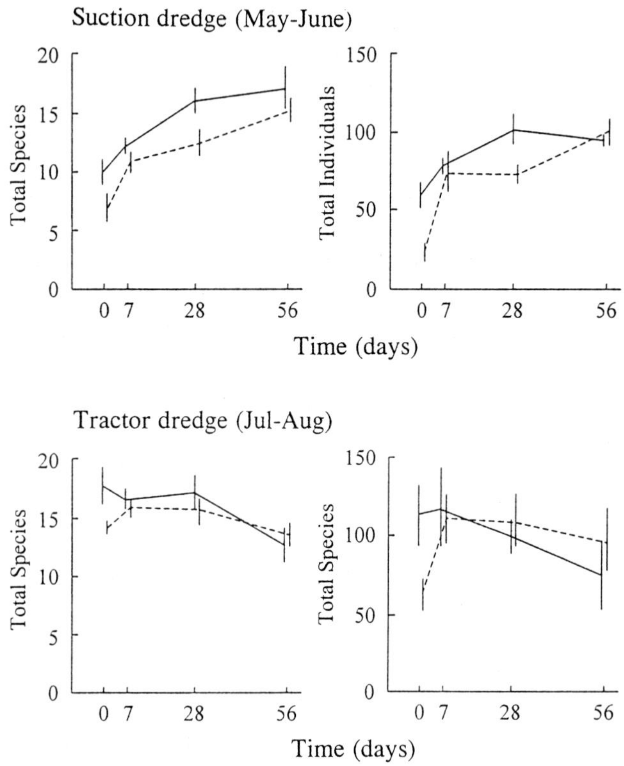

Fig. 4.3 Summary of trends in total numbers of individuals and total numbers of species in core samples for two experiments on the effects of mechanised harvesting of cockles. (Adapted from Figures 4 and 9 of Hall & Harding, 1997.) (—) Control; (---) disturbed.

other started in the summer (July), but some general features emerged. In the case of the suction dredge experiment, statistically significant effects on the total abundance of individuals and species were found. Disturbed sites had significantly lower mean abundances (up to 30% reductions for numbers of species and 50% reductions in numbers of individuals). This decrease occurred against a background of consistent seasonal increases in the abundance of many taxa in both dredged and control areas. For the tractor dredge experiment fewer effects were found and recovery from disturbance, occurred against a background of general seasonal decline in the abundance of the fauna.

The important point about this study is that, although both experiments revealed some statistically significant results, taken as a whole the faunal structure in disturbed plots recovered (i.e. approached that of the undisturbed controls) within 2–3 months. In this case, effects on non-target fauna were probably not a major cause for concern.

There are only two ways in which densities in a disturbed area can increase –

reproduction within the disturbed site, or immigration from outside, either as larvae, post-larvae or adults. In inter-tidal sites with sediments that are susceptible to movement by wind and waves, movement of post-larvae and adults is strongly correlated with wind–wave induced sediment disturbance (Emerson & Grant, 1992). This transport mechanism is likely to be a major factor determining density change. In the cockle study described above, the fauna associated with cockles was very variable. One could expect, therefore, that quite large sub-sets of the population of any infaunal species in the area would occupy sediments that do not support densities of cockles that are worthwhile harvesting. Unlike most other fishing methods, dredge fishermen get instantaneous information on reward rates and rapidly cease to fish any area where the cockle catch falls below an economically viable level. Thus, areas with low densities of cockles are subject to little exploratory fishing. It seems likely, therefore, that there will almost always be areas that remain unfished to provide a source of colonists for disturbed areas, either through passive transport of animals with bedload, or through the provision of larvae. Moreover, although mortalities from harvesting were statistically significant, the numbers of survivors in dredged areas remained considerable, and recovery processes operating from within a disturbed patch probably made a substantial contribution to the recovery.

The question is how general is a rapid recovery from disturbance? I suspect it may be rather common and one can certainly find other examples of small-scale experimental studies that show recovery from disturbance within a few months in (e.g. Smith & Brumsickle, 1989). Kaiser *et al.* (1996) also give another good fisheries example for the effects of harvesting manilla clams. These clams are cultivated under netting for about two years at inter-tidal sites before being removed by suction dredge. Harvesting altered the sediment composition by removing the larger sand fractions down to the underlying clay substratum and there was a large reduction in the density of all individuals and total number of species. Seven months later, however, there was no statistically significant difference between communities in the dredged area and control sites, and sediment structure had been restored.

Most of the studies cited above were undertaken in areas where sediments were fairly mobile and, in general, more mobile sediments are likely to see relatively rapid recovery under most circumstances. But what about more stable sediments? A possibility that has received relatively little attention is that the first pass of fishing gear may remove species that act to stabilise sediments. Once lost, the increased sediment mobility may greatly reduce the probability that stabilising species can re-establish. In other words, the system may flip into an alternative state dominated by mobile sediments from which it is more difficult, or perhaps impossible to return. One obvious habitat type for which this is possible is inter-tidal seagrass. Seagrass beds support much richer faunal communities than surrounding areas and may find it difficult re-establish in mobile sediments once they are removed.

Peterson *et al.* (1987) provides a good example of a fisheries effect study in an area stabilised by seagrass. These authors examined the effects of traditional raking and two mechanical methods for harvesting clams in shallow seagrass meadows and sand flats in an experiment which ran over four years. The mechanical methods involved suspending both sediment and clams in the water column by intense disturbance using modified outboard motors and then trawling the clams. In this comprehensive and quite large-scale study (plots were 1225 m^2) effects were examined with reference to clam recruitment, seagrass biomass, macro-invertebrate densities and scallop densities. Two harvests were taken in a single year and the effects were then followed over the next four years. With regard to scallop recruitment, effects were ambiguous, but there was little indication of a dramatic increase in settlement resulting from reduced numbers of adults in the area. As one might expect, there was a large effect on seagrass biomass (Fig. 4.4). All fishing methods had a detectable short-term impact, but recovery had occurred within a year for the less intense disturbances. For the more intense methods recovery did not begin for two years and was not complete after four years. This extended time period led the authors to suggest that perhaps the seagrass beds and sand flats exist as alternative stable states. Once seagrass has been removed disturbance effects are more intense, making it more difficult for the grass to re-establish. No effect was observed on small infauna and it was suggested that this was because the fauna was already dominated by small polychaetes.

Although seagrass is an obvious candidate for stabilising sediments, other taxa can also play this role. In a nonfishery related example, Thrush *et al.* (1996) examined the effect of plot size on recovery rate following defaunation on a sand

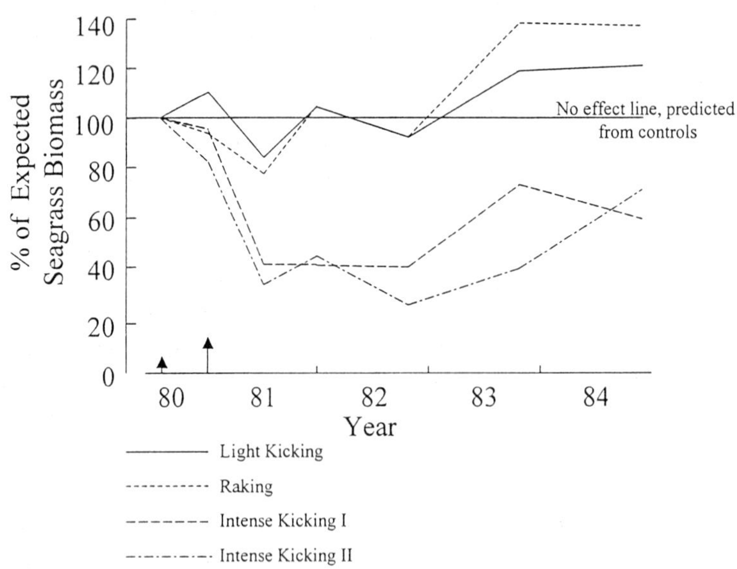

Fig. 4.4 Percentage difference between observed average biomass of seagrass in plots subject to different clam fishing regimes and expected biomass, based on biomass at control sites. The plot assumes that initial pre-fishing differences between control and treatment plots would be expected to remain constant over time. Expected biomass is then plotted as 100%, the 'no-effect' line. Arrows denote minor and major treatment applications. (Adapted from Figure 4 of Peterson *et al.*, 1987.)

flat. They found that sediment instability was higher in treatment plots owing to the removal of dense spionid polychaete tube mats. With increased sediment instability (which was also a function of plot size) the authors postulate that the rates of emigration from recolonising plots was higher. These changes in sediment dynamics are proposed as the explanation for recovery rates in excess of nine months, an unexpectedly slow rate for disturbed plot sizes ranging from 0.2–3.25 m^2. The possibility that fishing might alter sediment stability characteristics is one that deserves increasing attention.

Bait digging

Although exploitation of shellfish for human consumption is probably the most common form of fishery in inter-tidal areas, species are also exploited to provide bait for anglers in many parts of the world. Bait digging can either be by hand or with mechanical dredges similar to those employed for shellfish. Various studies have been undertaken to examine their impact. Perhaps the most comprehensive of these is that of Beukema (1995) who reports on the impacts of mechanical harvesting in the Wadden Sea. The strength of this particular study lies in the fact that fishing activity took place for four years between 1978 and 1982 in part of a 50 km^2 area that had been under long-term scientific investigation since about 1970. In this area, 12 permanent 1 km transects had been sampled each year and the commercial fishermen focussed intensely on the area around one of these transects using mechanical dredges to catch lugworms. During high water the dredges manage to dig about 3 gullies, each about 1m wide, 0.4 m deep and 300 m long. Sieved sediment and animals that are not retained are dumped back over the side leaving a 1.5 m wide ridge beside each gully. Dredgers work around an anchor at one location for a few weeks, until an area has been comprehensively covered and harvest rates fall. Not surprisingly the area they focussed on had historically supported higher densities of lugworm than other areas.

The prolonged stay by the dredges in an area that had previously been regularly sampled offered a unique opportunity for Beukema to study the long-term effects of intensive lugworm harvesting by comparing the changes in density at the fished site with many reference areas. Figure 4.5a shows how the lugworm biomass changed at the site compared to the other locations that were regularly examined. During the period of fishing, its rank as a site supporting lugworms fell from its traditional number one position to seventh after about four years. But, unlike in the cockle example described earlier, it was not just the target species that was affected; Fig. 4.5b shows how the density of the clam *Mya arenaria* changed over the same period. The size structure of the population also altered during this time with the removal of larger individuals by fishing. This latter aspect is particularly important when considering fisheries effects on many species, since alterations in the size or age structure of the population can have marked implications for its reproductive capacity. Several factors contributed to the decline in large clams, the biggest were retained and sold, shells were broken by the fishing process and individuals

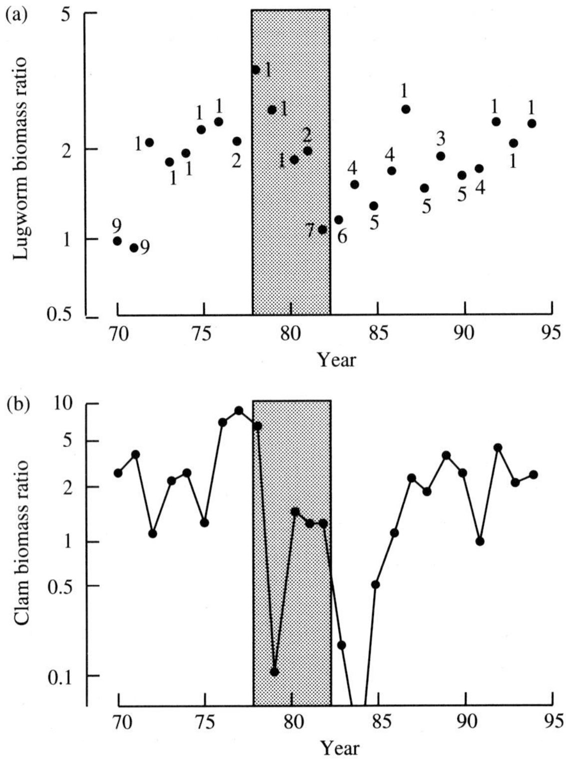

Fig. 4.5 (a) Relative values of lugworm biomass at a site in the Wadden Sea in a period before, during and after fishing. Values are expressed as the ratio of annual (March) values observed at the site, divided by a 15 station mean value for the wider area. The numbers near each point denote the rank this particular site took among the 15 stations, with respect to lugworm biomass. (b) Relative values of clam biomass for the same site as in (a). Shaded area denotes fishing period. (Adapted from Figures 2 and 5 of Beukema, 1995.)

deposited back on the sand were vulnerable to predation by birds and shore crabs. Changes in the densities of clams and lugworms accounted for most of the change in total biomass that was observed at the fished site, but some other polychaete species were also affected. Other studies of the effects of bait digging on tidal flats either mechanically or by hand show similar effects to those found by Beukema (1995) (e.g. McLusky *et al.*, 1983). The immediate effects of most harvesting processes in sedimentary environments include strong reductions in the abundance of many benthic species and of total benthic biomass in the areas from which sediment has been disturbed.

Van den Heiligenberg (1987) estimated that bait fishing disturbed approximately 0.23% of the total area of the western part of the Dutch Wadden Sea and that approximately 1% of the *Arenicola* population is removed each year. For the Wadden Sea, therefore, the scale of disturbance is relatively slight and the effects of this kind of fishery are probably of little concern for the system as a whole. This, of course, reflects the very extensive areas of mud flat which are available and the relatively limited market for fishing bait. This seems to be the key to the issue for most tidal flat systems. On less extensive sand flats in other areas, the ratio of demand to potential supply may be much lower and the proportion of the total area affected may be very high indeed.

4.3 Coral reefs

Perhaps more than for any other marine system, the apparently inexorable rise in the human population is taking its toll on coral reefs, the majority of which occur in countries whose populations are expected to double within the next 30–50 years. At a meeting in 1993 on the status of coral reefs, overfishing was deemed to be one of the three greatest threats to reef systems (Roberts, 1993). In many areas extreme poverty has led increasing numbers of people toward exploitation of the common property of fisheries resources. Unfortunately, some of the fishing techniques that are being adopted are among the most overtly destructive known. In global terms yields from coral reef environments are relatively trivial at about 0.5×10^6 tonnes per year, but this figure belies their importance as a contribution to the catch of low income fishers with few alternative opportunities for employment (Russ, 1991).

Pauly *et al.* (1989) identified a particular class of overfishing to describe a situation where such destructive practices are undertaken. Named after the Rev. I. R. Malthus, who in the eighteenth century considered the inevitable consequences of population growth coupled with finite resource availability, the term 'Malthusian overfishing' was adopted. This term refers to a situation where 'poor fishermen, faced with declining catches and lacking any other alternative, resort to wholesale destruction in their effort to maintain their incomes' (Pauly *et al.*, 1989).

An example of such a development can be seen in some fisheries, which traditionally have used poisoning as a method of capture. Indigenous coastal communities have used leaves, berries and roots to capture fish for centuries, but with increasing economic hardship, this activity has more recently developed into the use of bleaches in tide pools, and other shallow waters in some cases. Ultimately this has led to the use of sodium cyanide. For application to small areas, these poisons are applied either by tying tablets to sticks for application to fish hiding in corals, or by using squeeze bottles to apply the poison as a solution. For larger-scale application a solution may simply be ladled into the water, or mixed with baits. The effects of poisoning on fishes can be expected to be rather non-specific and alteration of fish community structure seems likely. The subject of greater concern, however, is the effect of poisons on corals. There is little specific information on the effects of poisons on coral mortality rates, but the tolerance of corals to sodium cyanide is believed to be low.

A more physically destructive method of fishing on coral reefs involves the use of stones, chains or poles to break up coral and drive fish out into nets, but perhaps the most obviously destructive of all methods is the use of explosives. Generally 'blasters' focus on clumps of coral and a single blast destroys an area of 1–2 m radius. Fish must then be collected by hand from among the rubble, leading to substantial waste and restricting the activity to days when water clarity is good, and to depths above 15 m. By 1988 the use of explosives had been reported in at least 40 countries or island territories, and poisoning in at least 15 (Wells, 1988). However, although spectacular, it may be that blast fishing is rather less problematic than had hitherto been thought. Analysis by Saila *et al.* (1993), for example, suggests that blasters tend to focus on valuable pelagic species, rather than coral fish and that the pressure wave from blasting in mid-water does not extend very far. This form of fishing tends to decline when coral cover falls below 20% in any given patch, and when fish concentrations fall to an order of magnitude below their unfished values. It may be the case that anchor damage from fishermen and poisoning are greater causes for concern, although this is not to say that blasting is an environmentally friendly pursuit!

The global scale of overfishing and habitat destruction in reef fisheries is difficult determine. However, Roberts (1995) suggests that in areas such as the Philippines, Haiti and Jamaica, fish population densities have declined to very low levels, adults of edible species have been almost eliminated from fished areas, and reef habitats have been severely damaged perhaps to a level where populations will be unable to recover. The prospects for recovery in such degraded systems seems highly questionable, particularly if the root cause of the problem – poverty and pressure by human populations on the coastal margins – is not alleviated.

Although such social factors leading to fishing as a last resort certainly seem to be important driving forces in the implementation of these methods, in some cases such fisheries are actually considered to be economically viable for more established fishermen. MacManus (1996) cites examples for Papua New Guinea where the perception of a limitless supply of coral leads fishermen to conclude that blasting is cost effective. Similarly, in offshore areas where long boat trips to a reef are required, blasting is perceived as an economically sound way to fill the ship's hold quickly and minimise the amount of time that needs to be spent at sea. With recovery rates for coral estimated at between 0.1 and 1% per year (Saila *et al.*, 1993), one can't help concluding that the perception of a limitless supply is flawed!

4.4 Conclusions

Rocky shores

The proximity of rocky shores and other inter-tidal areas to populations make them especially vulnerable to intense pressure from human activities. Fishing is clearly one of those pressures, but separating its effects from others will often be difficult. For example, the studies discussed above and others show clearly that trampling can have profound effects on rocky shores. In some cases, these effects are reversed relatively rapidly when the disturbance ceases, in others such as in the mussel-dominated habitat studied by Brosnan & Crumrine (1994), it may take much longer, perhaps up to a decade. In many areas it would be difficult to separate the contribution fishermen make to these effects from those by other visitors to the shore; and it is probably a pointless exercise to try and do so in most cases. Indeed, if one wants to protect rocky shores, simply preventing fishing may be insufficient if access remains for other sectors of the community. It should also be recognised that, at present, our appreciation of the effects of humans on these systems is largely based on studies at local scales. A good start in obtaining a wider perspective would be national assessments of the proportions of coastlines that are subject to human visitation indexed by the intensity of activity. Without such assessments an important (but by no means the only) scale on which to assess the effect is missing.

As with the sub-tidal systems discussed in Chapter 3, excluding (or restricting) human visitors from some sections of coastline can give valuable additional information and perhaps satisfy conservation objectives. As part of a programme of coastal zone management inter-tidal reserves have much to commend them. Examples of this approach are discussed in Chapter 5.

Tidal flats

For tidal flats, the issue is rather more focussed because human trampling is likely to be less extensive and its effects usually minimal. Pollution effects and other factors associated with coastal populations will also be an issue, but fishing activity which leaves highly visible tracks and pits on the sediment surface is often as prominent.

In the more developed countries the mechanised fishing methods are likely to continue to be of most concern, but it is the sustainability of the fishery and the potential competition between fishermen and other predators on the target species, that are probably most worthy of attention. Under most circumstances it seems that the populations of non-target benthic infauna are likely to be robust to the disturbance. This kind of conclusion seems to make many scientists uncomfortable. The doubt and qualifications we are apt to make leads to an understandable reluctance to say that something probably doesn't matter, particularly something as obvious as a device which ploughs sand flats. But if the data are supportive scientists have as much of an obligation to discount concerns as they have to raise them. In my view this happens to be one of those cases. Unfortunately, however, there are no 'yes' or 'no' answers. We cannot say that for all sand flats effects of dredging on non-target fauna will be trivial. Nor can we conclude the opposite about most other systems. We can say, however, that a problem is less likely to obtain on sand flats, although there will be occasions when effects will be more marked. When might such occasions arise? They are most likely in relatively stable sediments, either muds in sheltered areas or in coarser sediments which are stabilised by the resident fauna or flora. Importantly, however, the larger (older) individuals of bivalve species in particular will not actively migrate into disturbed areas or be normally passively transported as bedload. Since these older individuals contribute a great deal to the total biomass of the community, complete recovery often depends on their re-establishment through growth of juveniles. Thus, in this respect at least, the time-scales of recovery can be of the order of years.

Coral reefs

Although the data are relatively sparse and well-executed studies of effects of fishing on coral reefs are difficult to find, it seems self evident that practices which destroy the physical integrity of reefs will have profound consequences, both for fish populations and the other taxa of coral systems. To quote Bonshack (1993):

> 'Simplification of reef structure through direct damage leads to a decrease in branching structures, a filling of holes and a general increase in habitat similarity. These modifications reduce the number of fish that can find refuge, the number of distinct habitat types, and probably the suitability of settlement sites for the larvae of particular species. Experiments with artificial reefs have amply demonstrated the critical importance of habitat architecture in attracting fish and maintaining diverse assemblages, although factors such as reef size and location may be even more critical.'

Sebens (1994), quoting Sale also notes that

> 'The physical structure of coral reefs can enhance fish diversity by providing physical refuges from predators, microhabitat differences, and numerous sites for recruitment. [However,] although the expected relationship is an increase of fish diversity with structural complexity, not all studies of natural reefs have demonstrated a strong correlation.'

One cannot help but agree with these assessments and it is worth pointing out that the same effect is occurring in some temperate systems, such as the fishery for the rock boring clam discussed in Section 4.1. The real issue here is not that marked and undesirable effects ensue when such fishing takes place, but the extent of the fishing activity. Unfortunately, data on this is likely to be very difficult to obtain, particularly for reef systems where fishermen are widely distributed and can range over large distances, or where the activity is actually illegal. However, the consensus seems to be that the problem is widespread. I suspect measures are probably urgently required in some areas to address the problem before habitat degradation reaches a level and geographic scale that will make decadal time-scales for recovery optimistic. As a general proposition fishing methods that destroy habitat structure are undesirable, and alternative gear types such as pots, traps and lines are greatly preferred. The economic incentives and legislative instruments that might be used to implement shift are largely outside the scope of this book, but they are touched upon in Chapter 9.

Part 2
Species Interactions

Introduction

Even species that are not directly exploited by a fishery are likely to be affected by the removal of a substantial proportion of their prey, predator or competitor biomass. In this Part the evidence for such ecological interactions is explored for a variety of systems. The nub of the issue is how closely coupled are the dynamics of populations? If populations are largely independent of one another because other environmental factors limit their numbers, removing one species will have no effect on anything else. If they interact strongly, there will be changes, although whether they will be measurable, or predictable, is another matter. This section addresses two questions:

(1) With precipitous declines in some species, have there been concomitant changes in the abundance of others?
(2) If so, how much of the response can be attributed to ecological interactions with the harvested species and how much to other independent environmental effects.

There are certainly strong indications that interactions with exploited species should be strong enough to lead to population effects elsewhere. For example, in an analysis of the energy budgets for six major marine ecosystems Bax (1991) found that the major source of mortality for fish is predation by other fish. Predatory interactions may, therefore, be an important regulatory interaction for marine populations. This possibility is not lost on fisheries biologists trying to build multispecies fisheries models that take account of such effects.

One of the best ways to determine the role that a species has in a system is to do an experiment – manipulate the density of the species in question and measure the reactions of the other populations. To satisfy the idealised rigours of the scientific method, such manipulations should be replicated in any given place and the experiment should also be repeated in several areas, and ideally on several occasions too. Unfortunately, such degrees of replication can rarely be achieved and efforts to approach them constrain most experiments to relatively small spatial and temporal scales. Even then, the number of studies that adequately address the factors that might confound the interpretation of experimental results is surprisingly small. Underwood (1984) gives an excellent account of this issue with respect to competition experiments. The constraints of good experimental design also tend to lead experimentalists to focus on animals for which experiments are tractable, not necessarily those for which the questions most pressingly need to be answered.

In fairness, it is difficult to see how one could possibly do tightly replicated experiments on marine fish populations. But, while we cannot claim the experimental high ground, we do have experiments of sorts. After all, fishers have performed the density manipulations, even if the experimental design fails to reach the standards that one should aspire to. Indeed, Walters (1986) argues that fisheries regimes can be considered as unplanned experiments and develops the idea of adaptive management to make use of the information such experiments generate (see Chapter 9). Much more can probably be learnt from closer scrutiny of data that lie in fisheries laboratories around the world, particularly since fisheries provide data on changes in abundance over spatial and temporal scales that are unmatched in most other systems.

Caution is needed though, because some data can be misleading – especially catch data, which is the thing there is most of. Catch data can be particularly problematic when trying to estimate the abundance of pelagic species. For this group, fishermen usually concentrate on dense aggregations so that catch rates remain high even when overall stock abundance is low. Problems with catch data also occur in mixed species fisheries because species composition in a catch will vary according to the target species that is being sought. Catches from two fleets fishing in the same area can give very different impressions of species abundance – a problem that is especially acute in the tropical fisheries (see, for example, Gulland & Garcia, 1984). The development of a fishery is also relevant in this respect, because the typical picture for a previously unexploited area is of a progressive change in the spectrum of resources that are harvested. This effect is well illustrated by the Senegalese trawl fishery where very marked changes in landings occurred as export opportunities changed and other resources were discovered. Changes in discarding practices can also confound any relationship between catch rates and true abundance.

To avoid the problems with catch data, a much better index of abundance is one that scales catch by the amount of fishing effort used to obtain it. Catch per unit effort (CPUE) is a common currency for fisheries biologists and it is certainly better than catch alone. But deciding on the most appropriate measure of effort for a fishery and actually obtaining the data both present a challenge. What measure should one use? Number of hours fishing? Number of boats? Days at sea? How does one account for improvements in the efficiency of gear over time, the advantage gained from radio communication between fishermen or spotter planes? None of the answers are obvious, and CPUE data must also be viewed with a degree of circumspection.

Perhaps the best kind of data available to us is that gathered on research vessel surveys of fish populations. Although often collected for other reasons – to provide early indices of the number of young recruits in the next year, for example – biologists have a habit of measuring as many variables from their samples as they can think of. This often makes the data suitable for asking different questions from those that were originally envisaged. Such data are drawn upon in the next two chapters.

Much of what follows is devoted to the effects of fish species on each other – this is partly because the best data on the responses of populations to fishing comes from the fisheries scientists themselves, who rarely gather data on non-fish taxa. However, one might also expect that, since fish are fed on primarily by other fish (Bax, 1991), it would be other fish species which are most affected. If competition is important, this is perhaps also likely to be most intense between fish species.

Part 2 contains two chapters. The first takes a fairly broad sweep through the available literature which considers the role of biological interactions in marine ecosystems. In 5.1–5.3 data are drawn upon from a range of ecosystems and communities to consider the effects of harvesting predators, competitors and prey species. Then a single system, the Southern Ocean, is looked at in more detail. This region has been chosen because its relatively simple structure and the large-scale harvesting of whales that has taken place might make the occurrence of ecological effects mediated through species interaction more likely. A great deal has also been written about it. For the most part Chapter 5 ignores effects on demersal fish communities because this forms the basis for Chapter 6. Chapter 6 examines what has happened in three demersal fishery systems that have been subjected to intense fishing pressure. At the end of each chapter I have tried to synthesise the available information and draw some general conclusions about fisheries effects mediated through ecological interactions.

Chapter 5
The Responses of Predators, Prey and Competitors to the Removal of Target Species

5.1 Top-down control: the effects on prey of removing predators

There have been numerous experimental demonstrations that predators can control the abundance of their prey in a wide variety of terrestrial and marine systems (Pimm, 1991). In fisheries the descending limb of dome-shaped stock recruitment curves has often been attributed to cannibalism by older fish, and in recent years there have been efforts to explicitly model the effects of predation mortality in fisheries assessments. What is not clear, however, is how prevalent such top-down effects are in natural systems and whether there are any general principles that allow us to identify when top-down control might be expected. In temperate marine fish communities, for example, control of prey populations by predators may be very variable owing to the large, environmentally driven fluctuations in recruitment success. Consequent changes in the abundance of predators or prey may lead to marked variations in the importance of predator control on relatively short (1–10 yr) time-scales. Perhaps, for other communities, such as those on rocky shores or in the tropics, the pattern is more predictable. Before considering such possibilities, however, we should review the available information.

Mackerel and herring

In the pelagic realm, a number of possible examples of a release from predation have been suggested for sandeel (sand lance) in response to declines in mackerel and herring (e.g. Hempel, 1978). Sandeels are probably a very important component of many temperate marine systems. They often form a substantial component of the diet of both pelagic and demersal fish predators, seals, cetaceans and seabirds. Mackerel and herring are rather strong contenders for controlling sandeels and other small pelagic fish species because they not only eat them, but may also be competitors for smaller zooplankton prey such as copepods. It is tempting, therefore, to conclude that there is a functional link between observed declines in these

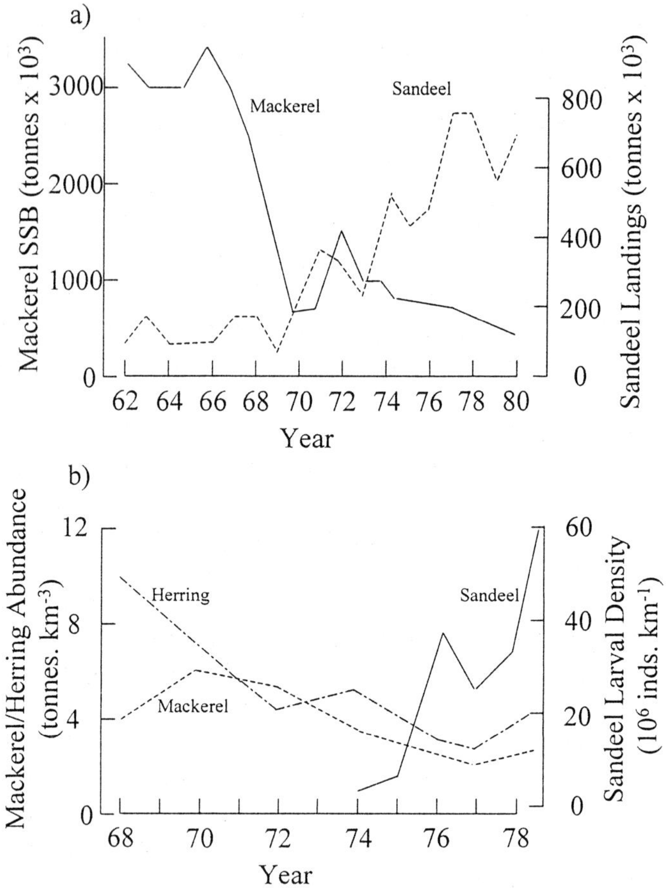

Fig. 5.1 (a) Trends in the spawning stock biomass of mackerel and sandeel landings in the North Sea (redrawn from Jones (1983) with permission from the Food and Agriculture Organization of the United Nations). (b) Trends in herring and mackerel abundance biomass and sandeel larval abundance for the northwest Atlantic shelf (redrawn from Sherman *et al.*, 1981).

species and coincident increases in sandeel abundance. Figure 5.1 shows two examples that would appear to fit this pattern, one from the North Sea, the other from the northwest Atlantic. Let us consider each of them in turn.

Fishing played an important part in the decline of North Sea mackerel stocks, which fell precipitously between 1965 and 1970, coincident with the development of the Norwegian purse-seine fishery. During this period, herring also collapsed, although the role that fishing played is less clear. In this case poor recruitment during the 1960s probably resulted from a combination of environmental events affecting larval mortality and low levels of spawning stock as a result of exploitation. Regardless of the cause, at the same time as the mackerel and herring stocks declined, landings of sandeels (and other small fast-growing pelagic species) increased (Jones, 1983). Anderson & Ursin (1977) noted the same thing and their analysis adds weight to the hypothesis that the events may be linked by showing a good negative correlation between the catch of sandeels, sprat and Norway pout, and the size of the adult stock of herring and mackerel three years earlier. This lag in response could be explained as the time required for the sandeel population to respond to changes in the abundance of adult herring and mackerel.

Although these patterns seem rather compelling, a number of factors hinder interpretation. First, landings data often provide very poor indices of stock size for pelagic species. Thus, much of the trend in the sandeel abundance could be a consequence of fishermen diverting their effort from the dwindling mackerel stocks and from the industrial fishery for juvenile herring and whiting to the abundant sandeels, which had previously been ignored. Unfortunately, we have no good fishery independent estimate of sandeel abundance, so we cannot be certain whether sandeel abundance in the North Sea has really changed, let alone whether harvesting other species caused it. What we do have, however, is one very good study of local sandeel population dynamics by Wright & Bailey (1993) that gives some additional insights. Wright & Bailey (1993) show that during the 1980s sandeel abundance has increased in parallel with an increase in herring in the northwestern part of the North Sea, suggesting that, for herring at least, the argument for top-down control may be weak. We also know that marked oceanographic changes occurred in the North Sea during the 1970s and that for some sandeel populations, oceanographic factors play a very important role in recruitment to the stock (Wright & Bailey, 1993).

The second example for the east coast of North America has also been subject to considerable analysis. Perhaps the most prominent of these is that of Sherman *et al.* (1981) who improved considerably on the North Sea situation by using larval survey data to provide an index of abundance for sandeels. These data are clearly consistent with a predation effect. For the same system, Fogarty *et al.* (1991) attempted to quantify the strength of the interaction of herring and mackerel on sandeel, using time-series data spanning the period from 1970–1986. Significant negative effects of a combined index of mackerel and herring abundance on sandeel recruitment were found. Stomach contents data also showed that sandeel formed a large proportion of the diet for both these species. It was not possible to formally separate the effects of the two predators and determine their relative importance. However, the timing of a previous sandeel outburst in the early 1960s, which coincided with low mackerel but high herring abundance, led the authors to suggest that it may be mackerel that are the dominant controlling influence. Analysis of the relative changes in the abundances of potentially interacting species cannot determine causal relationships, nor can they refute them. Nevertheless, the study by Fogarty *et al.* (1991) does suggest that harvesting mackerel and herring might lead to increases in the abundance of sandeel.

Top-down control in the Baltic

The majority of work on top-down control in open water systems has been in lakes. Here its importance is considerable (Carpenter & Kitchell, 1993). The closest to a lake in the marine environment is an enclosed sea like the Baltic, and there are good reasons to suppose that top-down control may also be important here. Rudstam *et al.* (1994) note, for example, that fish yield per unit primary production is 5–10 times higher in the Baltic (and in other semi-enclosed coastal areas) than it is in lakes (Nixon, 1982). This feature suggests a strong coupling between fish and lower trophic levels. The Baltic food web is also relatively simple (Fig. 5.2), owing largely to the intermediate salinity, which limits the number of species that can persist there. If top-down control is important it may, therefore, be easier to detect here than in more complex systems (Rudstam *et al.* 1994). Basing their assessment primarily on estimates of food consumption by the various species in the system, Rudstam *et al.* (1994) reviewed the evidence for top-down control.

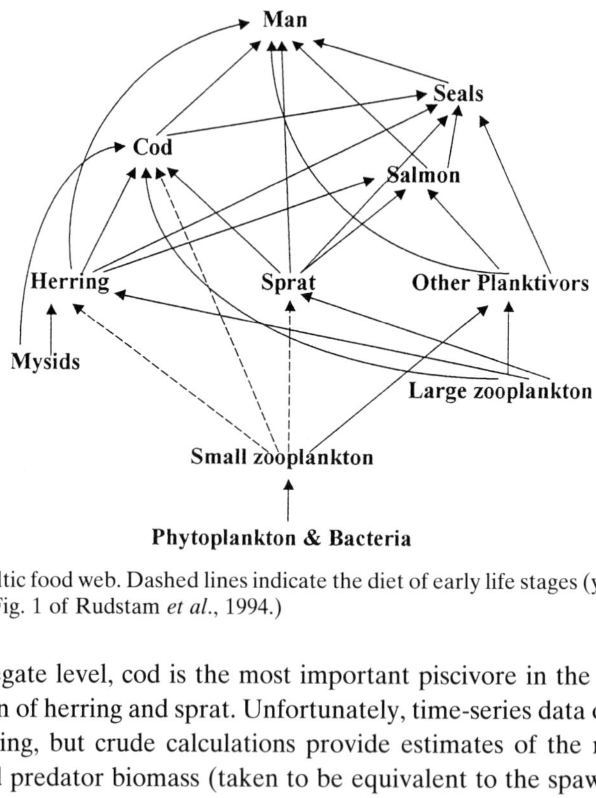

Fig. 5.2 The Baltic food web. Dashed lines indicate the diet of early life stages (young of the year). (Adapted from Fig. 1 of Rudstam *et al.*, 1994.)

On an aggregate level, cod is the most important piscivore in the Baltic, eating a high proportion of herring and sprat. Unfortunately, time-series data on consumption levels are lacking, but crude calculations provide estimates of the ratio of clupeid biomass to cod predator biomass (taken to be equivalent to the spawning stock biomass, since it is mainly mature cod which eat clupeids). This ratio was approximately 16:1 in the early 1970s, but by the late 1970s cod biomass had reached a peak, due to good recruitment a few years earlier, and the ratio fell to 4:1. By the early 1990s the ratio had risen again to 90:1. Growth rate of cod is highly correlated with this ratio, supporting the contention that the latter is a reasonable index of food availability.

Since cod is the dominant predator, clupeids should respond to changes in the cod stock if top-down control operates. There is some evidence that this may have occurred (Fig. 5.3); but it is weak and perhaps masked by environmentally driven

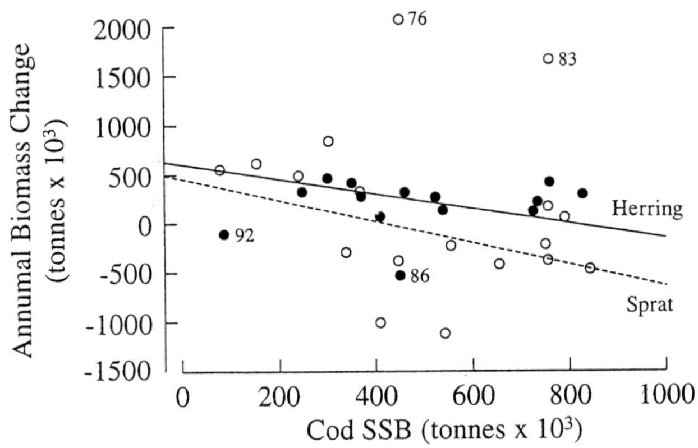

Fig. 5.3 The annual change in biomass of Baltic sprat and herring in relation to cod spawning stock biomass. Relationships are negative but not significant unless the two outliers in each data set are excluded. (Re-drawn from Rudstam *et al.*, 1994.) ● = Sprat; ○ = herring.

variability in clupeid recruitment. Although cod are, perhaps, the primary candidate for exerting a controlling influence, clupeids may also be important in this respect through predation on cod larvae and especially eggs. In support of this idea, Rudstam *et al.* (1994) notes a strong negative correlation between sprat and cod catches from 1960–1991, which could have been caused by sprat predation on cod eggs.

These various observations led Rudstam *et al.* (1994) to suggest two competing top-down control mechanisms, one leading to gadoid (cod) dominance, the other leading to clupeid dominance (Fig. 5.4). In the first (Fig. 5.4a) a rise in cod abundance leads to a decline in clupeid abundance through predation. This in turn leads to increased recruitment success for cod owing to reduced predation by clupeids on their eggs. An important limit on the ultimate size of the cod stock is cannibalism by adult fish on their young. The second (Fig. 5.4b) occurs with a rise in clupeid abundance which would increase predation on cod eggs, thereby reducing the number of adult cod. Since there is some evidence for a stock recruitment relationship for cod in the Baltic, it is to be expected that reducing the number of adults will itself contribute to lower recruitment in subsequent years.

Clearly if these mechanisms operate we have positive feedback systems which lead to dominance by either cod or clupeids. The question is what might cause a flip between the two states? Compared to clupeids, the natural variability in cod stocks is low; thus, in unexploited systems, a decline in cod numbers that would allow herring to increase is perhaps unlikely. Because the system is exploited, however, cod numbers have been reduced (although eutrophication and pollution effects are also believed to have contributed to this decline). Herring stocks are also high at present, which is of course consistent with the top-down control theory. Because cod are more commercially valuable than herring, a cod-dominated system is a more desirable state for the fishermen and if herring are indeed maintaining cod at low numbers, reducing herring densities may facilitate a recovery by cod. This would certainly be an

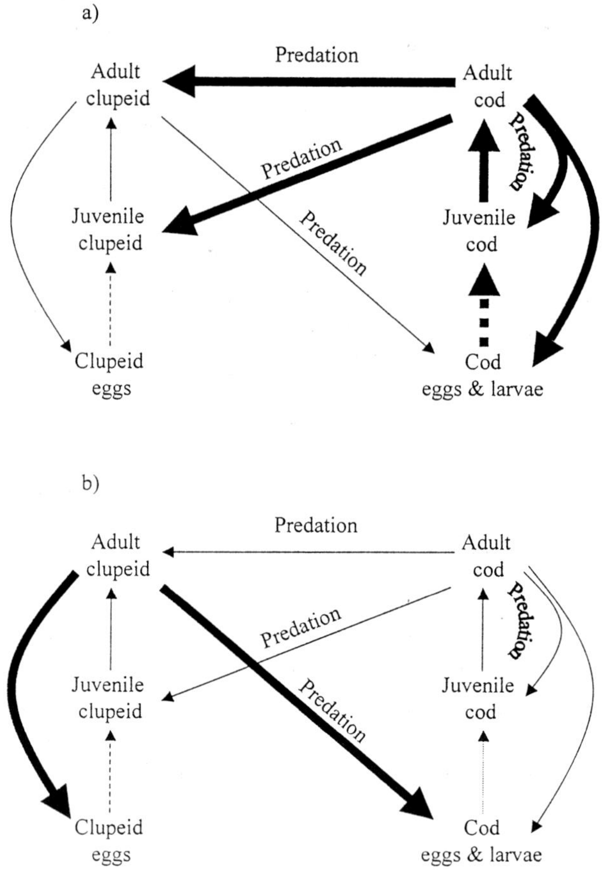

Fig. 5.4 Two competing top-down control mechanisms: (a) leads to cod dominance, (b) leads to reductions in adult cod abundance (for further explanation see text). Arrows not marked as predatory denote developmental (life-cycle) progression. Thick arrows denote stronger interactions, dashed lines denote strong external (hydrographic) influences. (Adapted from Fig. 7 of Rudstam *et al.*, 1994).

interesting experiment, but the low commercial value of herring means that fishermen would have to be heavily subsidised to undertake such an operation.

Sharks in Natal

A good example of the effects of removal of a top predator comes from a study by Van der Elst (1979) on the catch returns from the shore-based Natal sport fishery. Van der Elst (1979) showed that over a 20 year period there has been a decline in the numbers of large predatory sharks as a result of gill-netting to protect beaches for swimmers. In parallel with this decline there has been an increase in the numbers of the smaller-bodied elasmobranch species. These form the principal prey of the larger sharks (Fig. 5.5). It is postulated that this increase in small sharks has been

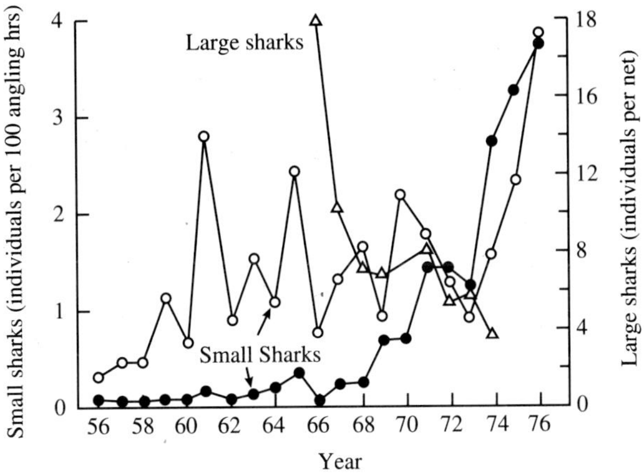

Fig. 5.5 Trends in the abundance of large shark species and of two smaller shark species on the Natal coast. Large shark abundance estimated from catches in the shark netting programme which began in 1964. (Adapted from Figs 6 and 12 of Van der Elst (1979) *Environmental Biology of Fishes* **4**, 349–62. With kind permission from Kluwer Academic Publishers.)

responsible for a decline in teleost fish stocks, which comprise their prey. These fish species form the basis for important coastal sport and commercial fisheries. Without removing the control measures for large sharks it is uncertain whether these changes are reversible. Hazarding a guess, they probably are, but it is always possible that predation on the juveniles of large species by the adults of the abundant smaller species might either prevent this or slow it down.

Responses in the Benthos

Perhaps the study on predation effects on benthic species that is most directly relevant to fishing effects is that of Witman & Sebens (1992), who examined regional variations in the intensity of fish predation on brittlestars and rock crabs occupying rocky sub-tidal ledges in the Gulf of Maine. Predation rates on tethered animals, obtained using time-lapse video and diver observation, showed that inshore sites experienced significantly lower predation by fish and supported larger populations of invertebrate prey than offshore sites. These differences in predation intensity were directly related to fish predator abundance, which in turn can be attributed to differences in fishing pressure. A fishery on inshore ridges and ledges for cod, haddock and other species has existed in the Gulf of Maine for many years. The authors speculate that fishing has reduced fish densities to such an extent that epibenthic crustacean prey populations have been released from predator control and have increased in size.

In a similar study, Aronson (1989) examined the predation intensity on dense brittlestar beds and on rocky reefs where ophiuroids were sparse. These results also suggested that fish predators were important in controlling prey populations because fish abundance and predation intensities were much lower in dense brittlestar beds. Aronson also offered a valuable evolutionary perspective, arguing that such predation effects were the most likely cause for changes in the structure of shallow-water benthic communities during the late Mesozoic. At this time dense populations of sessile and semi-mobile suspension feeders disappeared from most soft-bottom habitats and were replaced by infaunal suspension feeders and more mobile infaunal deposit feeders. This change coincided with the rapid diversification of crustacean and teleost predators, which led Vermeij (1977, 1987) to propose a causal connection.

Following from the above arguments it has been suggested that the numerous dense beds of the brittlestars *Ophiothrix fragilis* and *Ophiocoma nigra* found around the UK today are an anachronism, existing because fish and crab densities in these areas are low. In contrast to the Witman & Sebens case, however, it would be difficult to argue that low predator densities over UK brittlestar beds was a result of fishing, because some of the beds have been known to exist for over 100 years.

An important point about both Aronson's (1989) and Witman & Sebens' (1992) study is that they deal with the control of exposed non-cryptic invertebrate prey by predators and not on the cryptic species or infaunal taxa. In this regard, Aronson notes that 'It is significant that dense infaunal populations [of brittlestars] persist in the presence of fish predators, whereas dense epifaunal populations do not'. This suggests that fish predators may not exert a strong controlling influence on infaunal prey population densities, despite often high levels of sub-lethal damage (Bowmer & Keegan, 1983; Duineveld & van Noort, 1986). Such conclusions argue against the idea that reducing the numbers of fish predators in soft-sediment areas dominated by infaunal prey is likely to have marked positive effects on prey densities. An additional consideration here, of course, is the direct mortalities that demersal trawlers impose on benthic populations.

Predation and trophic cascades

In 1960 Hairston *et al.* published a paper in which they advanced the now famous 'The World is Green' hypothesis. The basic idea behind their argument was that herbivore species are not generally limited by the amount of plant material available and that if grazer abundance was not limited by food, it must be predators that served that function. Furthermore, since grazer populations must be predator limited, predator populations must be food limited. Developing on this theme a generic hypothesis emerged suggesting that organisms at the top of the food web are food limited and that at successively lower levels they are alternatively predator and then food limited (Schoener, 1989; Fig. 5.6). Although these ideas were developed with terrestrial systems in mind they were later adapted to aquatic habitats.

Accepting this idea of alternating mechanisms of control at different trophic levels allows the possibility that changes at the top of food chains with strong trophic links between them might cascade downward (McQueen *et al.*, 1986, 1992;

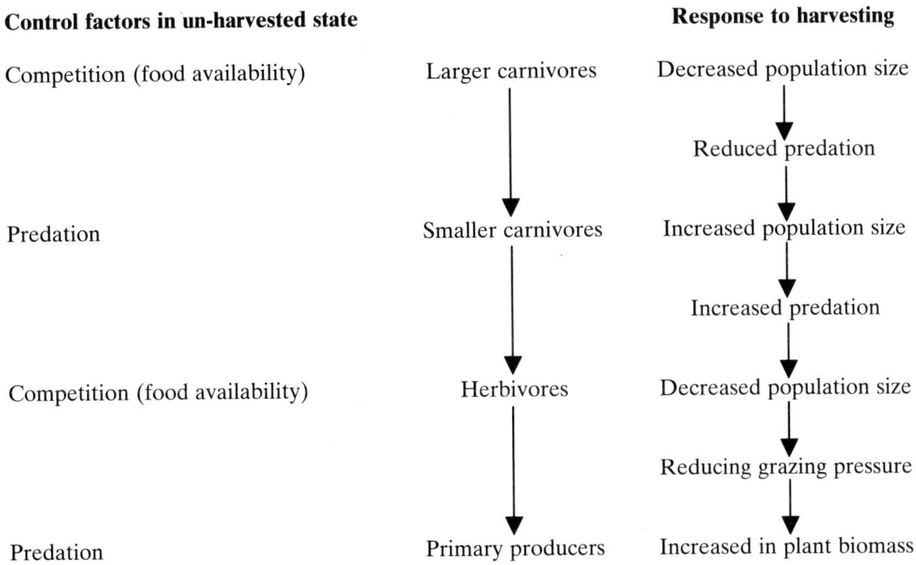

Fig. 5.6 The trophic cascade model for a 4-level system with harvesting at the top level.

Carpenter *et al.*, 1985). Thus, the trophic cascade model predicts that by their presence or absence, higher trophic levels will determine whether or not conspicuous accumulations of plant biomass will occur at the base of the system.

Pelagic systems

In a stimulating review of control mechanisms in pelagic systems Verity & Smetacek (1996) argue that trophic cascades might be more likely in aquatic than terrestrial systems. So, do we see cascades in marine systems, and if so, is fishing implicated? For the marine pelagos at least, it looks like the answer to the first question is basically 'no'. Verity & Smetacek (1996), found few in marine systems, and none of particular relevance to fishing activity. This finding is in marked contrast to the excellent examples of cascades in freshwater systems (Carpenter & Kitchell, 1993). These cascades form the basis for biomanipulative lake management programmes in which, for example, the harvesting of zooplanktivorous fish leads to an increase in water clarity through increases in zooplankton densities and consequent increases in algal grazing (McQueen, 1990).

Interestingly, in about three-quarters of the marine pelagic examples that do suggest cascades, the key predator was a gelatinous zooplankter. This observation led Verity & Smetacek (1996) to suggest that the key characteristics of organisms involved in pelagic cascades were: (1) the ability to rapidly increase population biomass, (2) a high weight specific ingestion rate and/or (3) large predator prey size ratios. The authors also speculated on the features of the ecology and biology of marine fish species that might make a role for them in trophic cascades unlikely. In particular, for a trophic cascade to work, the predator should be food limited. There is, however, considerable evidence that many pelagic fish species may often not be food limited. The typical lack of a density-dependent stock recruitment relationship, the occurrence of strong year classes emerging from weak spawning events, and the huge variation in fish recruitment among years with no concomitant large changes in the abundance of resources all support this idea. This contrasts sharply with the situation for fish in freshwater systems in which fish seem to be between 10 and 1000 times more abundant than in the oceans (Horn, 1972).

The only evidence I am aware of which supports the trophic cascade hypothesis comes from the Peruvian upwelling region that supported the anchoveta fishery. Rowe (1981) looked at the concentrations of organic matter in sediments below the upwelling zone and found that samples taken in 1969 before the collapse of the fishery had 3.3% organic matter. In 1972, after the collapse of the Anchoveta fishery, the values had risen to 5.7%. Rowe (1981) postulates that this increase may have occurred through an increase in the settlement of phytoplankton to the seabed, assuming that a large fraction of the primary production was previously consumed by anchovy. There appear few data to test this hypothesis, although very crude 'back of the envelope' calculations suggest that the magnitude of the change in organic matter could in principle be accounted for by the observed decline in the anchovy population.

Temperate sub-tidal benthic communities

While trophic cascades seem rare in marine pelagic systems, they may be more common in hard substrate benthic systems. In these habitats some of the best-known examples of strong predator effects have been experimentally demonstrated. Indeed, much of the theory about the role of the aptly named keystone species which act as control agents in communities stems from pioneering work on rocky shores (e.g. Paine, 1980). There are, perhaps, good reasons why trophic cascades might be prominent in such systems. First, a hard substratum, offers an attachment surface for macroalgae which, if grazers are under predator control, can dominate space and eliminate sessile animal competitors. Second macroalgae, unlike trees, do not have to invest in structural (woody) tissue owing to the support that the fluid medium supplies. If grazers are released from predator control, they therefore have the capacity to severely affect plant abundance because most of the plant tissue is palatable to grazers.

A good example of a cascade involves the sea otter (*Enhydra lutris*), which preys on the benthic invertebrates that inhabit shallow coastal waters of the temperate and boreal north Pacific region. Before the mid-1700s otters were probably extremely widespread, occupying the northern Japanese archipelago to the central coast of Baja California. Throughout the eighteenth and nineteenth centuries, however, the species was almost brought to extinction by fur-hunters and was eliminated from a large part of its range. When otters were protected in 1911 recovery of populations began and it appears to be almost complete in Alaska and the Soviet Union. However, there are still island and mainland sites that used to be populated by otters, but which have yet to be colonised. Evidence for the role of otters in the system is largely based on the comparison of these two situations. It would be stretching a point to argue that hunting for furs constituted a marine fishery, but examining the effects of reducing otter numbers is relevant when attempting to evaluate the role played by other predators, which are exploited by fisheries and which have similar diets to otters.

Sea-otters are very effective predators and appear to limit the numbers of their prey with remarkable success. It seems likely, therefore, that the high density of otters limited populations of abalone, clams, crabs and perhaps other invertebrates (e.g. urchins) in these regions. When the otters were removed it appears that no other species was capable of taking over their ecological role, so that by earlier standards dense populations of their invertebrate prey developed. Ironically, this increase in otter prey populations probably stimulated a number of commercial and recreational shellfish fisheries on the West Coast of the USA (Estes & van Blaricom, 1985).

Probably the most well-documented effect of the increase in prey species leading to a trophic cascade comes from studies in Alaska. The basic mechanism for the cascade is that otters control the densities of sea urchins, which, if their populations are unchecked, can have profound effects through grazing kelp forests. Urchin standing crop can vary by a factor of 9, due solely to the influence of otters (Fig. 5.7). With this range of biomasses production from kelp can range from 13 g carbon m^{-2} day^{-1} to unmeasurably low. So marked is the effect of high urchin numbers that features known as 'urchin barrens' can develop. In these barrens large canopy-forming algae have been completely eliminated. Estes & van Blaricom (1985) argue that such barrens could never occur if otters were present; they are quite widespread in areas where otters have yet to re-establish.

The evidence for a functional link between sea-otter decline, the release of algal-grazing urchin species and resultant changes in the productivity and species composition of the algal community is very strong indeed (Estes & Palmisano, 1974; Duggins, 1980). What characterises this example is the importance of a control on primary space occupiers – the algae. It may be that, for species-rich systems, cascading changes are most likely to occur in situations where one of the links involves a species with a firm grip on real-estate. For relatively species-poor systems such as those which occur in freshwater this rule does not seem to hold. In the case of the Pacific sea-otter an important point is that dramatic changes have reversed as populations have recovered.

Based on the otter experience the potential for trophic cascades in response to the removal predators in rocky sub-tidal communities seems quite large. But are there any harvested species which might play the same role? One candidate is the American lobster (*Homarus americanus*) which, like the otter, preys on sea urchins

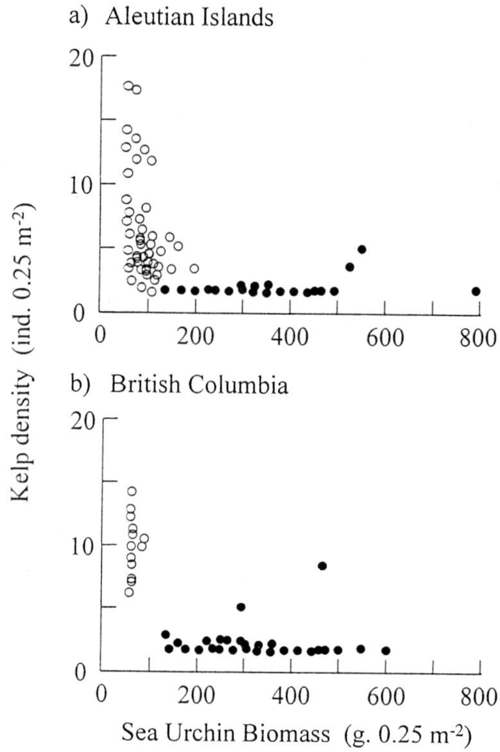

Fig. 5.7 The abundance of kelps and sea urchins in areas with and without sea otters. (Redrawn from Estes, 1996.) ● = Otters absent; ○ = otters present.

and might control their abundance. Data in support of such a hypothesis emerged in the early 1970s when there was a population explosion of sea urchins along the Atlantic coast of Nova Scotia, which was coincident with a dramatic decline in the commercial landings of lobsters. The resulting intense algal grazing converted macroalgal beds into coralline barrens (Mann & Breen, 1972). Over the following 15 years a large number of studies were undertaken to explain the alternation of macroalgal beds with sea urchin barrens. Although early papers strongly supported the idea that lobsters were indeed an important causative agent, more recent work suggests a temperature- and disease-induced 15–20 year cycle between these two ecological states. The approaches adopted and the conclusions drawn from the work conducted over this 15-year period is the subject of an illuminating analysis by Elner & Vadas (1990) which highlights some the shortcomings in the ecological research undertaken to address issues of this kind. This work is returned to in Section 5.5.

Rocky shores

In many regions of the world, human harvesting of the rocky inter-tidal has been going on for centuries. Recent studies in Chile, for example, suggest that humans could have exerted significant fishing pressure on inter-tidal invertebrates up to 12 000 years ago (Moreno *et al.*, 1986). As in all other systems, the best way to understand the role that harvesting might play is to do an experiment. Rocky shores offer opportunities for such experiments that are unrivalled in other marine systems. The most obvious experiment is of course to simply exclude humans from the shoreline over an extended period and observe how the system responds. This is the approach that was adopted in a series of experiments in Chile (e.g. Moreno *et al.*, 1984, 1986; Castilla & Duran, 1985). Moreno *et al.* (1984) report that in the first two years of the experiment there was an increase in the abundance of a herbivorous limpet species and a concomitant decrease in the cover of algae. As a result of the reduced algal cover, the densities of mussels and barnacles also increased. After about two years, however, effects of changes in the population of the predatory gastropod (*Concholepas concholepas*) began to occur. This species is an important harvested resource and in the protected area increased grazing pressure by *Concholepas* led to a decline in mussel and barnacle prey. Interestingly, this increase in grazing pressure was largely a result of an increase in the mean size of individuals rather than an increase in density. The change in the size of individuals compared to the unprotected areas was consistent with the patterns of size selectivity shown by the fishery. Similar effects of an increase in *Concholepas* grazing pressure were also observed in a separate exclusion experiment by Castilla & Duran (1985). This effect occurred more quickly through a numeric response by the *Concholepas* population.

In other areas indirect effects are less apparent. Keough *et al.* (1993), for example, examined the population structures of inter-tidal molluscs at eight sites in Port Philllip Bay, Victoria, Australia. Two of these sites had been protected for around 70 years because they were on a rifle range. Humans frequently visited the remainder of the sites. Surveys of human activity at these sites showed that, for those that were heavily visited, 25% of the people were actively collecting, despite regulations prohibiting the removal of inter-tidal gastropods. Analysis of the size structure and abundance of the mollusc populations showed that non-collected species did not differ between visited and protected sites, whereas three out of four collected species were significantly larger at protected sites and one was significantly more abundant.

Coral reefs

Fish serve important functions on coral reefs. Herbivorous species remove benthic macroalgae which compete for space with corals and other invertebrates, and schooling fish which take shelter in coral heads can increase coral growth rates by supplying nitrogenous waste products (Sebens, 1994). In addition, large predatory fish may control the populations of benthic invertebrates, and of smaller reef fish, thereby potentially enhancing the diversity of both groups by limiting competitive dominants.

The effects of fishing on the fish community in reef areas can be very evident. At heavily fished sites in Jamaica, for example, it is rare to see a fish over 30 cm long (Sebens, 1994). Such reductions almost certainly affect the dynamics of the rest of the system, but it is often in conjunction with other factors. For example, in the Caribbean there was a disease outbreak, which led to the destruction of the algal grazing sea urchin (*Diadema antillarum*). This decline led in its turn to increased growth of benthic algae (Hughes *et al.*, 1987) which threatens the survival and recruitment of corals and has slowed the recovery of these reefs from hurricane damage (Liddell & Ohlhorst, 1993). In addition to reduced densities of urchins, however, there were also low numbers of herbivorous reef fish, which probably contributed to the problems. These low numbers were almost certainly a consequence of fishing activity. Contrasts between Jamaican reefs and those in the Cayman Islands which are less than 100 miles away and subject to much less fishing activity are very marked (Sebens, 1994) (see also Koslow *et al.*, 1988). Figure 5.8 summarises the basic relationships that obtain in a Caribbean reef community; it illustrates how fishing and these other factors can interact.

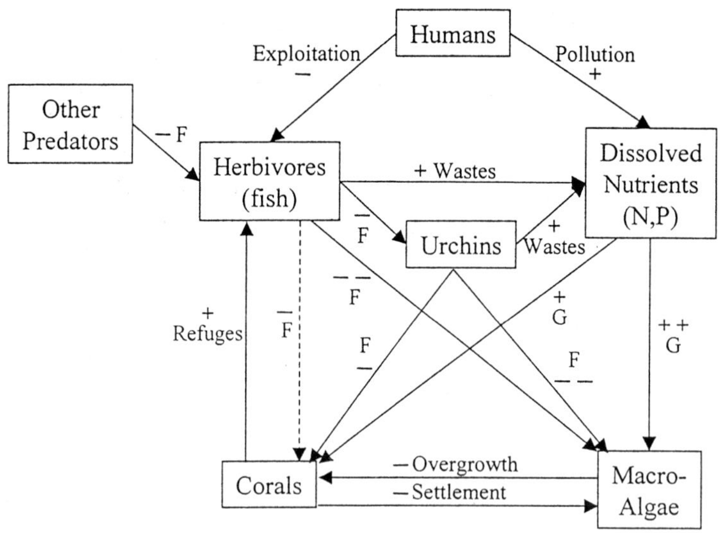

Fig. 5.8 Relationships between corals, macroalgae, nutrients, and consumers. Signs of interactions denote positive or negative effects. (Redrawn from Sebens, 1994.)

Although pathogens have reduced urchin densities in some areas, the large populations that existed previously, may themselves have occurred through the actions of fishing. For example, many well-studied parts of the Caribbean were near human population centres. This may explain the high densities of D. *antillarum* populations with their consequent controlling influence on algal cover (Hay, 1984). Evidence for this hypothesis came from assessing fish and urchin grazing on strips of seagrass placed on reefs designated as 'overfished' or 'less-fished'. A hypothesis was developed which stated that *D. antillarum* is the dominant herbivore in areas subject to heavy fishing pressure, but herbivorous fish dominate in more pristine environments where urchin densities are controlled by fish predators. This hypothesis has been supported by Levitan (1992) who correlated historic and regional changes in the mouth parts:body size ratio in urchins (a measure of food limitation) with changes in human population size and fishing pressure. Since the mouth parts and urchin test are hard, museum collections could be used to examine historic trends. Levitan showed that urchins did appear to be more food limited in areas of high human abundance supporting the ideas that populations had been released from predator control. If fishing had led to a competitive release through a reduction in the numbers of herbivorous reef fish, this should have led an increase in the food available to urchins and the opposite trend in the ratio of mouth parts to body size.

Of course, effects are not restricted to the Caribbean. For example, McClanahan & Muthiga (1988) compared Kenyan reefs with different historical and observed levels of fishing exploitation. They showed that more exploited reef lagoons had greater sea urchin densities and sizes, fewer and smaller fish and less coral cover. In the most highly exploited lagoons the sea urchin biomass increased 5-fold over a 15-year period. McClanahan & Muthiga (1988) postulate that fishing leads to a reduction in sea urchin predator and competitor densities which leads to a release of urchin populations. This increase in urchin densities in turn leads to a reduction in coral cover and calcium carbonate accretion and an increase in rates of bioerosion, compared to fish-dominated communities. The authors also develop an hypothesis regarding competitive interactions between urchin species in which the dominant competitor, which normally has to occupy crevices to escape predators, excludes the others in the system.

McClanahan & Muthiga (1988) address the possibility that their results might be due to intrinsic differences between sites rather than an effect of fishing, but a number of factors argue against this possibility. First, there are anecdotal reports that urchins used to be found in an area where fishing activity is now prohibited. Second, in an area where fishing has occurred consistently, urchin densities have increased. Since all sites are subjected to the same current systems and are separated by relatively short distances, it is unlikely that stochastic differences in sea urchin, fish or coral recruitment, the spread of disease, or physical differences between sites could explain the results. Interestingly, these results suggest differences in the response of reefs in the Indo-Pacific region, compared to the Caribbean, mainly due to the differences in the diets of the urchin population. It would appear that the important effect on Indo-Pacific reefs is an increase in bio-erosion caused by coral grazing by omnivorous urchin species. In contrast, algal grazing dominates on Caribbean reefs.

Speculations on the Black Sea

The Black Sea fishery from 1960–1970 comprised 26 species that were commercially exploited, many of which were larger predatory species. In the period from 1980–

1990 this number had declined to five species all of which were relatively small-bodied – a familiar response to heavy fishery exploitation. During this same period the total landings increased from 300 000 to 600 000 tonnes per annum. Concomitant with these changes was an increase in the abundance of jelly plankton from 10^5 to 10^8 tonnes (FAO, 1994), particularly of *Mnemiopsis*, a ctenophore, which is not endemic to the region, but was probably introduced in ballast water.

The *Mnemiopsis* explosion is incontrovertible, but what is not clear is whether the species would have obtained such a foothold if the fish populations had not been reduced. There is even the possibility that a positive feedback loop now operates in which *Mnemiopsis*, having found a niche through reduced densities of fish, further facilitated their decline through increased predation on larvae. The possibility of such interactions is intriguing, but at present they remain little more than educated speculations.

5.2 Harvesting lower in the food web: the effects of prey removal

So far this chapter has focussed primarily on the effects of removing predators, but of course some harvested fish species also constitute important prey resources. One might reasonably expect that reductions in the abundance of an important food through fishing will have major consequences for its predators.

The effects of fishing on the food supply to birds is probably one of the best studied questions. ICES, for example, which has traditionally focussed almost exclusively on the management of fish stocks, now supports very active study groups examining seabird/fishery interactions. Montevecchi (1993) provides numerous examples of correspondences between the collapse of fish stocks and breeding failures or population declines in seabirds. These effects usually occur when both the fishery and the birds exploit the same sized individuals of a prey species (usually pelagic fish). Perhaps the best known example is the collapse of the anchoveta fishery in Peru. At its peak around 1970 the Peruvian anchoveta fishery was by far the largest single-species fishery in the world, but by 1973 landings had collapsed to 16% of their peak value. This collapse was probably partly as a consequence of heavy fishing, but also due to the effects of an El Niño event in 1972. From 1970 to 1975 the harvest declined from 12 million tonnes per year to 2 million tonnes and it has stayed low ever since. In turn there was a dramatic decline in seabird populations, which have also yet to recover to their earlier levels. By the late 1970s the anchovy eating guano bird population had fallen from 25 million to 5 million individuals.

Although many studies have demonstrated a relationship between seabird prey harvests and fishery catches, for the most part, these correlations have been found only at scales equivalent to the foraging range of the predators, and have been thought to reflect local fluctuations in prey abundance. More recently, however, correlations at larger scales have also been found. For example, Montevecchi & Myers (1995) showed significant correlations between the catches of mackerel and short-finned squid by the Newfoundland inshore fishery and the prey harvest of northern gannets (*Sula bassana*). These correlations obtained for fish abundances estimated at a range of different spatial scales, from the local abundance around the colony, the abundance over the foraging range of the birds (approx. 200 km), and for the entire Newfoundland region (thousands of kilometres). These observations suggest that birds can be useful bio-indicators of the state of fish populations.

The Shetland sandeel debate

The lesser sandeel (*Ammodytes marinus*) is one of the commonest fish species on the continental shelf of northwest Europe and accounts for somewhere between 10 and 15% of the total fish biomass of the North Sea (Sparholt, 1990). As noted earlier, sandeels are of considerable ecological importance owing to their central place in the diet of many fish, seabird and marine mammal species (Harwood & Croxall, 1988; Daan, 1989; Furness, 1990). Sandeels are also commercially important, supporting the largest single-species fishery in the North Sea. It is not surprising that these factors have led to concern over the potential impact of sandeel harvesting on prey availability to marine predators.

Up until 1990 an important sandeel fishery in the northwestern North Sea operated close to the Shetlands. Exploitation started in 1974 and rose to a peak in landings of just over 52 000 tonnes in 1982. During the 1980s the landings from the fishery fell, due to a decline in both the stock and the economic value of sandeels, and in 1991 the fishery ceased to operate. Coincident with this decline in the fishery in the mid- to late-1980s was a general decline in the foraging and breeding success of kittiwakes, fulmars, Arctic skua, puffin, great skua, razorbill and black guillemot. In particular, there was an almost total breeding failure for Arctic terns in 1984 (Fig. 5.9). Perhaps inevitably, the decline in sandeels on which the seabirds rely so heavily was blamed on the fishery (Avery & Green, 1989). Indeed, public opinion was so strong on this matter that a number of environmental organisations jointly funded a programme of research to determine the extent of the link between the two.

The sandeel fishery might have affected the birds in three ways: (1) through effects on the total sandeel stock, (2) by slowing stock recovery following poor year classes, or (3) through local depletions near seabird colonies. All of these hypotheses were examined during the jointly funded study, the findings of which are summarised by Wright (1996). Based on fisheries stock assessments and research survey data, Wright concluded that it was variability in the survival and immigration of pre-recruit stages into the Shetland area that determined the changes in sandeel abundance for birds. In other words, the adult stock, which the fishery exploits, was large enough to produce sufficient larvae, but effects during a life-stage over which the fishery could have no influence determined availability. A parallel situation

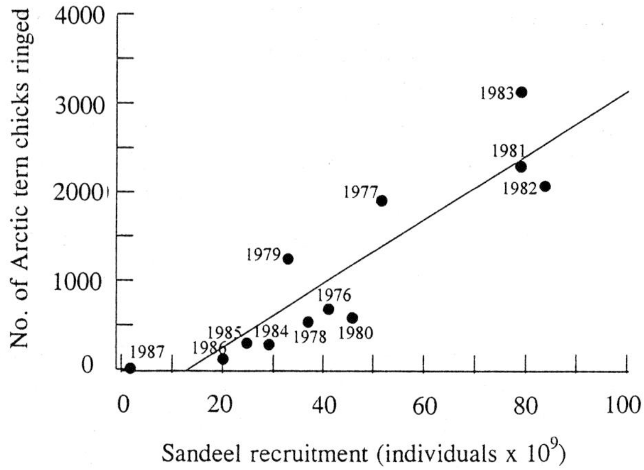

Fig. 5.9 The relationship between the number of Arctic tern chicks ringed in Shetland each year and the recruitment of O-group sandeels. (Redrawn from Monaghan *et al.*, 1989.)

appears to apply in the Bering Sea (Collie, 1991). Other data also suggest that the fishery was not responsible for the decline in the stock. For example, an initial decline in recruitment preceded any change in the size of the spawning stock (Kunzlik, 1989). Following the closure of the fishery, the apparent lack of a relationship between recruitment and spawning stock has been further highlighted. Although the spawning stock biomass in 1990 was one of the lowest recorded since surveys began in 1984, in 1991 juvenile abundance was very high, leading to a marked increase in spawning stock biomass by 1993. In this case at least then, it seems unlikely that the fishery played a major part in either the decline or recovery of the total stock.

What about local competition for food resources? Monaghan (1992) postulated that direct local competition for sandeels between the fishery and seabirds near colonies could occur, even though fishing may not affect sandeel stocks as a whole. A key point in this argument is the higher energetic costs of foraging at greater distances from nest sites. Local sandeel depletion that might lead to competition could occur under two conditions (Wright, 1996). First, if there is little exchange between sandeels occupying different local areas, high fishing pressure could lead to localised depletion on fishing grounds. This hypothesis is unlikely to be true because sandeel are a low value species. Thus, fishers tend to stop fishing when catch rates become uneconomic and grounds are usually abandoned before sandeel abundances fall to levels low enough to cause a problem for birds. Alternatively, if sandeel densities on fished grounds are maintained by a continual influx of individuals from less preferred areas that are not fished, local depletion at the less preferred areas could occur. When overall sandeel stock size is low due to recruitment failures, there does appear to be a density-dependent habitat selection which maintains densities on 'preferred' (fished) grounds, at the expense of less preferred areas. For example, there is evidence for a continuous recruitment to certain fishing grounds, even in years of low recruitment, and it may be that adults also move into these areas when there is space available. Thus, in years when there are few sandeels, it would be the less preferred areas that are depleted. Any birds that need to forage in these areas would suffer. However, even though non-preferred areas may become depleted, this is not to say that fishing is the cause. The same arguments about the determinants of overall year class strength discussed above also apply. When recruitment is poor (almost certainly an event unrelated to the fishery in this case) birds suffer in localised areas that tend not to be near traditional fishing grounds.

The important point to be learned from the Shetland sandeel experience is that large effects often have complex explanations, and often require detailed data collection and analysis. In this case, while reduced prey densities had effects on birds it is hard to sustain the argument that the fishery was responsible. This is not to say of course that the story might not be different in other areas. However, it is as important also to highlight examples of ecological effects where fisheries have been implicated but been shown to be relatively benign as it is to point the finger of blame.

5.3 Species replacements and other competitive effects

Species replacements have been a topic of conversation in fisheries circles for many years and it is often loosely invoked to explain trends in the abundance of fish stocks. Daan (1980) argued that one could not legitimately talk about a replacement unless it could be shown that the two (or more) species were functionally related by a comparable response to a common resource. In other words, the species must compete in some way. Daan (1980) also offered some further operational criteria, arguing that the decreasing stock should undergo a reduction of at least 50%, and that the compensating increases in another stock should also be of the same order. He also argued that the change should persist over a number of years. Below, I look at a number of cases that might be considered to be species replacements and ask the extent to which the competition hypothesis can be supported.

Sardine and anchovy

Of the candidates for species replacements, the sardine–anchovy species pair is perhaps the most compelling. Both genera (*Sardinops* and *Engraulis*) have a worldwide tropical and sub-tropical distribution and co-occur in many areas. The potential for food competition provides the mechanism for a functional replacement. Where populations of the two species have changed markedly in abundance, the changes have often (but by no means always) been out of phase (Lluch-Belda *et al.*, 1989). In these cases, the first requirement for a species replacement is clearly satisfied. But are the responses of one species a direct consequence of changes in the other, so that fishing might be implicated, or are they independent responses to something else over which fishing has no influence? The case for the latter appears to be the strongest.

The first line of argument is that, in general, historical catch records for different areas of the Pacific are statistically correlated; changes have either occurred simultaneously or with a short lag (Crawford *et al.*, 1991). For example, sardines off Japan, and in the California, Humbolt and Canary current systems were all high in the early 1980s after lows from the 1950s–1970s. This coherence in the responses of a number of widely separated populations strongly suggests a large-scale environmental effect as a primary driving force. A correlation between increasing sardine abundance (and decreasing anchovy abundance) and periods of warmer air and sea surface temperatures supports this view and suggests that changes in the patterns of ocean circulation are important (Fig. 5.10). Such large-scale changes in fishery systems that occur on decadal time-scales and are associated with hydrographic changes have been termed regime shifts (e.g. Steele, 1996).

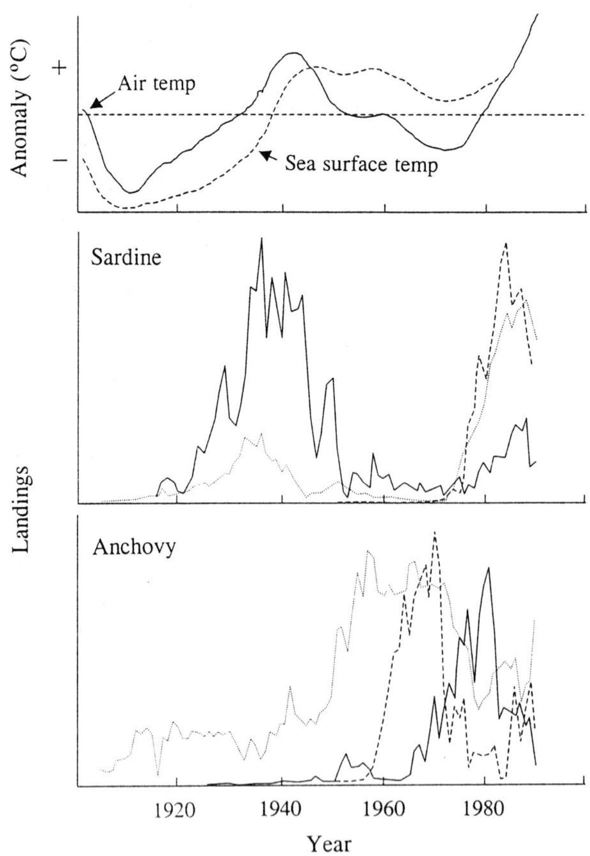

Fig. 5.10 Trends in landings of sardine and anchovy from different geographic regions and global air and sea surface temperature anomalies. (Redrawn from Lluch-Belda *et al.*, 1992.) (—) California; (----) Humboldt; (·······) Japan.

There is also the question of the timing of the abundance changes in the two species. Although the available data are often restricted to catches, which are difficult to interpret, and the pattern does not always hold, Lluch-Belda *et al.* (1992) argue that the first signs of a change in dominance is often an increase in the abundance of the subdominant species. This timing also suggests that the population changes are not functionally linked.

There is also some doubt about whether anchovy and sardine really compete for food. Speculation about the possible role of food competition as an explanation for the species replacement has been considerable, but clear evidence for such an effect has yet to be presented. Part of the problem is the high taxonomic diversity of planktonic prey that are taken by the two species and the plasticity of feeding mode that the species can adopt (filtering or biting, depending on prey size). Even when there is considerable overlap in the diet of the two species, however, this is by no means sufficient to show that competition for food is operating.

The dynamics of the California sardine fishery, probably one of the most well-documented fisheries stories available, also suggest that anchovy and sardine populations may be operating largely independently. Estimates of the spawning stock biomass suggest that sardine and anchovy were equally abundant in 1940 (about 4 million tonnes each) and that both stocks had declined to approximately 500 000 tonnes by the early 1950s (Fig. 5.10, Smith, 1972). This trend in both species appears to have been a response to a period of very weak upwelling. It is easy to imagine that the reduced primary production associated with weak uwelling reduced the carrying capacity of the California current system. Indeed the marked fluctuations in scale deposition in sediments over the past 150 million years that has been shown by Soutar & Isaacs (1974) would appear to support the idea of marked natural fluctuations in the pelagic fish stocks of the area. The important fact from our perspective is that when upwelling strengthened again in the early 1950s the sardine continued to decline, whereas the anchovy increased. Did heavy exploitation of sardine lead to prolonged recruitment failure, thereby allowing anchovy to reach levels that would otherwise be unobtainable? Daan (1980) concluded that the answer was probably 'yes' (Fig. 5.11). In the light of the more recent analyses, however, another possibility is that the trend from the 1950s onward might be explained by oceanographic factors other than upwelling. The kind of shifts in temperature regime discussed above may be significant in this regard. In the final analysis it must be admitted that the data are equivocal, and we certainly cannot reject the replacement hypothesis. Taking all the evidence into consideration, however, it is difficult to argue for a functional replacement. Fishing almost certainly contributes to the collapse of the dominant, but there is little to suggest that this collapse (on its own) allowed the subdominant to increase.

Pelagic species in the North Sea

Another potential replacement phenomenon concerns the dramatic increase in

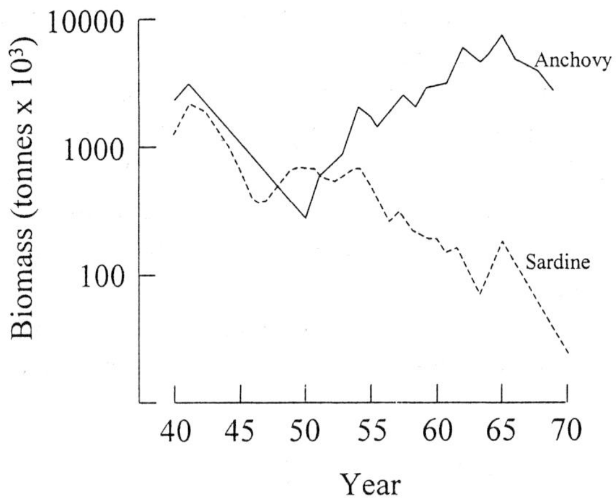

Fig. 5.11 Estimated biomass of California sardine and northern anchovy. (Redrawn from Daan, 1980.)

sprat abundance in the North Sea that coincided with the collapse of herring stocks and the subsequent recovery of herring and decline in sprat. These observations led to the hypothesis that the sprat boom occurred because the population was able to temporarily occupy the niche vacated by herring. Corten (1986, 1990), however, rejected this theory on the basis of discrepancies in the geographical patterns of decline and recovery of both stocks. The basis of Corten's argument was that herring first started to recover in the southern North Sea (1978) and only later in the central and northern North Sea. It follows that if returning herring were to effect changes in the sprat population it ought to have occurred in the southern North Sea first, the opposite of what actually happened. Analyses by McKay (1984) also suggest that the sprat boom in the 1970s was the culmination of a trend that started a decade before.

Corten shows that the data do not support a species replacement of sprat by herring, and his analyses also illustrate another important point – when data are aggregated on too coarse a spatial scale spurious trends emerge. It was only by examining the data with respect to broad regions of the North Sea that the holes in the replacement argument emerged. Too often we are tempted to draw conclusions based on data gathered at inappropriate spatial and temporal scales. Admittedly, there are sometimes occasions when data are only available on a coarse scale, but wherever possible the robustness of patterns needs to be examined over a range of spatial scales and the relevence of the scale of observations for the postulated mechanism of interaction needs to be carefully considered.

Sciaenids and sparids in West Africa

Ballistes carolinensis, a member of the family Spariidae, occurred in insignificant quantities on the continental shelf off West Africa in the early 1960s. In the late 1960s, as *Sardinella* stocks started to decline, it started to increase in abundance and by the 1980s two stocks in the area had grown from almost nothing to more than 10^5 tonnes.

Catches constituted 80% of the pelagic biomass in some areas. Is this another example of a fishery induced replacement?

No-one can say for sure why *Ballistes* increased, but Gulland & Garcia (1984), by drawing together environmental and catch per unit effort data (excluding *Ballistes*), build a sound case for concluding that environmental changes were the major instrument (Fig. 5.12). The authors suggest that a decrease in river output, an anomalous rise in upwelling and salinity and a decrease in plankton abundance led to a decrease in the suitability of the environment for the warm water, low salinity sciaenid Sardinella species which were traditionally exploited by the trawl fishery. But for *Ballistes*, which prefers cooler water of higher salinity, the environment had improved.

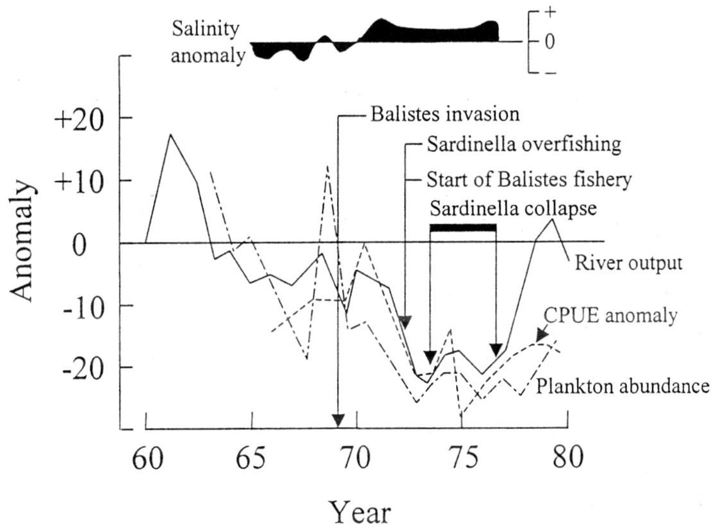

Fig. 5.12 Trends in river output, plankton abundance, salinity and CPUE anomalies in the West African trawl fishery. (Reproduced with permission from Fig. 7 of Gulland & Garcia (1984) In Exploitation of Marine Communities (R.M. May, ed.), pp. 155–90. Springer-Verlag, Berlin.)

While fishing may have contributed something to the observed replacement, there is one thing that suggests that it was not a major factor - *Ballistes* is also an exploited species. Without a favourable change in the environment it is difficult to see how it could have increased so dramatically. Interestingly, *Ballistes* is now caught fairly frequently off the west coast of Scotland and in the Irish Sea and has recently been reported in the northwest North Sea – an area where it was rarely recorded in the past.

5.4 Responses in simple systems – lessons from the Southern Ocean

When considering species interactions, one system that may be especially useful to study is the Southern Ocean. The central role of krill (*Euphasia superba*) in the diet of many large-bodied taxa in the upper part of the food web makes it a particularly tractable system for analysis. Since several species of whale have been reduced in numbers by more than 90%, surely here we can see how effects of fishing can manifest themselves through trophic interactions.

To try and see more clearly the kind of responses one might expect, Fig. 5.13 shows a food web diagram for the Southern Ocean. Of course, the situation at the lower levels is not really as simple as has been depicted. Although krill has received considerable attention, other non-krill species often comprise the majority of the biomass (Schnack-Shiel & Muijica, 1994). The true food web structure may, therefore, be as complex as many in lower latitude systems, especially for levels below krill. Nevertheless this simplified view is justified because it is not just biomass that is important – the scale at which resources can be exploited must also be taken into account. It is only because krill form dense swarms that large-bodied individuals such as baleen whales can obtain enough food from such a resource (Murphy *et al.*, 1988). It is this scale effect, coupled with its high population biomass, that makes krill such a central component of the Antarctic system.

Competitive release: the whale reduction – krill surplus hypothesis

Given the central position of krill in Fig. 5.13 an explicit hypothesis comes easily to mind:

Hypothesis: When baleen whales were at high population levels (i.e. prior to exploitation), other species populations which exploit krill were food limited as a result of competition.

Prediction: When baleen whale numbers were drastically reduced due to fishing, other populations increased as a result of the increase in food availability.

This hypothesis – a classic competitive release – has been termed the whale reduction–krill surplus hypothesis. It has received a good deal of attention. Table 5.1 gives us a crude estimate of the magnitude of the krill surplus that might have arisen. There are great uncertainties over the values in this table because estimates of whale numbers and their consumption rates are very difficult to obtain. For example, allowance must be made for the fact that a number of the species are migratory and do not spend all of their time feeding in the Antarctic. These data do, however, give an indication of the scale of the effect and which whale populations were most affected by fishing. The table shows that krill consumption declined by 147 million tonnes between 1940 and 1977. To put this value into perspective, in 1977 the entire Antarctic seal population was estimated to consume only 67 million tonnes, about one-third of the amount thought to have been consumed by the initial stocks of whales and 1.5 times their present consumption. From this perspective at least, there ought to be a lot more food available for the populations that remain after fishing.

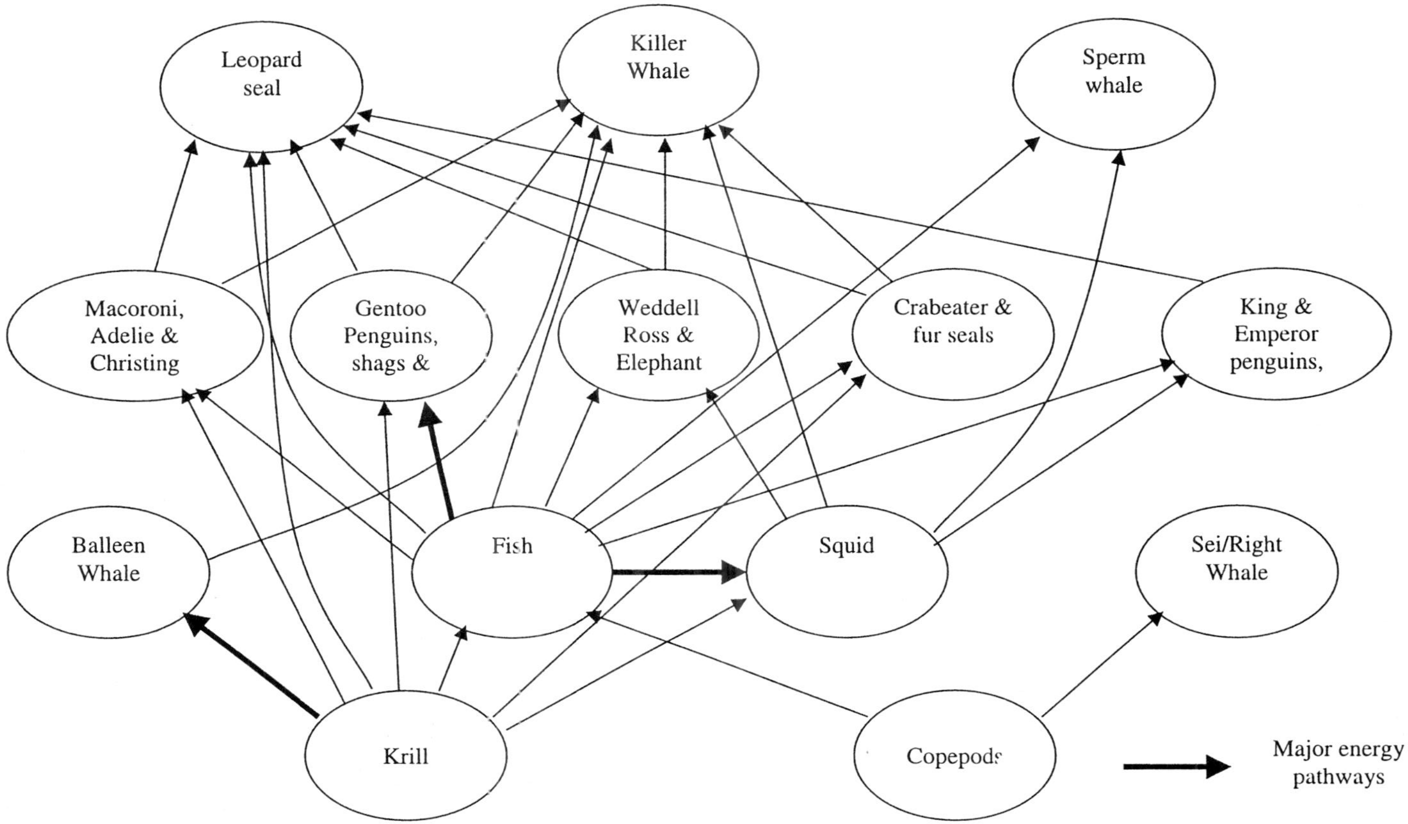

Fig. 5.13 The Southern Ocean foodweb. (Based on Table 6 of Kock and Shimadzu (1994) In *Southern Ocean Ecology* (S.Z. El-Sayed, ed.). Reprinted with permission of Cambridge University Press.)

Table 5.1 Crude annual estimates of changes in baleen whale numbers and biomass and the surplus krill that would be released as a consequence of harvesting. (Adapted from Table 2 of Laws, 1977.)

Numbers ($\times 10^3$)	pre-1940	1977	% decline
fin	400	84	79
blue	200	10	95
sei	75	40.5	46
humpback	100	3	97
minke	200	200	0
Total	975	337.5	65
Biomass ($\times 10^6$ tonnes)	pre-1940	1977	% decline
fin	20.00	4.03	80
blue	17.60	0.83	95
sei	1.39	0.71	49
humpback	2.70	0.08	97
minke	1.40	1.40	0
Total	43.09	7.05	84
Krill demand ($\times 10^6$ tonnes)	pre-1940	1977	Surplus
fin	81.48	16.43	65.05
blue	71.70	3.38	68.32
sei	5.65	2.89	2.76
humpback	11.00	0.32	10.68
minke	19.83	19.83	0.00
Total	189.66	42.84	146.82

But which populations in the systems were the most likely beneficiaries? The most obvious will be those species that overlap most in resource use and exploit their prey at a similar spatial scale. Thus, until about 1970, we might have expected the minke whale to be the most likely winner, partly because it was unexploited until that time, and also because it has the closest feeding affinity with the blue and fin whales which were so substantially reduced. The next most likely candidate are the seals, and then perhaps birds. Benefits for fish and squid would be less likely (Murphy *et al.*, 1988). What is the evidence for the krill surplus leading to increases in populations?

For minke whales, the answer is not much. There appears to be very little data available on population trends which would allow an assessment of either the direction or magnitude of change. Laws' (1977) estimate shown in Table 5.1 for example, suggests that the population was not substantially different in 1977 from that in 1940. Although minke had been exploited for about six years by the time of this estimate, with a total harvest in that period of about 37 000 individuals, the evidence for a population increase in response to increased food availability is nevertheless weak. Note this is not the conclusion drawn by Beddington & May

(1982), who concluded that of all species the evidence was most compelling for minke whales.

The next most likely beneficiaries in the system are seals. There are seven species found in the Southern Ocean, four of which (crabeater, leopard, Weddell and Ross) are truely Antarctic and confined to the pack- and fast-ice zones. Another two species (elephant and Antarctic fur seal) occur abundantly on sub-Antarctic islands and were themselves exploited during the nineteenth century – fur seals to especially low levels. The final species, the sub-Antarctic fur seal is a rare vagrant from north of the Antarctic convergence. Unfortunately, as with the minke whales, the overall status and population trends of the true Antarctic seals are rather poorly known, but this has not inhibited speculation. For example, based on data from replicate censuses in the late 1960s, the early 1970s and the early 1980s it has been suggested by Erickson & Hanson (1990) that crabeater seal populations have declined dramatically, leading them to argue for a causal relationship between this and the rebuilding of whale stocks. Kock & Shimadzu (1994), however, question the validity of the abundance estimates for Antarctic seals and robustly dispute this interpretation. In particular, the authors point out that changes in the dispersal of individuals between years could affect the censuses badly and that no estimates of variability in the survey data were presented to show the degree of precision. In the final analysis, Kock & Shimadzu (1994) concluded that they were unable to decide whether a change in abundance of crabeater seals has really occurred.

Some population data is also available for leopard seals which are seen each year on Macquarie Island, a location well to the north of the usual extent of pack-ice. The occurrence of leopard seals on Macquarie reflects a northward dispersal during winter. Peaks and troughs in the time-series apparently reflect juvenile dispersal rates in response to conditions on the pack-ice to the south. In view of these complications, relating the number of seals sighted at Macquarie Island to leopard seal population numbers is heroic at best. But, if the leopard seal population was increasing, we might expect there to be a trend in the data in addition to the oscillatory component related to changes in the environment. There appears, however, to be no such trend.

Since Weddell seals feed mainly on small fish and cephalopods, rather than krill, a population response to increased krill abundance seems less likely in this species than in others. There have been few suggestions that the dynamics of Weddell seals are related to the harvesting of whales. Environmental factors, however, have been shown to be important. For example, Testa *et al.* (1991) report that data on pup production in McMurdo Sound from 1970 to 1988 showed cyclical behaviour which was in phase with the Southern Oscillation Index (SOI), an indicator of the El Niño. Although Testa *et al.*'s data are most convincing for Weddell seals, they also show some degree of coherence between the SOI and population parameters in leopard and crabeater seals. Once again a multi-factorial interpretation will probably be required to adequately account for observed change.

Another possible competitive release response in seals concerns the recovery of fur seals from intense exploitation in the nineteenth century. On Bird Island, South Georgia, fur seal numbers remained at a low level for 70 years from about 1885–1955, despite little or no hunting. Interestingly, the population only started to increase from the mid-1950s when the whale stocks around the island were heavily depleted (by the mid-1960s there were insufficient whales to support land-based fishing operations). Another interesting observation is that, from 1958–1972 the rate of seal population increase was about 17% per year, a value that is much higher than is normal for the recovery of a marine mammal population (it is usually 6–10%; Laws, 1985). Kock & Shimadzu (1994) simply note these as interesting observations. Nevertheless, krill are a very important component of the diet of fur seals (Doidge & Croxall, 1985) and the observations are certainly consistent with the krill surplus hypothesis.

Trends in population sizes for penguins are only really detectable in rookeries that are regularly surveyed and censused. But the data may not be consistent from colony to colony or region to region. There is considerable temporal and spatial variation between data sets. For many bird species the effects of fisheries activities may be dominated by the incidental mortalities associated with fishing operations, or by the increased food availability from unwanted catch discarded by fishermen (Chapter 2). Nevertheless, there are some trends for which the cause has been attributed to competitive release. There seems little doubt that both chinstrap and Adélie penguin rookeries have increased in abundance since fishing began and there is evidence that the breeding range of chinstrap penguins has also increased. (Fig. 5.14). Although it has been argued that these increases are due to increased

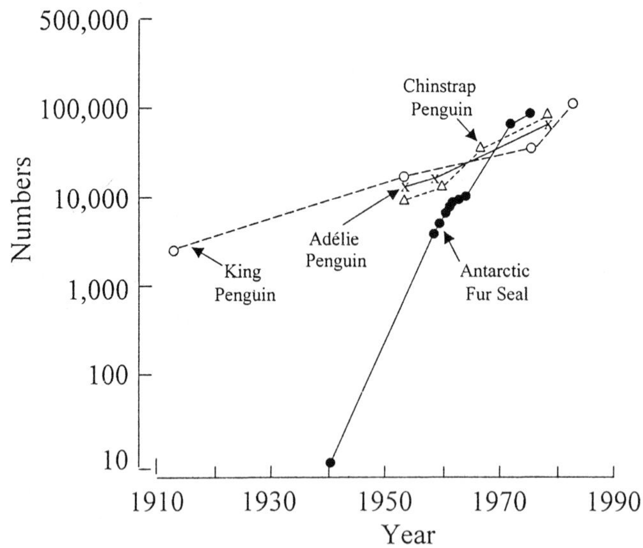

Fig. 5.14 Trends in population size of penguins and fur seals. (Adélie and chinstrap penguins: breeding pairs on Signy and South Georgia island; King penguins: South Georgia; Antarctic fur seals: pups, South Georgia). (Reproduced with permission from Croxall *et al.* (1988) In Antarctic Ocean Resources Variability (D. Sahrhage, ed.), pp. 261–85. Springer-Verlag, Berlin.)

food availability caused by a reduction in whale numbers (e.g. Sladen, 1964), there seem to be growing doubts as to its veracity. Croxall *et al.* (1988) point out, for example, that growth in penguin numbers has continued for twenty or so years since whaling ceased and that this trend is apparent, even in King Penguins, which do not eat krill. With respect to the recovery of King penguins, Rounsevell & Copson (1982), conclude that the krill surplus hypothesis is conjectural. For other species Croxall *et al.* (1988) also argue (more from a position of faith than data) that a krill surplus was very likely to explain the initial increases in penguin numbers, but that something else must now be responsible for the continued growth. Taylor *et al.* (1990) and Fraser *et al.* (1992) offer evidence that increases in food availability for chinstrap colonies in South Shetland Islands and Adélie penguins in the Ross Sea region may have been the result of decreasing ice cover (due to progressively warmer winters), which allowed the species greater access to their food resource.

Prey removal: effects from harvesting krill and fin fish?

Although the krill surplus hypothesis has been the most widely examined fishery effect, as commercial fisheries have developed for resources lower in the Southern Ocean food web, there has been some consideration of the effects these activities might have, or will have. Croxall (1987) notes that, to date, krill harvests have been very small in relation to the food requirements of seabirds and seals and, although fish harvesting has been much more intensive, thus far, no species except perhaps the southern Elephant seal and male Antarctic fur seals take significant quantities of commercially exploited fish.

With regard to Elephant seals, populations appear to be declining over most of their distributional range, with the exception of the population on South Georgia (McCann & Rothery, 1988). Several hypotheses have been proposed for the decline of these stocks and increased competition for prey with commercial fin-fish fisheries is one of them. However, despite some reservations about the quality of the data on the relative importance of squid and fish in the diet of Elephant seals, Kock & Shimadzu (1994) conclude that there is very little evidence for a fishery interaction. In particular, an apparently stable seal population in South Georgia, despite the presence of a large-scale fin-fish fishery for over more than two decades, supports this view. A clear candidate to explain the decline in population numbers at other sites has yet to be found.

Croxall *et al.*'s (1988) considered taxa which might act as indicators of the state of the system, perhaps through changes in reproductive performance that could be related to food availability. They concluded that even if harvests of krill and fish were to increase markedly, and the rate of increase in seabird and seal numbers were to diminish, it would be difficult to exclude the possibility that the populations had reached their carrying capacity independently of krill fishing. The authors also note that as populations reach carrying capacity they may be expected to fluctuate appreciably, adding further to the difficulties of detecting and interpreting population trends.

Whaling and deep-sea biodiversity?

Organisms in the deep sea are highly food limited, relying primarily on organic matter falling from above. Recent research suggests that whale carcasses may be particularly important because they sink fast enough to have sufficient tissue intact for deep-sea organisms to exploit (Butman *et al.*, 1995). Lipid rich whale skeletons have been found to support an animal community nourished by sulphur reducing chemoautotrophic bacteria which was previously thought to be restricted to hydrothermal vents (Smith, 1992). Thus whale skeletons may act like islands in the deep sea and provide critical stepping stones for organisms between vent communities, which are themselves insular and temporary habitats.

Clearly, if whale carcasses do play such a role in the deep sea, then the whale fishery, which will certainly have substantially reduced the supply of carcasses to the sea floor, may have had a profound effect on the possibilities for the dispersal of fauna between reducing habitat communities. This in turn has the potential to have marked effects on the biodiversity of hot vent and cold seep regions. This is of course an impossible effect to quantify, not least because we have been alerted to it a century after the event. However, it does highlight the potential for fishing activity to effect changes far removed from the initial perturbation.

5.5 Conclusions

On the basis of the information reviewed above, what conclusions can we draw about the importance of species interactions with respect to fishing? The key to answering this question and finding out whether there are any general principles that can be applied lies in understanding the mechanisms responsible for observed effects. Hence, the effort to structure this chapter with respect to the position of harvested species within food webs and the consequent candidate mechanisms of interaction. Effects have been considered only in terms of changes in population sizes, so what we are asking is whether predation and competition leads to increases or decreases in population size, or to changes in the variability in population size. Unfortunately, while our ideas about how a particular set of species could interact are legion, the actual data to support one alternative over another is often weak. Moreover, there seem to be many cases where even the very existence of the change that is being explained is open to question; explanations for changes in abundance based on CPUE data provide, perhaps, the best examples of this kind of problem. These weaknesses in our data leave ample opportunity for the development of 'just-so' stories to explain observed changes. Without studies directed specifically at the processes underlying the population dynamics of specific groups of species, we will remain ignorant of the true importance of the effects of fisheries acting through species interactions in marine systems. Indeed, a distinguished body of academics that met to discuss these matters concluded that:

> 'clear and unambiguous evidence of interspecific interaction in major marine ecosystems, which could be used directly to achieve a significant improvement in single-species assessments, as yet hardly exists. This does not, of course, mean that such interaction does not occur or even that it is unimportant; rather limitations of data and the influence of other factors have so far rendered the empirical approach alone of little diagnostic value.' (Beverton, 1984)

In the intervening twelve years it would be difficult to argue that very much has changed, although there are some general conclusions that can be drawn.

Top-down control and trophic cascades

Hard substrata

By far the best evidence for top-down control in marine systems comes from hard substratum communities. It is also in these habitats that trophic cascades seem to be most prevalent. The evidence that fishing has reduced predator abundances and that this has led to trophic cascades is also quite convincing in a number of cases. Coral reefs and sub-tidal hard substrate communities both provide good examples. In most cases cascades occur where prey species compete for space (i.e. mussels and algae) or themselves graze on sessile species (sea urchins).

Given the compelling evidence in some systems it is tempting to assume that all predators which occupy a similar trophic position will determine system structure in the same way. Such an assumption can lead to pressure to alter fisheries management regimes to limit or mitigate cascading effects. A concrete example of this comes from the lobster fishery in eastern Canada where adherents to the keystone species paradigm strongly urged fisheries managers to consider this interaction when deciding between management options (Elner & Vadas, 1990). Unfortunately, there is a tendency in ecology (as in other areas of science) to be somewhat uncritical of studies which appear to support the established wisdom of the day. Put simply, papers which suggest that a cascade does not occur in a particular system are less likely to be published that those that claim it does. Elner & Vadas (1990) argue that this tendency was readily apparent in the literature on lobster–urchin interactions. Many of the studies which purported to support the keystone species paradigm for lobsters were uncritical, and failed to specify adequate null hypotheses or experimental tests with appropriate controls. Moreover, evidence which ran contrary to the paradigm tended to be discounted. It was only gradually over a 15-year period that the hypotheses shifted away from predator control (Fig. 5.15).

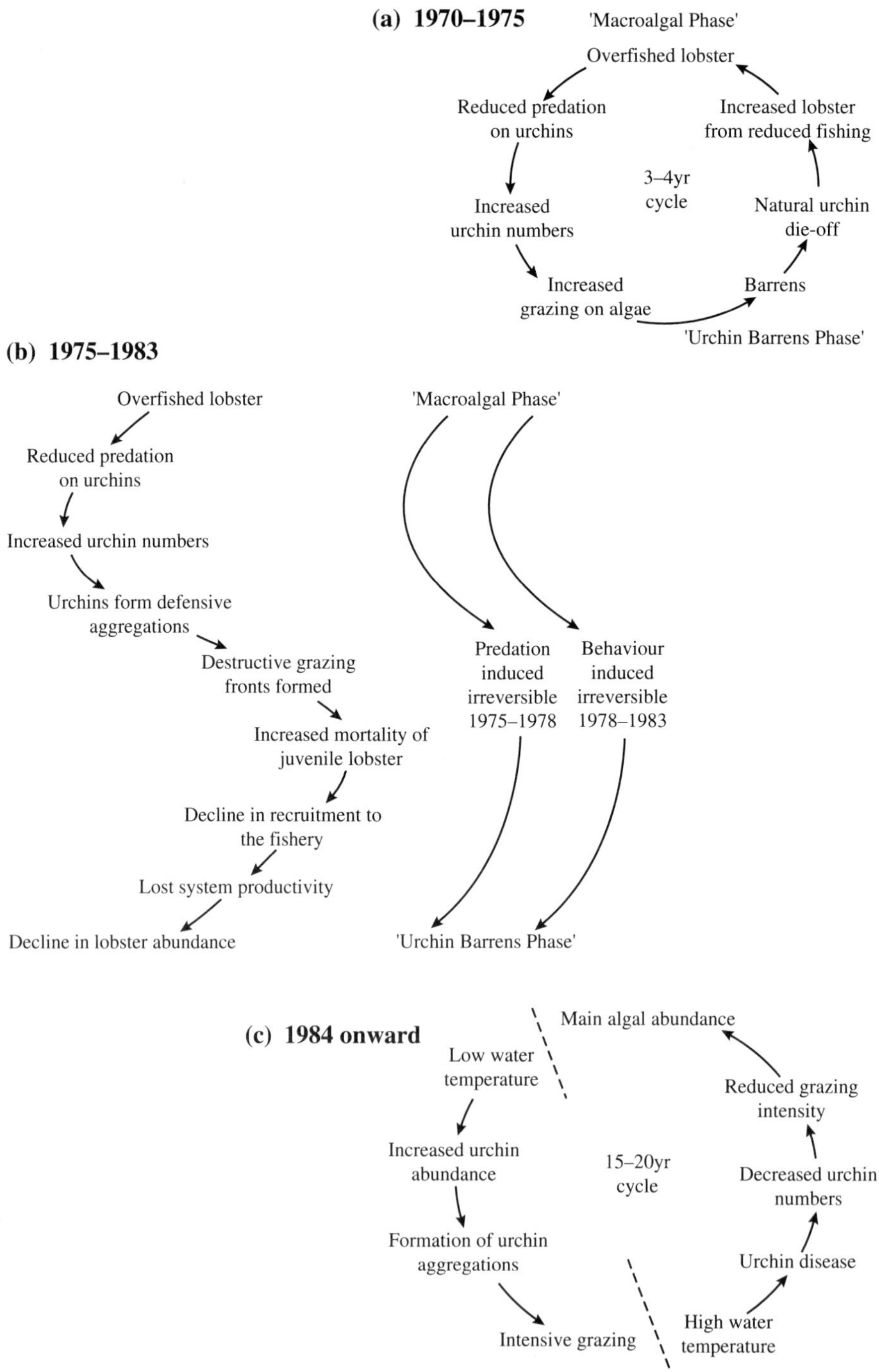

Fig. 5.15 Three periods in the scientific assessment of lobster–urchin interactions showing how the explanation of changes in community structure has changed. (Adapted from Figs 1–4 of Elner & Vadas, (1990) *American Naturalist* **136**, 108–25. With permission from the University of Chicago Press.)

This lobster–urchin story highlights how easy it is to confirm prior expectations, even in a system for which experimental manipulation and detailed observation is possible. In the Canadian lobster fishery, conservatism and/or inertia by fisheries managers meant that no steps were ever taken to account for predator effects, and eventually the climate of opinion about predator control by lobsters started to shift.

It seems certain that keystone predators and trophic cascades can be important in marine systems with hard substrata, and that fisheries effects from removing such predators can and do occur. However, expecting all species which seem to have the required attributes to control communities in this way is a mistake. Moreover, with the possible exception of very well documented rocky shore systems where many experiments have been performed (Menge, 1995), there seems to be no way of determining *a priori* which species might be keystone predators. Only very careful observation and experiment can provide such information. Of course, it may be that mitigating potential trophic cascades by reducing exploitation rates is desirable regardless of the uncertainty about whether a cascade will actually occur. Over-exploitation of fisheries resources is all too common and reducing fishing effort will often pay dividends in terms of long-term yields. However, arguing for such effort reductions solely because a cascade may occur would be unreasonable, unless experimental evidence is available to support the hypothesis.

Pelagic systems

Clear examples of strong predator control are much less easy to find in pelagic systems than they are in hard substratum communities. This perhaps suggests that predator control is less important in the pelagos. Alternatively, the lack of evidence may simply reflect our weak powers of observation – it is much harder to get data that would support the predator control theory in the pelagic than it is on a rocky shore. With respect to the evidence, it must be acknowledged that some would disagree with this assessment. Parsons (1995), for example, is a strong advocate for top-down control in both the North Sea, the western North Atlantic and the Southern Ocean, all regions that were discussed in this chapter. Parsons argues that the kind of responses observed in lake systems are also occurring in the sea. Clearly different weight has been given to the available information. The interested reader is urged to return to the original sources to make up their own mind.

While there is usually insufficient evidence to draw strong conclusions about predator control in pelagic systems, some examples are indeed compelling. The events that followed the reduction in shark numbers in Natal are a case in point. In this case, two factors are notable. First, both the predator and the prey (large sharks and small sharks, respectively) invest heavily in each individual offspring, either brooding their young, or laying a few well protected egg cases. There is likely, therefore, to be rather strong density dependent recruitment with a good relationship between the size of the adult stock and the number of recruits in the following year. Second, dramatic environmentally driven fluctuations in predator and prey stock size are likely to be rare. In the absence of such fluctuations the potential for a close coupling of predator and prey dynamics becomes greater, and increases in prey numbers in response to the removal of predators becomes more likely. This situation contrasts with many pelagic species where stock recruitment relationships are highly variable with only weak evidence for a density dependent recruitment relationship. Given these often high levels of variability in the stock recruitment relationship for pelagic species it is difficult to argue that predation is a consistent controlling influence. This is not to say that predator control on pelagic populations does not occur, but rather that identifying when it does occur will be difficult and adapting fisheries management to account for it even more so.

Removing prey

There is little doubt that fluctuations in the abundance of prey resources can affect a predator's breeding success. Thus, if prey population collapses are sustained over the longer term this will translate into a population decline for the predator. Examples can be found particularly for bird species, but also for other taxa such as seals. Since many people have strong emotional attachments to such taxa, there is often intense interest when breeding failures or population declines occur. In the search for a culprit fishing activity is often readily offered as an explanation for the prey decline, or at least as an important contributory factor. In assessing the effect of prey removal, however, one must consider whether the fishery and predator compete for the same portion of the population, either in terms of spatial location or stage in the life cycle. For example, if the predator eats juveniles whose abundance is unrelated to the abundance of the fishable stock, the potential for interactions is greatly reduced. Such a feature seems rather common and probably needs to be examined closely in cases where a fishing effect is implicated.

It is also important point to bear in mind that competition is usually for pelagic prey species whose populations are intrinsically variable. Indeed, on the available evidence, it often seems to be rather difficult to sustain the argument that fishing plays the dominant role in effecting large-scale changes in recruitment and spatial distributions. Rather it appears that it is the vagaries of water movement that are most strongly implicated in the pelagic fishery collapses that have led to effects on many bird species. In the final analysis though, there can be little doubt that unrestrained exploitation will increase the likelihood of fisheries collapses and this is turn will take its toll on predator populations.

Competition

No one would deny that competitive effects can occur in ecological systems, and from an evolutionary perspective, they may well have shaped the communities we now see. Demonstrating that two species are currently competing in the real world, however, is notoriously difficult. Indeed, without the ability to conduct carefully controlled field experiments on the species in question, it is impossible. It will be no surprise to learn then that unequivocal demonstrations of competition in most marine systems are rare. Perhaps the only exception to this is for communities occupying hard substrates where competition for space has been demonstrated and can be important in determining community responses to predators (see above). For other systems (e.g. the pelagic or soft-sediment benthos), we can only offer opinions, based on our assessment of the importance of other factors (e.g. predation, low quality food, environmental conditions). My prejudice is that the latter are more important in most cases (but see section 6.1).

Even in the Southern Ocean, where we might expect the signals to be strongest, the evidence for competitive effects is equivocal. While a number of observers have linked apparent changes in the population size, reproduction or growth of baleen whales and other taxa to extensive reductions in whale populations caused by commercial whaling (see, for example, Beddington & May, 1982; Kawamura, 1994), a recent assessment by Clapham & Brownell (1996) has even cast doubt on this interpretation. These authors conclude that there is little evidence that populations have responded to an increase in available resources resulting from a decline in competitor densities. Indeed while the authors accept that the idea of competitive release is intuitively reasonable, not only is the data fragmentary, but many of the purported changes which support the hypothesis are in question.

The conclusion that the data are often insufficient to draw firm conclusions about fishing effects on ecological interactions will no doubt be frustrating to many readers. Why devote the bulk of a chapter to arguing the point? The reason is that many of the examples used here have been considered seriously as examples of fishing effects. For this reason alone, it would be remiss not to explore the issue fully, regardless of the final conclusion. As with the issue of predator control, the importance of competition will remain a key issue for ecology in the coming years. For fisheries systems the difficulties of obtaining data and effecting critical tests of hypotheses are particularly formidable, but as we will see in Chapter 6 when comprehensive data are available good progress can be made.

Chapter 6
Demersal Fish Populations: Three Case Studies

It is probably fair to say that some of the best data on the effects of fisheries on both target and non-target fish populations are for demersal taxa in temperate waters where there is a long history of fishing. This is largely because relating catch data to abundance is less problematic than for pelagic species. Demersal taxa do not generally form dense aggregations so fishermen cannot hunt for schools and maintain catch rates until most of the stock has been caught. The population dynamics of demersal species are usually also less volatile so that trends in fishing effort are less variable. Another advantage is that a long history of trawl survey data for some areas provides additional fishery independent estimates of population sizes. These surveys provide some of the best data of all on the long-term trends in fish populations during periods when fishing effort has been especially high.

In an effort to draw together what is known of the trends in fish populations from demersal systems, this chapter concentrates on data from three areas for which there is good information. Two of these, the Georges Bank/Gulf of Maine area and the North Sea could in some senses be viewed as comparable systems from different geographical areas. The contrast between these two will illustrate how differently, ostensibly similar systems can respond. The third example is from the Gulf of Thailand, which allows differences between temperate and tropical fishery systems to be explored.

6.1 Georges Bank and the Gulf of Maine

The shelf seas of the North American east coast are among the best studied marine systems in the world and much of the available data for the area has been gathered in support of fisheries management objectives. Within this region the data for Georges Bank and the Gulf of Maine (Fig. 6.1) are, perhaps, the most comprehensive and the rate of fish production in the Georges Bank region in particular, is among the highest in the world (Grosslein *et al.*, 1980). This high production arises largely because the relatively shallow topography of the bank, on to which deeper

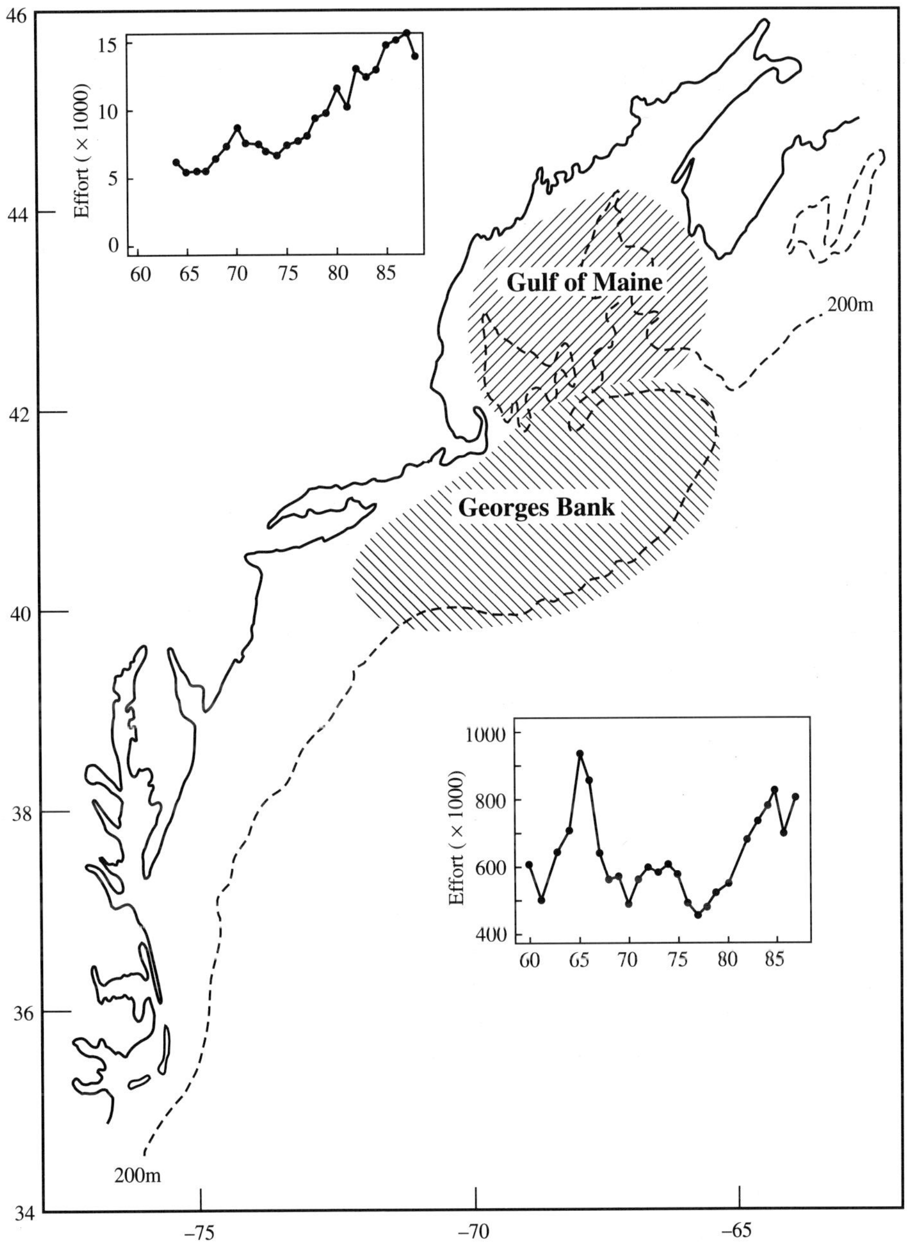

Fig. 6.1 The Georges Bank and Gulf of Maine region and time-series of nominal effort data for the demersal fleets in each region. Nominal effort data are simply the total hours fished by all vessels with no account taken of the relative sampling efficiency of different types of gear, power of boat etc.

nutrient-rich water is advected, allows almost continuous primary and secondary production from April to November.

The area has supported a rich fishery since the eighteenth century but, it was in the early 1960s that it began to develop substantially. By 1965 foreign fleets had started to fish in the area and in the following decade the level of exploitation rose dramatically. Brown *et al.* (1976) estimated that during the 1961–1972 period standardised fishing effort increased six-fold, while landings more than tripled. At its peak in the late 1960s and early 1970s landings of fish and squid combined exceeded the 400 000 million tonne estimate of maximum sustainable yield (Brown, 1987). Clarke & Brown (1977) estimated that this level of effort led to a decline in the total fish biomass in the area in the order of 37% between 1963 and 1974 – a substantial effect on the fish assemblage by any standards. During this period, the range of exploited fish species also expanded from the early concentration on the principal groundfish species to include a wider variety of groundfish species and pelagic and semi-pelagic taxa (Fig. 6.2). In 1975 when the foreign fleets were excluded, the level of fishing effort fell, but they have gradually risen again to approximately 60% of the 1965–1975 average.

Georges Bank is an extension of the shallow submerged sandy moraine of Nantucket Shoals, separated by a deep channel to the south. The Gulf of Maine comprises deep-water basins and ledges and is essentially an extension of the deeper offshore habitats. The overall level of fishing effort has been much lower in the Gulf of Maine, but has shown a general rising trend over time. Clark & Brown (1977) estimate that, between 1963 and 1974, the overall finfish biomass in the area declined by 41%.

Population trends

A number of analyses of the spatial and temporal trends in the fish populations in this area have been undertaken (see, for example, Clark & Brown, 1977; Overholtz & Tyler, 1985). One of the more recent efforts is by Gabriel (1992) who used cluster analyses to identify characteristic groups (assemblages) of species which tended to occur together. Gabriel showed that the membership of the major species groups was fairly persistent over time and that many groups were also spatially persistent, although some species did show variable group affiliations (Table 6.1).

Using the species groupings shown in Table 6.1, Gabriel examined the changes in assemblage structure in different sub-regions. For the Gulf of Maine, the overall pattern with respect to total biomass was of an initial decline between 1963 and 1974, a resurgence between 1975 and 1981 and a secondary decline from 1982–1988. These trends were apparent for most members of Groups A and B. The increase in biomass in the mid-1970s – contrary to the trend in fishing effort – was largely a consequence of good year-classes for the Group B species (haddock and Atlantic cod) and to the presence of pollock. This is a good example of how environmental variability can introduce large population signals. Of course, by definition,

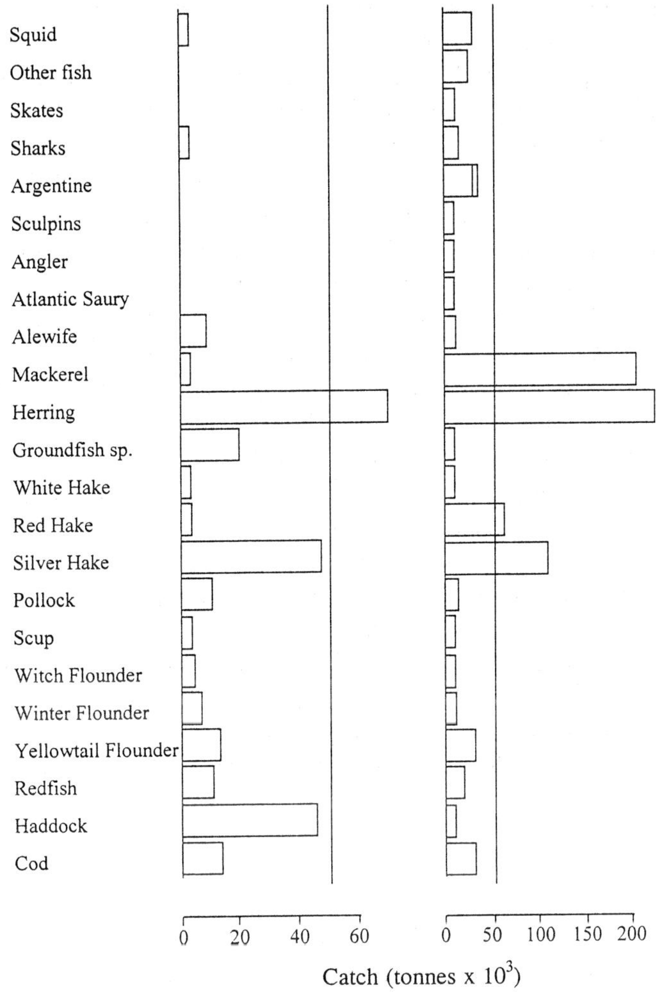

Fig. 6.2 The nominal catches reported for 1962 (left) and 1972 (right) for ICNAF area 5 (Georges Bank and the Gulf of Maine) showing the increasing total catch and range of fish caught. (Data from Table 2 of Brown *et al.*, 1976.) The vertical line on each panel denotes landings of 50 000 tonnes.

aggregate trends must reflect the changes for individual species; the interesting question is which ones do not fit the pattern? For the decrease in biomass between 1963 and 1974 a notable exception to the trend was white hake, which showed a slight initial increase in abundance. Along with the other members of Group A, however, white hake declined in the 1982–1988 period. The major exceptions to the trend have been in the transition species red hake and spiny dogfish. Until 1980, the biomass of spiny dogfish declined, but both this and red hake have shown increases in abundance from the 1980s onwards.

For Georges Bank the species in Group B and the flatfishes in Group C showed a similar initial decline, resurgence and subsequent decline, a pattern largely

Table 6.1 The major species groupings present on Georges Bank and the Gulf of Maine from 1967–1988. From Gabriel (1992)

Group	Species	Area
A	Thorny skate American plaice White hake Redfish Witch flounder	Deep water Gulf of Maine and Georges Bank
B	Atlantic cod Haddock Pollock	Gulf of Maine–Georges Bank transition zone
C	Winter skate Little skate Windowpane Winter flounder Yellowtail flounder Longhorn sculpin	Shallow water Georges Bank
Transition species	Silver hake Red hake American goosefish Short-finned squid Spiny dogfish Cusk	Various group affiliations, including groups not shown here, but in the mid-1970s most species were associated with Group A. The exception to this pattern was the spiny dogfish which was associated with group C from 1978–83 and with the Groups A and B in the mid- to late-1980s.

consistent with the trends in fishing effort. The most striking feature of the Georges Bank picture, however, is the rise in elasmobranchs, which began with increases in spiny dogfish numbers in 1973 and continued with rising numbers of winter skate, which began in 1977. Expressed as a proportion of the community biomass or numbers this increase is particularly striking (Fig. 6.3), but even in absolute terms there can be no doubt that skates and rays have increased dramatically. We will return to the elasmobranch question later, but first we should consider other ways of looking at what the survey data can tell us.

Other ways of looking at assemblages

With time-series data on the biomass or abundance of many species, what other ways are there of examining trends? Looking for simple correlations in the time-series of many species may not be especially revealing. For example, Rothschild (1991) showed that for the 16 most abundant species on Georges Bank there were 22 statistically significant correlations in the time-series from 1963 to 1988 out of the 120 possible species pairs. Eighteen per cent might seem like a lot, but with so many comparisons, some will be significant by chance alone, and if we correct for this possibility none appear to be significant. A similar, although somewhat more comprehensive effort to quantify the strength of interactions between populations in the area was undertaken by Sissenwine *et al.* (1982) who fitted a simple model to

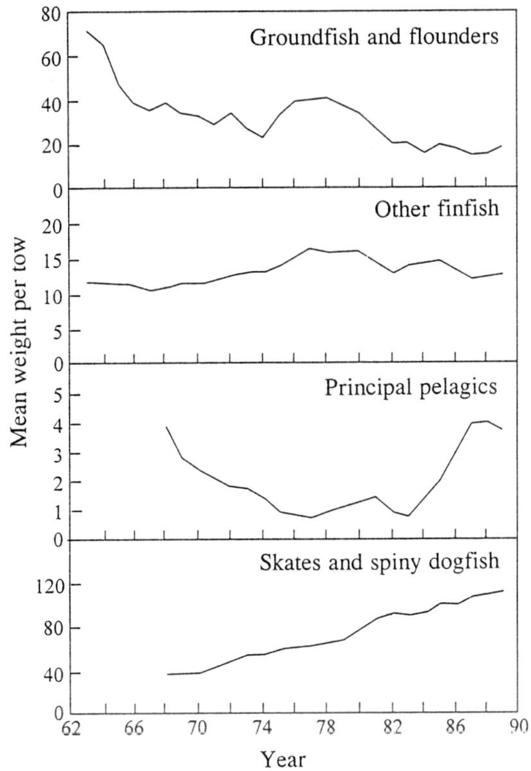

Fig. 6.3 Trends in the relative contribution to total biomass (numbers) made by major fish groups. (Adapted from Mountain & Murawski, 1992.)

estimate interaction strengths for data from 926 species pairs. For some of these comparisons data for the predator abundance was lagged by one year. This was because effects on the prey population may well occur some time after the predator population reaches a high level. Although the signs of the interactions were generally correct, Sissenwine *et al.* (1982) found little evidence of statistically significant effects of one population on another – in this case, as in others, the correlative approach seems likely to be a poor route to understanding.

But there is an alternative approach where, rather than focus on the mean abundances, some measure of overall structure, or correlations between the populations, we might instead focus on the variability of the system. Such variability is, of course, a consequence of the variability in the populations of all the component species. The system can, therefore, be considered as having many dimensions, each one represented by a different species. A number of methods exist to explore such multidimensional data and to characterise variability and dynamics. The basic idea underlying all these methods is to summarise the multidimensional variability in terms of a smaller number of values.

Rothschild (1991) used principal components analysis (PCA) for his study of the behaviour of 16 species on Georges Bank. In a completely random system where all species abundances behaved independently of one another one would need as many variables as are there are species in the system to describe the variation. Rothschild (1991) found, however, that including only three variables could account for nearly 98% of the overall variance. With each of these three variables, or eigenvalues, is a vector of numbers (an eigenvector) which represents the associated variability in the biomass of the 16 examined components of the system. Each successive eigenvalue accounts for the greatest possible proportion of the unexplained variance, and each associated eigenvector represents the relative importance of the species elements for that variation. Thus, all eigenvectors are mutually orthogonal. Figure 6.4 a shows the magnitude of each element in these first three eigenvectors.

The figure shows that the first eigenvector (the one that accounts for most of the variability in the system) reflects a high positive correlation with dogfish and winter skate and a negative correlation with haddock. The second eigenvector reflects a positive correlation with haddock and spiny dogfish, and the third reflects a negative correlation with winter skate. From this analysis it is clear that (in biomass terms at least) the systems response over the 25-year period of study can largely be captured by considering the variations in these three species.

Rothschild (1991) terms the three dominant eigenvectors 'eigen-assemblages' and, in any one year, each assemblage makes a variable contribution to the total biomass (Fig. 6.4b). What this figure shows is that there was a fairly steady increase in the dominance of spiny dogfish and winter skate, but that the increase became more substantial and less variable from the mid- to late-1970s. For the dominant haddock and spiny dogfish eigen-assemblage we see decreases during the early-1970s then a sharp increase in the mid-1970s followed by increasing variability and a declining trend. For the third eigen-assemblage (lack of winter skate) we see increases in the early-1960s, then an increase in variability in the mid-1970s followed by a decline.

How does the PCA approach adopted by Rothschild (1991) differ from approaches, such as that adopted by Gabriel (1992) where assemblages of fish are defined using cluster analysis or similar techniques? The difference lies in the treatment of time; Gabriel (1992) examined the species composition in different areas for each year and concluded that, on the whole they were spatially consistent. She then examined the temporal trends in those assemblages over time. In contrast, the PCA approach does not define an assemblage that is fixed, but explores the variability in the data to defined eigen-assemblages which contribute in varying degrees over time to the overall variability. In effect, what Rothschild did was to define two clusters: a variance-dominant cluster containing spiny dogfish, haddock and winter skate, and a noise-dominated cluster which contained the other 13 species he included in his analysis.

Both of the approaches described above suggest one thing – there have been dramatic changes in community composition in the region, which appear to be closely related to the heavy exploitation of gadoids and flounders. On Georges

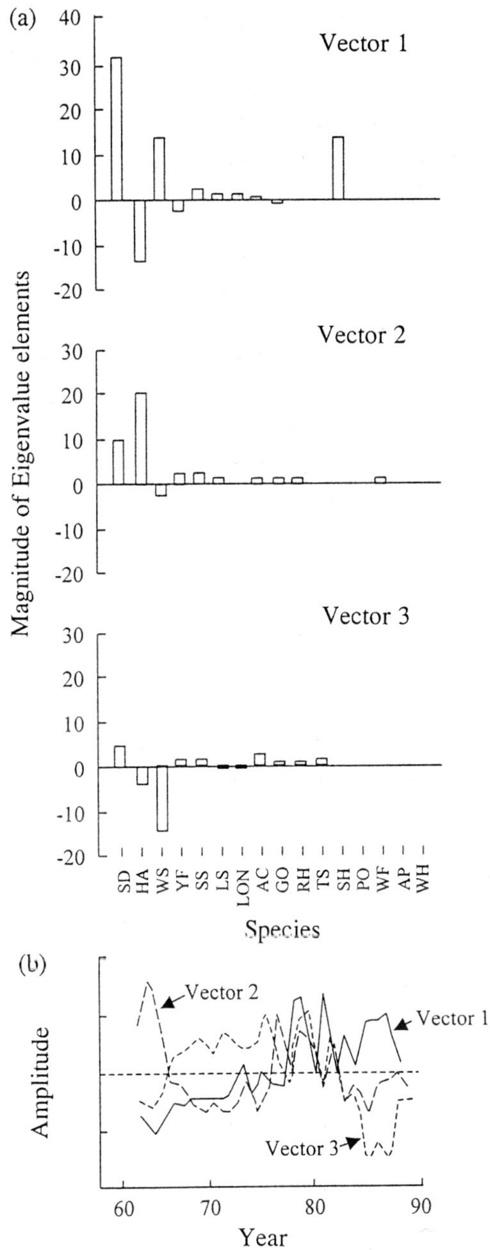

Fig. 6.4 (a) Time-series associated with the magnitudes of the elements (corresponding to each of the species) for each of the first three eigenvectors. (b) Time-series of amplitudes associated with the three eigenvectors. (For further explanation see text). (Redrawn from Figs 1 and 2 in Rothschild, 1991.)

Bank in particular, data from trawl surveys indicate a change in species composition in which the dominant gadoid and flounder components are replaced by dogfish and skate, previously only a very minor component of the system. As ever, the question then is the degree to which these changes are an indirect consequence of fishing mediated through population interactions or independent responses to environmental conditions.

Simulation modelling: a route to understanding?

We have already seen in Chapter 5 that the empirical evidence for population interactions is often poor, but we are in no position to discount its possible importance. This is partly because the approaches that correlate time-series data depend on good estimates of population sizes, which may not be available, but also because of the spiky population data that obtains when recruitment success is variable from year to year. The time lags in response between predators and prey that may occur might also make the correlations in abundance a poor indicator of effects. How can we overcome such difficulties and come closer to understanding how important interactions between species might be? One approach is to try and formally quantify our understanding and develop mathematical models of interacting populations, which attempt to explain some of the observed fluctuations in species abundance.

Building sensible models is a formidable challenge, and to adequately represent the effects of competition and predation it is particularly important that account be taken of the temporal and spatial patterns of species occurrence. At its simplest, if a predator and prey species are not in the same place at the same time, one cannot feed on the other. We are often left guessing, therefore, when and where the circumstances that would permit significant interaction to take place occur. However, if the important spatial and temporal data that is required is available for any system, it ought to be for Georges Bank. It is on this system that an informative modelling effort by Overholtz & Tyler (1986) focussed.

Overholtz & Tyler (1986) consider the period from 1963–1978 when, as we have seen, major changes in the fish assemblage occurred. Rather than try and create fully predictive models of the fish community, which would almost certainly be wrong, these authors lowered their sights and built a simulation model to simply pose and test hypotheses about the ecological processes that might operate in the system. In other words, they built a model that described how they believed competition and predation between species in the system might work and then looked to see if the consequences of their assumptions (the model output) matched the observed data.

The dynamic part of the model was restricted to an assemblage of species that had previously been identified from cluster analysis – essentially the same as the shallow-water Georges Bank assemblage from Gabriel's (1992) analysis (Group C, Table 6.1), but including juvenile haddock. During the period considered, three distinct changes in the assemblage occurred: in the early 1960s juvenile haddock dominated the biomass, but from 1968 to 1972 the haddock declined and a mixed group comprising yellowtail and winter flounders, winter skate, and longhorn sculpin became important. Thereafter, winter skate increased in importance while little skate and windowpane predominated. From the early 1970s there was also a marked increase in spiny dogfish which occur in the summer and which may be an important predator on the rest of the assemblage. It is these changes – a general shift from an assemblage dominated by commercially important species to one where species of little or no commercial interest were ascendant – that the model was built to try and understand. Overholtz & Tyler (1986) developed the following overall hypothesis summarised in Table 6.2.

Table 6.2 The hypothesised mechanisms to explain changes in the fish assemblage in Georges Bank. From Overholtz & Tyler (1986)

Time	Response	Mechanism
1960–1968	Juvenile haddock dominate	Competition by haddock keeps growth rates of other species at relatively low levels.
1968–1972	Haddock decline due to a combination of high fishing effort and poor recruitment, yellowtail flounder and longhorn sculpin increase in importance.	Competitive release of other species, allowing them to expand in numbers and biomass. Rapid response by yellowtail flounder and longhorn sculpin allowed them to dominate initially. Skates and windowpane also experience release, but respond more slowly.
1972–1978	Skates and windowpane dominate	Competitively superior species eventually dominate those species which responded more quickly to the decline in haddock. Fishery on yellowtail flounder accelerated the response. Additional predation effects by resident and seasonally dominant piscivorous species, such as spiny dogfish, also important.

The model was designed so that the consequences of various combinations of competition, predation, predator preference, and prey turnover could be investigated in simulation experiments. As a null hypothesis it was assumed that the physical environment played no role in the dynamics of the system, so that efforts to explain the historic trends in the data were based solely on the interactions between populations. Most importantly, the authors state that once the parameters of the model had been set, they were not modified to make the model output fit the observations better.

The performance of the model was evaluated by comparing the model output with the actual data for the 1963–1978 time period. Best fits were obtained with a combination of both predation and competition, rather than competition alone. Importantly, however, the model output did not match some of the prominent features in the data, notably the decline in yellowtail flounder, the increase in windowpane in the early 1970s, and all of the time-series for longhorn sculpin.

The principal way for competition to affect species in the model was via growth, and the results suggested that the decline in haddock allowed invertebrate prey biomass to increase such that the size at age of some of the model species also increased in response to the greater food supply. When the model was set up so that competition was moderately strong, the size at age of three of the species continued to increase after 1970 when haddock had fallen to a low level, but one species (windowpane) experienced a decline in growth rate during this same period. In contrast, when competition was intense, prey resources fell and the size at age of all species declined. In this latter case an increase in fish predation served to increase growth rates by decreasing competition through the increase in food available to the survivors. Importantly, however, the predators led to decreases in the biomasses of yellowtail flounder and longhorn sculpin during the latter part of the time-series with concomitant increases in skate and windowpane.

The model results, are in essence formal expressions of the commonsense notion of how competition and predation ought to work. Unfortunately, these ideas have yet to be tested, although data on changes in size at age, food habits and other related information might be used for such purposes. Importantly, however, the simulations do show that synergistic effects of competition and predation between species could have played a part in the change from demersal assemblage dominated by haddock and flatfish to one dominated by elasmobranchs.

Environmental influences

As with population interactions, evidence for environmental effects through correlations between fish abundances and environmental data is often (if not always) equivocal. This is partly because a changing environment may influence stock abundance (through variable recruitment), but also through changes in the spatial distribution of the stock. The effects of both of these may, in their turn, be confounded by the influence of fishing on stock abundance. Nevertheless, a number of data sets are consistent with the idea of a functional link between changes in recruitment patterns for a number of species in the northwest Atlantic and a changing environment (e.g. Koslow, 1984). How much of a role could such effects play in the observed changes on Georges Bank and in the Gulf of Maine?

One of the most interesting studies of the effect of the environment on fish distribution is that of Mountain & Murawski (1992) who examined the relationship between water temperature and abundance and distribution data obtained from fisheries surveys. Figure 6.5 shows the temperature anomaly about the overall mean value for Georges Bank. These data show that the periods around 1970 and 1980 were generally cool, whereas the middle of these two decades was warm. A broadly similar temperature pattern occurred for the other areas examined. The inset panel on Fig. 6.5 shows the trends in aggregate abundance for the four major species

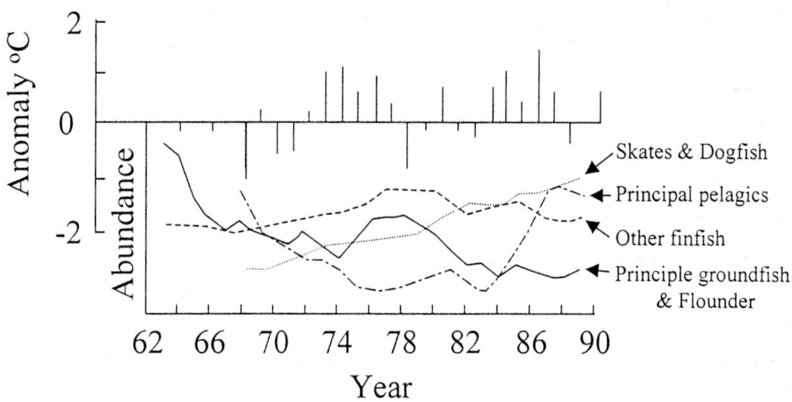

Fig. 6.5 Temperature anomaly about the overall mean values for Georges Bank and the trends in the abundance of major taxonomic groups. (Adapted from Mountain & Murawski, 1992.)

groups. Clearly the trends in temperature and abundance do not correlate very well. It seems unlikely, therefore, that a temperature effect played an important part in determining abundance. Mountain & Murawski (1992) do show, however, that distributional changes associated with temperature are quite marked for most species. Indeed, it is only because surveys have covered almost the entire habitat for most of the species of interest that we can be fairly certain that abundance estimates have not been confounded by populations moving out of the surveyed area.

The rise of elasmobranchs: a true species replacement?

The available evidence points to the responses of the fish assemblage in the Georges Bank/Gulf of Maine area as one of the clearest examples of a species replacement caused, at least in part, through fishing pressure. During the 1980s the principal groundfish species, flounders and other finfish, declined markedly in abundance after modest increases in the late 1970s. These increases were associated with the eviction of foreign fleets and some good recruitment events that were probably related to favourable hydrographic conditions. It seems almost certain that the subsequent decline was a direct result of over-exploitation by the fishery. In contrast, the elasmobranchs (skates and spiny dogfish) continued to increase during the 1980s.

It would appear, therefore, that the elasmobranchs have responded opportunistically to the decline in the other species in the system, perhaps by being able to exploit food resources that were no longer removed by target species. It should be stressed, however, that we have no data to support or deny the hypothesis. Moreover, the increase in dogfish numbers in the early 1970s may have been the result of a shift in the distribution of the population in response to hydrographic changes which brought warmer water into the area (Gabriel, 1992). It is difficult to believe, however, that the continued high abundances can be explained in this way.

Two other features of this example are noteworthy. First, there is no market for dogfish in North America, and second, the species has high discard survivorship. Both of these may be significant factors to explain the success of these taxa in the region. Interestingly, the decline in gadoid stocks in the North Sea (described in the following section) has not been associated with a similar outburst of elasmobranchs, perhaps because a market for the species exists in Europe.

6.2 The North Sea

The North Sea provides 50 plus species of fish and invertebrates of commercial interest, of which a dozen or so support major fisheries. For the most part fleets exploit single species or a narrow range of species, and one might imagine, therefore, that fisheries management would be a reasonably simple affair. Unfortunately, this is far from the case, the main reason being that, unlike many other systems, no one nation has sole control of the resources. All of the nations bordering the North Sea jointly manage the fisheries, largely through the action of the Common Fisheries Policy (CFP), an instrument of the European Union. The essence of the problem for the CFP is that it is usually difficult to take the hard decisions which are often required to conserve stocks when it is against the interests of only one or a few of the nations. For a fishery under the control of a single state, forcing one sector of the fishing industry to take short-term losses to achieve long-term gains is sometimes politically acceptable (although probably always difficult). However, when such action is perceived to disadvantage a particular state, nationalist sentiments come into play and make life more difficult. For an excellent account of the strengths and weaknesses of the CFP Holden (1994) is recommended.

The North Sea has been intensively fished for over a century and some of the oldest fisheries laboratories in the world are situated in the countries that surround it. Despite this history it is an unfortunate fact that, with a few exceptions, systematic efforts to obtain fishery independent survey estimates of fish abundance for the North Sea have only been made in the last two decades or so. For the most part, therefore, we are forced to piece together the picture from available catch and fishing effort data. This, of course, restricts data to those species that are commercially fished. Perhaps the dominant feature of these data is the 'gadoid outburst', for which this account draws heavily on Hislop (1996).

Interactions between target species

The gadoid outburst was a large and sudden increase in the abundance of several gadoid species which occurred at the beginning of the 1960s (Cushing, 1980, 1982). Haddock and whiting were the first to increase, with very strong year classes in 1962. Cod also produced above average year classes from 1963 to 1966 and a very strong year class in 1969 (Daan, 1978). Over the next 20 years haddock, whiting, cod and

Norway pout all produced one or more outstanding year classes although, for the most part, these were not synchronised (Fig. 6.6). The maturation of strong year classes led to a general increase in the spawning stock biomass of cod, haddock and whiting in the mid-1960s and throughout the 1970s. These increases were by factors of 13.5, 5 and 2.3 for haddock, cod and whiting, respectively, but they were rapidly removed by the fishery.

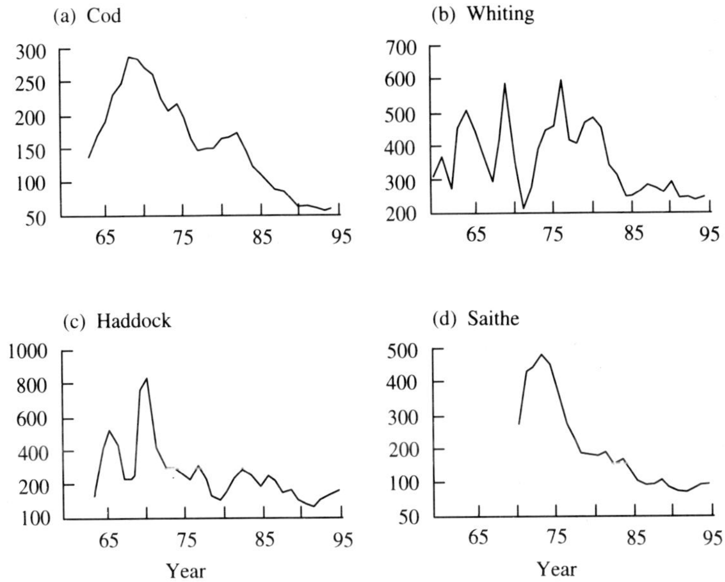

Fig. 6.6 Trends in abundance for four major whitefish species in the North Sea. (Adapted from Fig. 3 of Hislop, 1996.)

The extent to which species interactions played a part in the gadoid outburst has been the subject of considerable speculation. This speculation has focussed particularly on the relationship between demersal and pelagic species, largely owing to the collapse in pelagic stocks, which occurred at around the same time as the outburst. If gadoids do interact with pelagic taxa it will be when they are in their own pelagic larval and juvenile phases, and both predatory and competitive interactions could play a part. Pelagic species are known to prey on the eggs and juveniles of both gadoids and flatfish (pleuronectid) species, and there is considerable overlap in the diet of larval and post-larval gadoids and adult herring and mackerel.

Could the large decrease in the abundance of herring and mackerel that occurred in the 1960s and 1970s (in which fishing certainly played a part) have opened a 'window of opportunity' for the gadoids? Hislop (1996) argues that the evidence for this is rather weak, largely because a closer look at the timing of events reveals holes in the hypothesis. First, the abundance of herring in the central North Sea was already low by the mid-1950s, but there is no indication that the recruitment of cod and whiting, which have important spawning populations in this area, increased at this time. Indeed, the responses in these species occurred after those of haddock, which spawns mainly in the northern North Sea. Second, a truly remarkable year class of haddock emerged from the 1962 spawning, an event that cannot be explained by a reduction in herring and mackerel because it had yet to occur in the northern North Sea where haddock are found (Cushing, 1980). Although Daan (1980) is right to suggest that criticisms based on the differences in the timing of events are often not very strong, owing to the time lags that are likely to be inherent in a complex system, events which precede the action of the hypothesised causative factor cannot be ignored.

In addition to the numerous attempts to analyse and explain trends in the abundance of commercial taxa, there have also been attempts to construct mathematical models to explore phenomena such as the gadoid outburst. One of the earliest and most prominent of these efforts is that of Anderson & Ursin (1977) who, under an assumption of constant total food supply, concluded that the observed increases in stock abundance could be explained by the food released with the reduction of the herring and mackerel biomass. In effect, their model supported the idea that there might have been a large-scale shift in the pattern of energy flow through the system. This result is certainly intriguing, and is similar to the conclusion drawn from a very general energy flow model constructed by Jones & Richards (1976). However, as Daan (1980) points out, in a model where food supply is assumed to be constant and mass is conserved, if some stocks go down, others must by definition go up. Hislop (1996) concludes that 30 years after the onset of the gadoid outburst we are no closer to understanding its causes. One cannot help but agree, but it would be difficult to argue for a fishery-induced shift in community dominance over one largely controlled by changes in environmental conditions which altered the probabilities of recruitment success.

There is certainly abundant evidence of environmental variations during the last 40 years, and there appear to have been parallel shifts in phytoplankton, zooplankton, herring and kittiwake breeding success (Aebischer *et al.*, 1990). A number of factors have been suggested as important determinants of gadoid recruitment. Salinity anomalies (Dickson *et al.*, 1988a), the prevalence of westerly (Aebischer *et al.*, 1990; Corten, 1986; Turrell, 1992) or northerly winds (Dickson *et al.*, 1988b), wind stress and winter cooling (Svendsen & Magnusson, 1991) have all been implicated in the gadoid recruitment process to varying degrees. However, convincing explanations for the gadoid outburst remain elusive, particularly since, although conditions for recruitment showed a general improvement during the 1960s and 1970s, conditions in any given year only favoured one or a few species. Understanding recruitment remains one of the greatest challenges for fisheries science.

One important point, however, is that we appear to have seen a return to the pre-outburst state, suggesting that effects are reversible. From a fishery perspective this is unfortunate, since abundant gadoid species are more commercially valuable, but such a change over decadal time-scale may be another example of the so-called regime shifts in marine systems that seem to be driven by broad-scale changes in hydrography.

Although, it is tempting to discount biological interactions as the primary cause of events during the gadoid outburst, this is not to say that there is no evidence that important species interactions occur. Indeed, the recent efforts to model the interaction between the major resource species in the North Sea described below suggest the opposite.

More models of species interactions

Of the efforts to tackle the question of species interactions from a theoretical perspective, the work of the ICES Multispecies Assessment Working Group (e.g. Anon, 1988) is perhaps the most prominent in fisheries circles. This group developed a multispecies extension of virtual population analysis (VPA), the assessment methodology upon which the bulk of the management advice for North Sea fisheries is founded. The basic objective of single-species virtual population analysis (SSVPA) is to reconstruct the history of fishing mortality and numbers for each cohort in a stock using catch data (see Pitcher & Hart, 1982, for a description of the method). The multispecies extension of the method (MSVPA) differs from its single-species counterpart in the way natural mortality (i.e. death by anything other than fishing) is handled. In SSVPA, natural mortality is described by a single term M, which accounts for all non-catch mortality and is provided as an input to the model. The total mortality rate is given by $e^{-(F+M)}$ where F is the fishing mortality. In contrast, MSVPA partitions M into two components M_1, the residual natural mortality that is not explicitly attributed to any particular source and M_2, the predation mortality imposed by species that are explicitly included in the model. Note that M_1 includes both predation and other sources of mortality. For North Sea MSVPA M_2 accounts for predation by 11 commercially exploited fish taxa. Admittedly this is a very small sub-set of the species in the North Sea food web, but it does include the main exploited species. An excellent account of MSVPA methodology is provided by Magnusson (1995).

A key feature of SSVPA and MSVPA is that, not only can they be used to reconstruct the history of a stock, they can also be run forward in time (if one makes assumptions about the pattern of recruitment to the fishery) to examine the development of the population under differing patterns of fishery exploitation. Thus, the potential for species interactions to affect management actions can be explored. For example, at its 1988 meeting (Anon, 1988), the ICES Multispecies Assessment Working Group explored the effect of increasing the minimum legal mesh size for the roundfish fishery from 70 to 120 mm. Figure 6.7 shows the results of the comparison between SSVPA and MSVPA predictions of the spawning stock biomass of cod, haddock and whiting. For cod and whiting an increase in mesh size produces an increase in spawning stock biomass, but the predicted increase is less

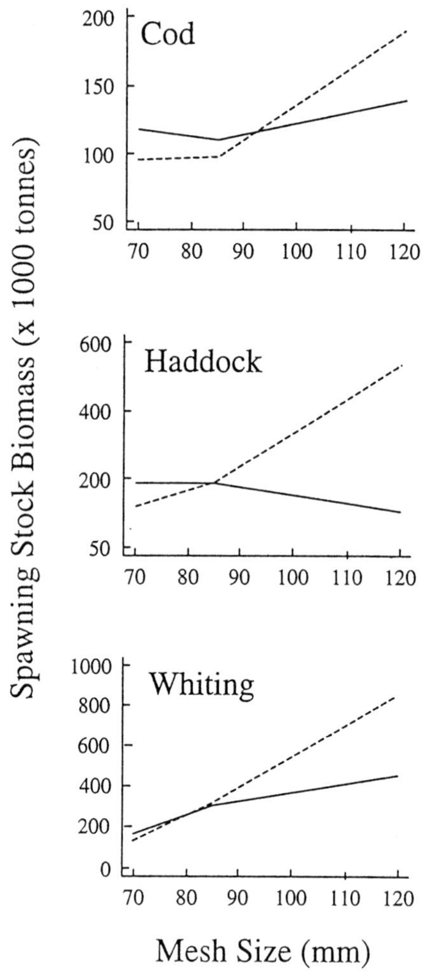

Fig. 6.7 Contrasts between MSVPA and VPA predictions of the spawning stock biomass of cod, haddock and whiting under different minimum mesh sizes regulations. (---) Single species; (—) multispecies.

with an MSVPA assessment; this occurs because the larger fish escape capture with the greater mesh size and continue to consume some of the production. In the case of cod, this appears to be largely a result of cannibalism. For haddock, the increase in predation mortality resulting from the greater spawning stock biomass of cod and whiting leads to an actual decrease in the abundance at the largest mesh size under MSVPA. The results of the analysis are quite clear – accounting for multispecies interactions substantially alters our perception of the effects of a mesh size change. Interactions between target species look like they might matter.

On the basis of these results it seems possible, therefore, that the reduction in predation mortality that occurred when fisheries removed large numbers of predatory fish may have more than offset any fishing mortality generated on smaller

North Sea fish species. In contrast to the conclusion regarding the interaction between gadoids and pelagic taxa, the interactions between some of the gadoid species probably are important for much of the time. Fortunately, for the North Sea at least, efforts to take such interactions into account in fisheries management are under way.

One question that is often raised with respect to the North Sea is the effect of the cessation of fishing during the Second World War. There seems to be a generally held belief that populations of exploited fish species were in decline in the years leading up to the war and that these recovered markedly during the war period when little or no fishing was possible. After the war stocks are considered to have decreased again with the resumption of fishing. One might reasonably ask whether there is any data to show what the effects of this were. Unfortunately, however, I think there appears to be rather little evidence to support the hypothesis that demersal stock did increase during the war years. For most groundfish species there is little to suggest a wartime population recovery (Hempel, 1978; Cushing, 1980), or a rapid post-war decline (Daan *et al.*, 1990). The exception to this may be the haddock, as suggested by Jones & Hislop (1978), but their conclusions are strongly influenced by one datum; the very low stock estimate obtained in 1939.

Non-target species

The foregoing discussion centred on interactions between target species, but there are many non-target fish species in the North Sea that might also be affected by changes in those that are fished. Is there any evidence for population changes here? As mentioned earlier, there are relatively few survey data for the North Sea prior to the 1970s, so following trends over extended periods is rather difficult. Nevertheless, earlier data do exist and can be used, provided great care is taken in interpreting the result of analyses. For example, it has been possible to piece together data for three areas in the northern North Sea, from trawl surveys taken between 1929 and 1953 and compare it with more recent survey data from 1980 to 1993 (Greenstreet & Hall, 1996).

The principal objective of this work was to establish whether community level changes in the groundfish assemblage have occurred over the last 60 years by comparing the two time-series. Data were stratified on an area basis, first to investigate possible habitat effects on assemblage structure, and second to determine whether any between-area differences are modified by different, and varying, fishery regimes. Available data suggest that, over the period, fishing effort in two areas has increased by between 50% and 100%, while in the third it has declined by at least 50%. To evaluate the possible role that fishing might have played the structure of the complete fish assemblage was contrasted with one in which species for which there are directed fisheries are excluded. Ordination and clustering of these data revealed highly significant differences in species composition, both between time periods and between areas – differences that are apparent, not only within the whole groundfish assemblage, but also when only non-target species are considered (Fig. 6.8).

Although highly significant differences in species composition between areas and surveys were identified, these were generally rather subtle in nature. The most obvious change was an increase in both the relative and absolute abundance of Norway pout. This was the only change detected for a comparatively abundant

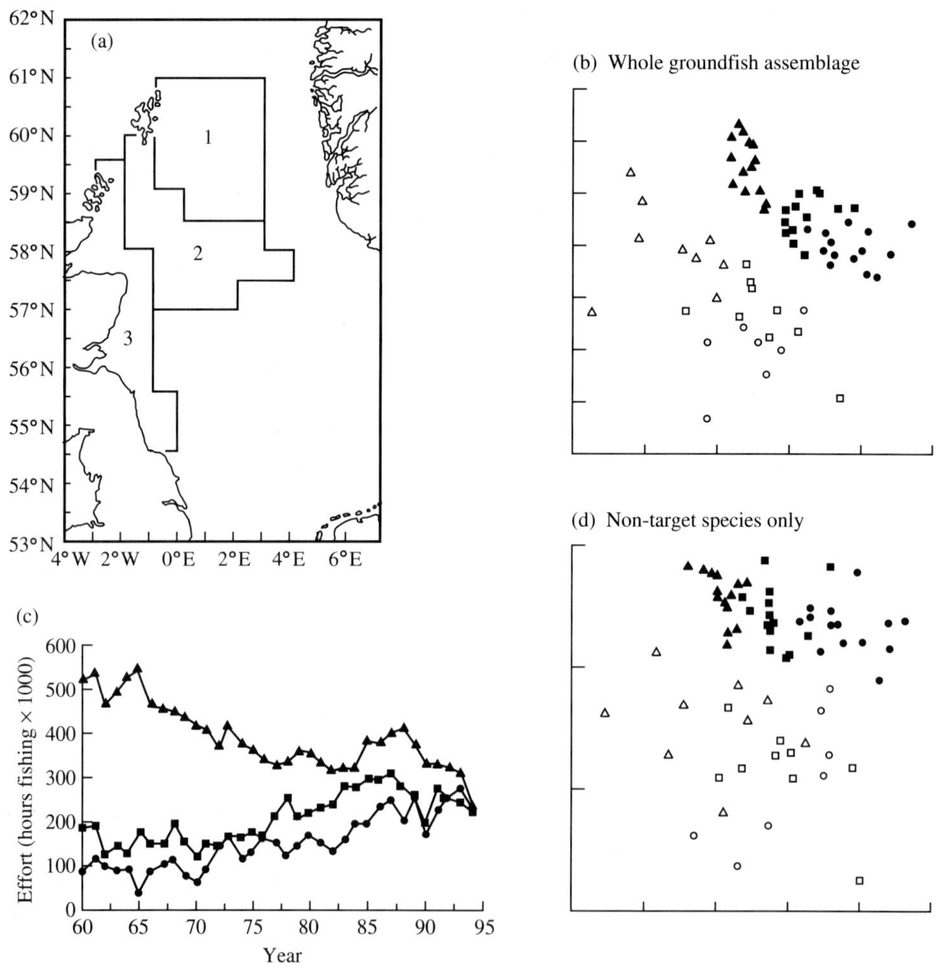

Fig. 6.8 Trends in fishing effort and the relative composition of the groundfish community for different survey years and areas as revealed by non-metric multidimensional scaling. On the MDS scatter plots points which occur closer together are more similar to one another. Each point represents data for a single year/area combination. Open symbols denote data from surveys between 1929 and 1956, closed symbols from 1981 to 1990. For all plots, circles = Area 1; squares = Area 2, triangles = Area 3. (Adapted from Hall and Greenstreet (1996) *Nature* **383**, 133. Copyright 1996 Macmillan Magazines Ltd.)

species. However, this increase in the northwestern North Sea, which has also been noted elsewhere (e.g. Hempel, 1978; Daan *et al.*, 1990), is difficult to attribute to a direct effect of fishing. Norway pout have been increasingly targeted by the industrial fishery sector so, if anything, one would have expected a decline. It also seems highly unlikely that the increase in Norway pout abundance could have occurred through reduced competition from the juveniles of other gadoid species, such as whiting and haddock, which appear to have been more or less equally abundant in both periods. Although there are doubts, it is possible that, as with the other gadoids, the increase in abundance might be due to reduced competition associated with the decline of pelagic species – the principal planktivores in the North Sea. An alternative and perhaps more likely hypothesis that is also consistent with these data is that reductions in the abundances of large gadoids have reduced predation pressure on pout.

With respect to other non-target species, two (long-rough dab and grey gurnard) showed declines in relative abundance. In absolute terms, however, only grey gurnard appears to have declined. Both these species are comparatively rare when the whole groundfish assemblage is considered, but they are two of the most dominant species in the non-target assemblage. These declines appear to have been balanced by increases in both the relative and absolute abundance of lemon sole and especially common dab (Fig. 6.9). Such population changes are consistent with those found in other studies (Hempel, 1978; Richards *et al.*, 1978). It is also perhaps worth pointing out how strongly flatfish dominate the non-target component of the groundfish assemblage, a trend that seems to be on the increase. Between 1929 and 1953 flatfish made up just over 70% of the non-target species assemblage, while between 1980 and 1993 this percentage had risen to just under 90%. With this increase in mind it is worth noting that fishing effort is rarely directed at flatfish in the northern North Sea and that this group may have higher discard survival rates than other species (Hudson & Furness, 1988).

While fishing might conceivably have played a role in the overall increase in flatfish species, it is difficult to conceive of any mechanism by which fishing might be responsible for declines in the populations of long-rough dab and grey gurnards and increases in the populations of common dab and lemon sole. It is also hard to imagine an environmental factor that might have the same specific impact. In cases such as this we should, perhaps, recognise the possibility that the patterns are a result of stochastic drift in population growth (Pielou, 1969; Krebs, 1978). Such a possibility is especially likely when the communities under scrutiny contain many species and the changes affect only a few, of which a high proportion are rare.

These analyses suggest that changes in the structure of the groundfish assemblage have occurred over 60 years, but that these have largely been confined to the species directly targeted by the fishery. It would perhaps be surprising if the large-scale removal of fish from the ecosystem did not have some structural consequences for those species on which efforts are most focussed. It is perhaps surprising, however, that the non-target assemblage does not appear to have been affected to a larger

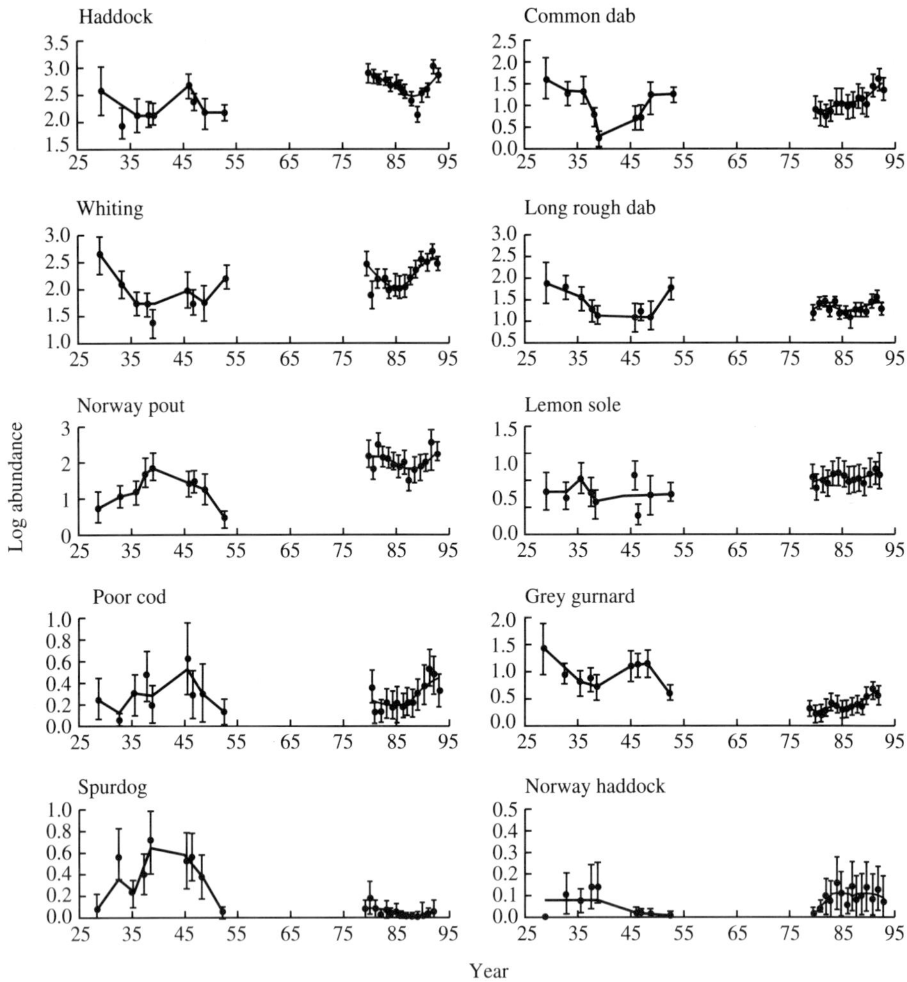

Fig. 6.9 Trends in non-target species abundances in the North Sea. Adapted from Greenstreet & Hall (1996).

extent. For two of the North Sea areas examined, the structure of the non-target assemblage appears to have been robust to increases in fisheries activity over the last 60 years. Structural changes were only apparent in the non-target component of the Scottish east coast groundfish assemblage. Perhaps this is because fishing effort for groundfish species has historically been much higher in this area than in the two more offshore regions that were looked at; high levels of exploitation have also occurred over a much longer time period. Such intense and prolonged perturbation may have been sufficient to have a knock-on effect on the non-target component of the groundfish assemblage. If this is the case then the trends of increasing fishing effort in the northwest, central and east Shetlands regions may lead to similar effects in future.

Despite the changes detected one is forced to conclude that the structure of the non-target groundfish species assemblage of the greater part of the northwestern North Sea appears to be highly resilient. It has remained relatively unchanged despite a century of intensive fishing disturbance. Differences in species composition are certainly apparent, although largely restricted to one abundant and a few comparatively rare species. Overall, one is left with the impression that the script remains the same, while a few of the minor players have swapped parts.

6.3 The Gulf of Thailand

The Georges Bank and North Sea examples discussed above suggest that the demersal fish communities in two temperate systems have responded in rather different ways to fishing activity. Might tropical systems react differently again? The fishery in the Gulf of Thailand provides data to examine this question. The Gulf of Thailand covers an area of approximately 300 000 km^2, 55% of which can be classed as 'inshore', lying less than 50 m deep. A further 45% lies between 50 and 85 m. The history of the trawl fishery in the area is well documented and tells a familiar tale: from a more or less unexploited stock and high catches in 1960, there has been a marked decline in total catch rates ever since. Before considering the responses of the fish community to fishing pressure, however, it is worth examining some of the key features of tropical fisheries systems.

Features of a tropical demersal fishery

Two aspects of tropical fish fisheries stand them apart – the taxonomic diversity of the catch and its size-distribution. With respect to taxonomic diversity, the current estimate for the number of fish species in the Indo-Pacific region is between 6000 and 7000 (Carcasson, 1977). This compares with estimates for the North Sea of 224 (Yang, 1982). In a brief review of published trawl survey data from the Indo-Pacific, Pauly (1979) reports that the total number of species recorded during a survey ranges between 173 and 341. Intensive surveys of the North Sea record between 50 and 60 species at most. Another distinguishing feature is that for tropical systems the line between target and non-target species is very blurred because there are few species which can be fished for without catching many other species at the same time, and most species are landed and used.

As for size, the majority of the species that comprise the landed catch in tropical demersal systems are small-bodied, compared to temperate fisheries. In shallow tropical waters, the slip mouths (*Leiognathidae*) comprise the bulk of the catch and these have a mean maximum length of about 12 cm. In deeper water the similarly small-bodied *Gerridae* often make up the bulk of the catch. For the pelagic and demersal North Sea fisheries, fish of this size would rarely be retained in the net. Only industrial fisheries would harvest fish this small. Another important feature of these tropical fisheries is that highest densities of commercially exploited fish occur in shallow waters. The slip mouths, for example, have their biomass maximum at 25 m (Pauly, 1979). In common with most other fish species, there is a clear trend for larger individuals to occur in deeper waters (Macpherson & Duarte, 1991), but these tropical species do not appear to make large-scale migrations. An important consequence of the size distribution in tropical fisheries is that, in catching small-bodied species, a significant proportion of the catch can also consist of the juvenile stages of valuable larger-sized fishes. Pauly (1979) argues that this feature makes it highly likely that both recruitment overfishing and growth overfishing could occur.

Trends in abundance

One valuable feature of the Gulf of Thailand fishery is that, in parallel with its development, independent surveys of species composition were undertaken. Thus, we have a rather good picture of how the fish community changed as the fishery developed. Pauly (1987) summarises the trawl survey data from 1961 to 1982 for major taxa. These data have been used to construct Fig. 6.10 which shows the smoothed trends in abundance over the 20-year period. Because the intention of this figure is simply to illustrate trends, the axes labelling has been left off to reduce clutter. One feature of this figure is striking – over about the first half of the time-series, the majority of the taxa (24 out of 36) showed marked declines in abundance. Of more interest in the present context are the remaining species, most of which show a humped shaped curve. These data suggest that, during the period when most species were declining, some species (mainly benthic invertebrates, cuttle fish, and some flatfish) were increasing in abundance. Eventually, however, all the species which increased during the early stages of the fishery succumbed to the pressure.

By far the most marked positive response was that of squid (*Loligo* sp.) which, having reached a high abundance by the early 1970s, only experienced a slight decline into the 1980s. As a proportion of the total catch the squid increase was even more dramatic. From being a trivial component of catches when the fishery started, it rose to comprise about 25% of the catch from 1973 to 1975, and remained at 15–20% thereafter.

Understanding patterns of decline

In an effort to explain the underlying mechanisms for the responses of the community during the first decade, Pauly (1979) undertook a detailed analysis of the trawl data. First, he calculated the rate of change in stock size for each of the component species and examined the speed of response in relation to the biological characteristics of the species. Pauly showed that, on the whole, it was the small-

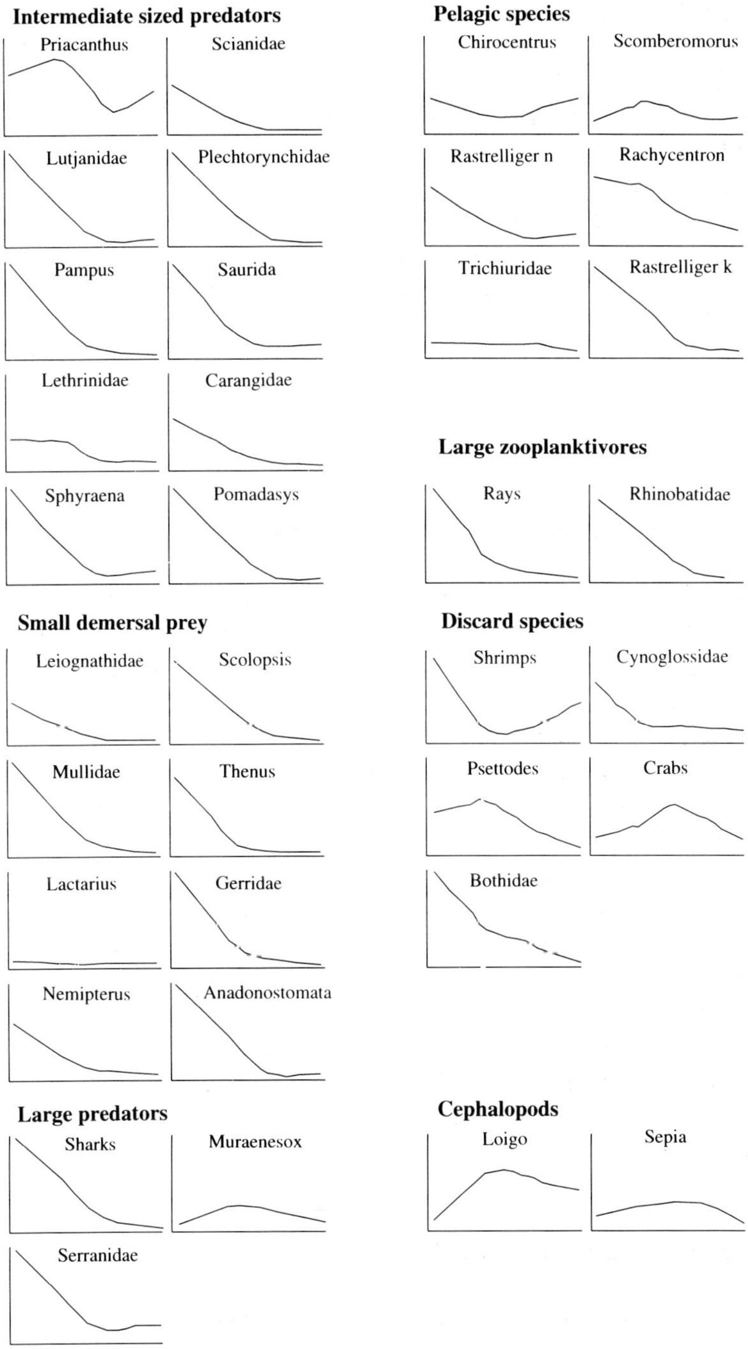

Fig. 6.10 Smoothed trends in species abundance from 1960–1983 for taxa in the Gulf of Thailand trawl fishery. Taxa are grouped according to the functional classification used by Pauly, 1979. (Note: Y axis scales differ between panels, and labelling has been removed for clarity). Data are from Pauly (1987).

bodied prey species which declined most rapidly, usually faster than the species which preyed upon them. In general these species tended to be specialist feeders, adapted to exploit a limited range of prey. This same type of response was also noted for another Thai fishery prosecuted in the Malacca Strait, implying that the response in the Gulf of Thailand may be a general one for a tropical demersal fishery in the region. Part of the explanation for this pattern is probably that species such as the *Leignathidae* occur in shallow water and are, therefore, vulnerable to both artisanal and trawl fishermen. They also occur together with shrimps and are therefore subjected to a much higher fishing mortality than other species in the area. However, using the concepts of maximum sustainable yield and simple yield assessment models, Pauly also argues that small prey fishes declined fastest because they were already being harvested at their maximum sustainable rate by their predators. Their reproductive biology probably also made them more vulnerable to poor recruitment when stock sizes fell. Thus, further mortality imposed by fishing led to a precipitous decline in abundance. This decline led in turn to the collapse of the predator stocks.

Understanding patterns of increase

With respect to those fish and invertebrate species which increased in biomass, Pauly argues for the following scenario. In the virgin unexploited system, small demersal prey fish were competitively dominant owing to their specialist feeding adaptations. These taxa kept populations of benthic invertebrates (i.e. crabs etc.), cuttle fish, eels, squid and some flatfish species at low levels. In their turn, these small prey species were controlled by top-down predation by larger species. With an increase in fishing pressure and the removal of the small demersal fish, predation on benthic invertebrates declined and their populations consequently increased. In its turn this increase provided increased resources for the more generalist species which also saw a decrease in the abundance of the predators which eat their eggs, larvae and juvenile stages. It is difficult to escape the conclusion that species interactions were responsible for the observed population increases. Indeed, with respect to the most dramatic increase of all, Beverton (1984), reporting on the conclusion of a workshop on multispecies dynamics concluded that:

> 'The increase in squid in the Gulf of Thailand following depletion by fishing of their fish predators is almost the only convincing evidence known to the group of predator–prey interaction at the population level in a major sea area.'

At the time of Pauly's 1979 analysis, the second half of the story had yet to emerge, but Fig. 6.10 tells it clearly. Even those species whose biological characteristics enabled them to exploit the reductions in competitor biomass, could not escape the inexorable rise in fishing effort.

6.4 Conclusions

The North Sea versus Georges Bank

Although there have been considerable fluctuations in demersal fish populations, the available data suggest that changes in the relative abundance of North Sea species have been minor, although flatfish do appear to have fared better than most. Given the current extent of fishing activity and cries from fisheries managers about the parlous state of North Sea fish stocks, it may come as something of a surprise to learn that, neither non-target species, nor heavily exploited gadoid species such as haddock and whiting have shown marked changes in relative abundance in fisheries survey data between 1929–1953 and 1980–1993 (Greenstreet & Hall, 1996). Indeed, these data suggest that even absolute abundance of exploited gadoids may be slightly higher during the later time period. Although one might have some reservations regarding the derivation of measures of absolute abundance from these data, this conclusion is supported by other work (Jones & Hislop, 1978; Hempel, 1978; Richards *et al.*, 1978; Daan *et al.*, 1990). However, it is important to realise that these data do not contradict the view that North Sea fish stocks are in a poor state because demersal fisheries are generally characterised as suffering from 'growth overfishing', where fish are taken too young. Greater overall yields could be obtained if fish are left in the sea to grow to a larger size before harvesting. There is no doubt that there are fewer large fish than there used to be.

The important point with respect to the North Sea is that the above observations seems to be in marked contrast to the situation on Georges Bank where the exploited gadoid species have been more or less replaced by non-exploited elasmobranch species. To further highlight the difference between the two regions, the relative and absolute abundance of the spur-dog in the North Sea (the only elasmobranch species abundant enough to influence community analyses), has declined between the two time periods. One possible explanation for this difference is that spur-dog caught in the North Sea are often landed, as are many of the skates and rays. In contrast, on Georges Bank these species are generally discarded and have high discard survivorship. The low fecundity of these species must render their populations particularly vulnerable to fishing mortality, so the exploitation of these species in the North Sea may have prevented them from replacing the heavily fished gadoid species. There is some evidence, however, that starry rays have increased in the North Sea (Heessen & Daan, 1996).

One worrying prospect is that the dominance by elasmobranchs on Georges Bank may not be reversible. One would like to think that we will eventually find ways to control fishing effort more successfully and allow stocks of desirable fish species to rebuild. It would be tragic indeed if one of the most productive fisheries in the world could not be returned to its former glory. Before such a question can be answered, however, fishing effort on gadoids and flounders would have to be reduced. As Sissenwine & Cohen (1991) point out, the short-term economic losses associated with such a reduction will be difficult to justify to fishermen, and it seems unlikely to occur in the near future. One obvious question is whether reducing the dogfish biomass can facilitate the rehabilitation of commercially valuable stock. Although the idea is quite seductive, the cautionary notes of Sissenwine & Cohen (1991) regarding the predictability of the outcome and the practical difficulties of actually achieving the necessary reductions in elasmobranch biomass are important.

Temperate versus tropical demersal systems

Differences in catch profile explain a lot

As noted earlier, important differences between temperate and tropical fisheries, are the taxonomic diversity and the size distribution of the tropical catch. Both of these features make fishing on only one or a few species almost impossible in the tropics. This means that a far wider range of taxa will be directly affected by fishing – there are few species which are truly non-target. Since most species are actually deliberately fished, it follows that fishing will have more widespread effects on the fish taxa in the system. Moreover, because much of the fishery is prosecuted in shallow water, both artisanal and commercial fishers can impose substantial fishing mortality. Effects on the juveniles of the stocks may be particularly important in this respect, because nursery grounds are in the shallower water and demersal catches consist, to a significant extent, of the juveniles of valuable larger-sized fish. Thus, much of the difference in the responses of temperate and tropical systems may stem directly from these basic differences in the nature of the fisheries. But might there also be more fundamental differences in the nature of the relationship between species in the two systems, which partly determine the response to fishing? There are good reasons to suppose that there might. The reason relates to one of the most discussed contrasts between temperate and tropical systems – the relative environmental stability that the tropics experience.

It is argued that, because the tropical environment is more stable, environmentally driven variation in recruitment to demersal fish species is less likely and species abundances will fluctuate less than in temperate systems. Conventional wisdom states that this relative stability has led to greater trophic specialism in the tropics. Thus, the coupling between populations of predators and their prey should be stronger, and indirect effects consequently more important. This seems to be a view held by Pauly (1979) who states that ‘It seems generally accepted that tropical fishes interact most strongly with the biotic components of their environment, while temperate fishes seem to be more strongly affected by the abiotic components’. Pauly further argues that a greater preponderence of specialists in the tropics should mean that when the food web is disrupted by fishing, it is generalist species (squid and heterosomatid flatfishes) which benefit. This is because these species can better adapt their behaviour to changing conditions and are less dependent on the continuing availability of one or a few prey species. Moreover, Pauly’s arguments to explain the observation that small-bodied specialist fish declined fastest in the tropics, also implies that trophic interactions may be more important in these systems. (The argument goes that: (1) these small specialists were already being harvested at their maximum sustainable rate by their predators, and (2) their reproductive biology made them more vulnerable to poor recruitment when stock sizes fell.)

The above response contrasts with the rather simpler-minded view that it should have been the larger predatory species that were affected first and most markedly by fishing. In contrast, for temperate systems where the size of landed fish is generally much larger, larger individuals are, of course, more likely to be affected first, but given the arguments above one might also expect there to be fewer specialist taxa in temperate systems. Thus temperate species might, in general, be less likely to be affected by any changes in the abundance of prey or predator species that might be caused by fishing.

Although the above arguments are fairly compelling, there seem to be few data to support (or refute) the idea that fluctuations in fish species abundance are lower in the tropics, or that diets are more specialised. However, it is certainly true that the annual production cycle shows much less seasonality, and reproduction in many species is prolongued or semi-continuous, so there seems to be less selective pressure for life-history traits to evolve to exploit seasonal windows of opportunity for successful reproduction. Pauly (1979) states that there is little to suggest that for multispecies tropical fisheries highly variable recruitment occurs, but the possibility cannot be ruled out for particular taxa. Of course, for temperate demersal fisheries systems year class failures are well documented, but in contrast to pelagic systems, there are generally more year classes present to buffer against failure in any one year. That said, the prevalence of growth overfishing in temperate demersal fisheries is reducing the number of year classes, with a consequent reduction in buffering capacity.

Species replacements and alternative stable states?

One lesson that comes from the above comparisons is that depletion of valuable fishery resources can lead to concomitant increases in species of lower value, regardless of the latitude of the system. From a commercial viewpoint, dogfish are much less desirable than the groundfish they have replaced. Similarly, although markets for cephalopods are certainly burgeoning around the world, abundant groundfish are generally preferred. The question that is repeatedly asked of course is whether the changes are reversible; in general, they probably are. Although there is evidence for multiple stable states in some marine systems (see Chapter 8), there is little basis for supposing that reductions in fishing effort would not lead to the eventual rebuilding of most stocks owing to some regulatory control by other species which have come to dominate. It must be recognised, however, that the vagaries of recruitment make predicting the time-scale of any recovery very difficult. Some demersal stocks, such as the northern cod in Canada which collapsed in 1989 have yet to show signs of rebuilding, despite a moratorium on fishing.

With respect to tropical fisheries, one might ask whether domination by squid is a general feature of an intensely fished demersal system? I don't think we are in a position to answer that question with any certainty, but there is certainly evidence that cephalopods have also increased in other areas. For example, Gulland & Garcia (1984) describe the circumstances surrounding changes in the structure of the fish community off the coast of northwest Africa, where catch data extend back to the 1940s. Until the 1960s, the community had traditionally been dominated by sparid fish species of various kinds. Although some cephalopods, especially cuttlefish (Sepia) and squid, have always been present, octopus was rarely mentioned until 1962 when it was first recorded as a significant component of the community. Some reports suggest that, as a proportion of landings, cephalopods had risen from almost nothing in 1960 to 80% by 1967, although others put the figure at about 30%. By the early 1970s, however, all agreed that the cephalopods constituted some 90% of the catch, a large proportion of which is octopus.

Unfortunately, the data with which to establish the true extent of a replacement of fish by octopus are simply not available, and catch data are always open to criticism because, as we have seen earlier, changes in the behaviour of the fishing fleet can mask the true picture. However, the survey data that are available support a replacement hypothesis (Fig. 6.11). Most scientists who are familiar with the fishery seem inclined to believe that octopus have indeed replaced fish as the dominant component of the system. Other regions with a long history of demersal fish exploitation have also seen an increase in the proportion of cephalopods in the

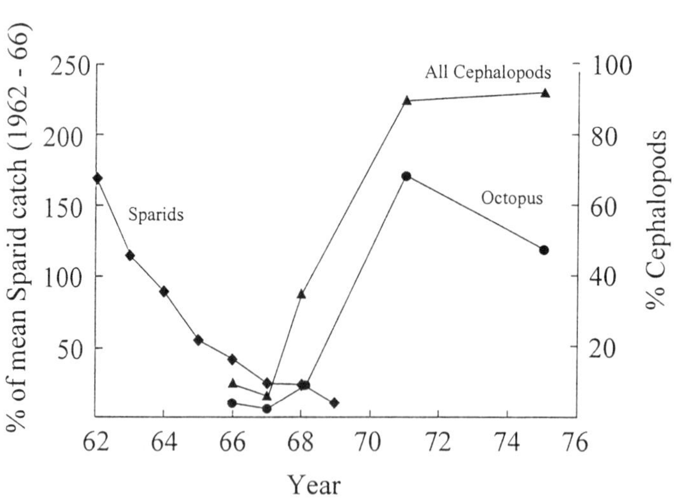

Fig. 6.11 Trends in relative abundance of sparids and cephalopods, based on fishery independent survey data and commercial catches. (Redrawn from Caddy (1983) with permission from the Food and Agriculture Organization of the United Nations.)

catch, but whether this reflects an increase in absolute abundance or marketability is unclear in most cases because the data do not go back far enough.

The transition in dominance shown in Fig. 6.11 appears to have occurred over a 5–6-year period – a rate of change that is difficult to explain on the basis of simple equilibrium population dynamics. Yet, similar abrupt changes have also been noted in other cephalopod fisheries (Caddy, 1983). As ever, the relative importance of changes in environmental conditions and fisheries exploitation is an issue. Caddy (1983) notes changes in the upwelling regime off west Africa, for example, that are associated with changes in the frequency and intensity of offshore winds. Drought in the Sahara seems to be another manifestation of this change. Similarly, there is a striking relationship between ocean climate and 100-fold changes which occurred in the *Illex* squid fishery off eastern Canada over a 10-year period (Fig. 6.12).

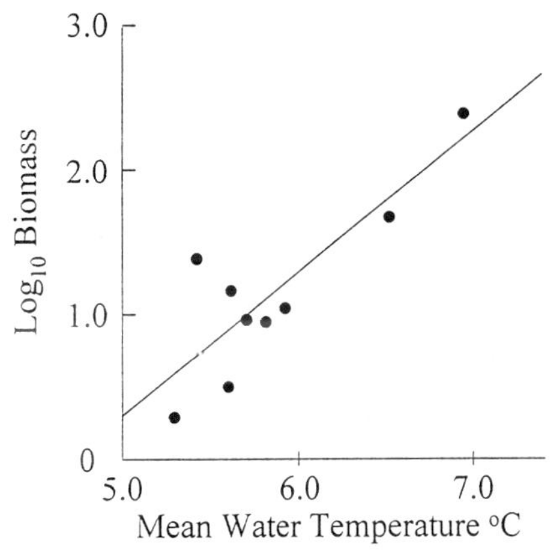

Fig. 6.12 The relationship between annual biomass of the squid (*Illex illecebrosus*) on the Canadian east coast and mean seawater temperature. (Redrawn from Caddy (1983) with permission from the Food and Agriculture Organization of the United Nations.)

In 1983 Caddy (1983) noted that hydrographic conditions off west Africa appeared to have been returning to those which obtained during the earlier period of sparid dominance. He suggested that if there was no decline in cephalopod dominance, one would be more inclined to believe that it was the reduction in sparids, rather than climatic change that was the main factor governing increase in cephalopod stock size in the late 1960s. As far as I am aware, there has been no such return, and it would appear that the case for a strong ecological interaction is reasonably compelling.

The mechanism of such an interaction can only be speculated upon, but one could imagine that in the unexploited state sparid populations comprised many age classes and were buffered against the vagaries of environmental fluctuations. From this dominant position, predation on adult and juvenile cephalopods might act as a controlling mechanism. With the onset of fishing on sparids, the life-history characteristics of cephalopods (e.g. shorter life-spans, high fecundity) offered the potential for large and rapid increases in populations given favourable environmental conditions. Having achieved such dominance, controls by cephalopods on fish recruitment through predation on juveniles may have become significant.

Flatfish seems to do better?

One seemingly common feature of the data discussed above is the relative performance of flatfish. Pauly (1979) notes that flatfish seemed to do well when other fish declined (although they eventually declined also) and the North Sea data also show a general increase in dominance by flatfish. It would appear that there has also been a large increase in flatfish in the Gulf of Alaska and eastern Bering Sea (pers. comm). Since trawl gears are generally less efficient at catching flatfish, responses in the Gulf of Thailand and the North Sea demersal fisheries are perhaps unsurprising, but it is possible that biological interactions may also play a part. Interactions related to changes in habitat structure may also be part of the story. If, for example, trawling the seabed increases the area of bare sand, it may be more suitable for flatfish than for other species. Increasing dominance by flatfish may be a common feature of exploited systems that is worthy of further investigation.

Data quality

For fisheries systems, the difficulties of obtaining data and effecting critical tests of hypotheses about ecological interactions are formidable. These difficulties are evident in many of the examples used in Chapter 5, but also in this chapter where the data were rather better. This lack of a clear signal also makes taking account of ecological interactions in management difficult. We know that predation and competition will often be important, but without knowing when, where and how, it is not clear what you do about it. This question is returned to in Chapter 9, but part of the problem may lie in the level of organisation at which the foregoing analyses were focussed. It may be that the signals at the species level are just too hard to detect. Variability in the population dynamics of individual species, lags in response and shifts in the relative importance of predation or competition over relatively short time-scales may all conspire to make unravelling the key process such a daunting task. Perhaps instead we should look at responses at a more aggregated level and ask how 'higher order' attributes of ecosystems and communities respond to fishing. This is the topic of the next chapter.

Part 3
System Level Perspectives

Introduction

Most would accept that an ecological system is rather more than the sum of its parts and that there are 'higher order' or emergent properties possessed by communities and ecosystems that are worthy of study. The terms 'biodiversity' and 'ecosystem function', for example, abound in the recent ecological and conservation literature and although precise definitions of the terms are often difficult to pin down, their use signals recognition that system level properties should be accorded some weight when trying to manage our environment responsibly. In an assessment of the effects of fishing then, it seems sensible to try and take such emergent properties into account. The next two chapters adopt this wider viewpoint and ask how the harvesting of one or a suite of species might alter the structure and function of the community or ecosystem as a whole.

Chapter 7 describes some of the approaches that have been adopted to try and evaluate marine community structure and the changes to it that fishing activity might have imposed. What each approach tells us (or may tell us in the future) about fisheries effects is summarised. In Chapter 8 related questions are asked: what might constitute ecosystem health or stress? What diagnostics of such states might one look for? What changes in structure might mean for the functioning of the system? This latter question is of fundamental importance, because if fishing affects how systems function, rather than just how they look, there may be even more pressing reasons to take action to mitigate effects.

Chapter 7
Ways of Looking

A continuing difficulty with community analyses is deciding what measures are most useful for characterising the state of the system in revealing and useful ways. A bewildering array of approaches is now available, but at present we have no real basis for choosing one approach over another. Different authors have adopted different techniques, but all ask the same basic question – have the emergent properties of communities actually changed over time? After all, it is only if we have evidence that changes at the community level have actually occurred is it worth trying to determine the extent to which fisheries activity can be implicated in the change.

In Sections 7.1 and 7.2 a number of the approaches that have been used to describe changes in fish species assemblages are examined for regions where fishing activity has been intense. Despite the extensive data sets held by fisheries institutes in some parts of the world, fisheries data were rarely examined from a wider community level perspective until recently. Single-species of commercial interest were the almost sole focus for analysis, even though data on entire fish community composition were often collected during fishery surveys. Happily, this deficiency is now being rapidly rectified with the publication of a growing number of community level analyses. Thus, our understanding of whether and how community level properties have changed in response to fishing has improved dramatically in recent years.

The reasons for treating the responses of the fish assemblage in isolation from the rest of the system are partly practical – there are more time-series data available on the abundance of fish species than there are for anything else. However, one might also expect structural effects to be more profound in the fish community than elsewhere since it is this community that experiences the most direct effect of fish removals. Fish assemblages seem like a sensible place to start looking. Later in the chapter, a rather wider focus is taken, which includes the non-fish fauna in an examination of how estimates of energy transfer between various animal and plant groups might be used to gain a better appreciation of the pressures fishing activity imposes on the system.

7.1 Species diversity – a blunt tool or informative measure?

One way to summarise the composition of a multispecies community is to use a diversity index. Trying to capture the diversity of species in the system in a single number has many pitfalls, but it remains a very common approach. Indeed, for many, analysis of differences in a diversity index between sites or over time is one of the first steps in unravelling community trends. There is a vast literature on the use of such indices and Magurran (1988) is recommended for an excellent introduction to the subject.

Perhaps the most commonly used diversity measure is the Shannon-Wiener index (H'), which is defined as:

$$H' = -\sum_{i=1}^{S} P_i \ln P_i$$

Where P_i = the proportion of the total number of individuals comprising the ith species. A commonly used variant on this measure is obtained by calculating *exp* H.

How does the diversity of a fish community respond to fishing pressure? Figure 7.1 shows the trends in *exp* H for two of the areas analysed in Chapter 6, Georges Bank and the Gulf of Thailand. In both cases the data are calculated from research surveys. For the Gulf of Thailand, the time-series begins more or less at the inception of the trawl fishery, whereas for Georges Bank, while fishing had occurred for many years

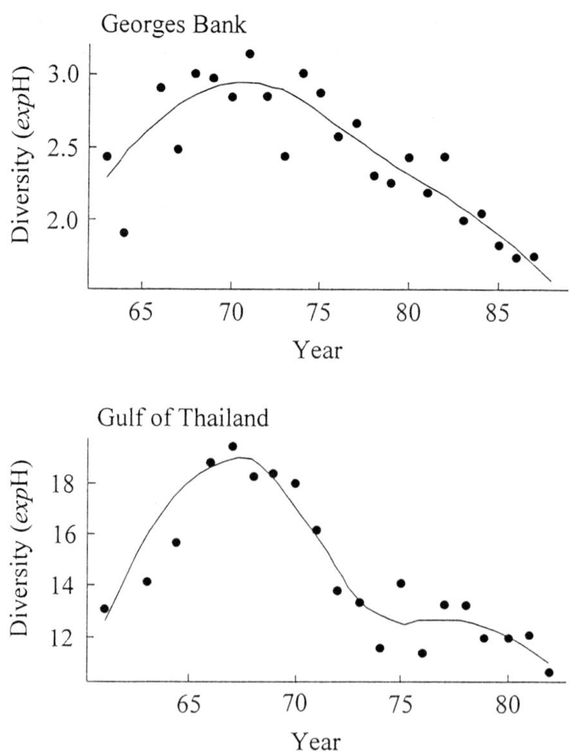

Fig. 7.1 Trends in *exp* H for the Gulf of Thailand and Georges Bank. Data for Gulf of Thailand is taken from Pauly (1987); data for Georges Bank taken from Fig. 2 of Solow (1994).

previously, the time-series covers the period when there was a massive increase in fishing effort. The trends in these two graphs are remarkably similar, with an increase in diversity in the early stages of the fishery and a progressive decline thereafter.

Bearing in mind that a diversity index captures aspects of both the species richness (the number of species in the system) and the evenness (the distribution of relative abundance among species), these trends make intuitive sense. Since species are unlikely to be added to the system by fishing, the rise in diversity in the early stages of the fishery must be due to an increase in the evenness of species relative abundance. It seems reasonable to suppose that the commonest fish species were the initial targets of the fishery and their decline led to the an initial rise in evenness. Over time, however, overfishing of these species may have led to further reductions in the population sizes of the initial target species, with a consequent decline in diversity. New species may then have been exploited until one reaches a point where commercially undesirable species dominate the community and diversity is low. One might call such an explanation the sequential cropping hypothesis.

Although such an explanation for the observed trend in diversity seems plausible and the pattern itself seems clear, other studies where the effects of fishing on diversity have been examined do not show such convincing patterns. For example, Watson *et al.* (1996) examined the effects of artisanal fishing on the abundance and diversity of coral reef fish assemblages in Kenya by contrasting the assemblage in a reserve where fishing was permitted and a park where it was not. While more fish were found in the protected area, there were no effects on diversity, a result that Watson *et al.* (1996) tentatively attributed to migration of adults from the protected area or increased larval supply. In temperate systems too, effects on diversity in response to increased fishing effort are not always apparent. In the northern North Sea, for example, species diversity (as revealed by indices such as Shannon's *exp* H′) varied between years, but trends over 60 years were generally inconsistent with a fisheries effect (Greenstreet & Hall, 1996). Given the fact that expressing diversity in conventional terms is not always informative, one might reasonably ask whether there is another approach?

Taxonomic diversity?

Assemblages that have the same value for a conventional diversity index may comprise species that are all closely related to each other taxonomically, or be of species that are only distantly related. Conventional diversity indices such as the Shannon Index H′ take no account of taxonomic relatedness. Noting this fact, Warwick & Clarke (1995) recently proposed two new measures of community diversity Δ and Δ^* which incorporate taxonomic relatedness into their calculation. The authors describe Δ as an index of taxonomic diversity because it is empirically related to the Shannon's species diversity H′, but has an added component of taxonomic separation. In contrast, Δ^* can be thought of purely as a measure of taxonomic distinctness, without the contribution from species diversity.

These indices behave in intuitively reasonable ways when small example communities are constructed. When used in an analysis of a marine benthic invertebrate assemblage around an oilfield, a continuous decrease in taxonomic distinctness could be detected along a gradient of increasing environmental contamination. In contrast, conventional diversity indices showed no trend along the same gradient. The authors also examined the effects of sample size on the behaviour of the indices and showed that, unlike most other diversity measures there was no systematic bias at low sample sizes – a highly desirable property for any index.

Based on these results it would appear that including taxonomic relatedness into diversity measures may provide more robust and sensitive indices of community perturbation. It can also be legitimately argued that such measures more truly reflect our notions of 'biodiversity'. A community of many closely related species would be viewed by most people as being less diverse than one where the same number of species is more distantly related.

To explore the value of this alternative perspective for examining fishing effects, Hall & Greenstreet (1998) calculated Δ and Δ^* for North Sea data to examine whether there have been shifts in the taxonomic relatedness of the assemblage. Contrary to expectations, the conventional diversity indices were lightly correlated with these new ones (Fig. 7.2). At first sight then, it would appear that taxonomic indices give us no additional insight. However, it must be borne in mind that these new indices measure another component of community structure that is of intrinsic

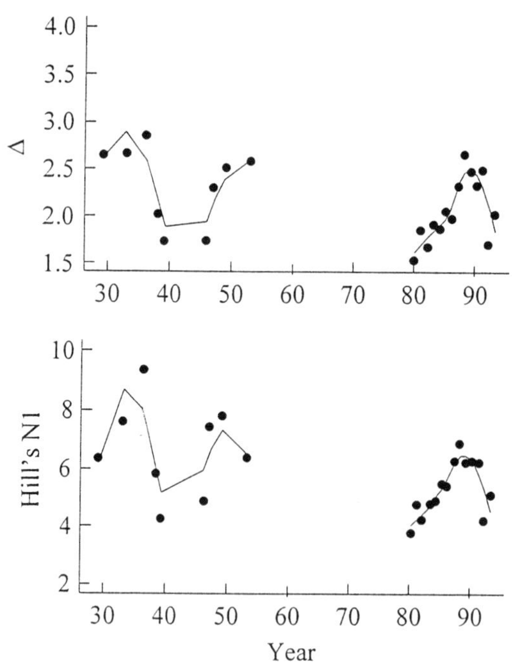

Fig. 7.2 Time-series plots for the standard diversity index *exp* H (Hill's N1) and an index incorporating taxonomic relatedness Δ. Adapted from Figure 1 of Hall & Greenstreet (1998).

interest. Their use with these data suggests that, as with conventional diversity, while there were trends over time, there is little indication for this system of marked changes in the taxonomic hierarchy ('biodiversity') that might be associated with, for example, increases in fishing effort. It will be interesting to see whether other fisheries data behave in the same way.

7.2 Ordination and clustering

How else might one examine community data? Another common approach is to use one of the various cluster or ordination methods. These methods calculate how similar in species composition different samples are and then classify them on the basis on their commonalities or differences. These are now commonly used and powerful tools for identifying shifts in community composition. To illustrate further how this approach can be used Fig. 7.3 shows ordination plots for the yearly biomass data given in Pauly (1987) for the Gulf of Thailand and Clark & Brown (1977) for Georges Bank. Both data sets are for a period when fishing pressure grew rapidly. To understand such graphs one simply needs to appreciate that samples that are more similar to one another in faunal composition (in this case data for different years) are plotted closer together. The graphs for both areas clearly indicate that in the early years of the fishery the community changed in a directed way and that this directional change became less apparent later in the time-series. Of course, without data from other areas that are not subject to fishing we cannot ascribe such directed change to fishing with certainty. However, the common trends in both the ordination and diversity plots for two regions subject to similar patterns of increase in fisheries exploitation are suggestive.

Although methods like MDS can detect changes in composition they give no idea of the nature of the changes or how distinctive or subtle they might be. For example, the MDS analysis presented here shows that change has occurred for Georges Bank and one might conclude that this change is ecologically significant in some way. An analysis of these same data by Grosslein *et al.* (1980), however, led them to conclude that there had 'been no drastic shift in species composition of the demersal finfish biomass and no large build up of other demersal finfish species to replace the principle groundfish and pelagic stocks'. A similar conclusion was drawn for data gathered from the North Sea, despite finding very clear long-term trends in community structure over time (Greenstreet & Hall, 1996, see Chapter 7.2). The root of the problem here is separating a detectable effect from a large and/or ecologically important one. Perhaps methods which include more obvious functional attributes like body size might be more informative.

7.3 Body size

Ecologists have long appreciated that information on the distribution of body sizes may help us to understand the functional basis for community organisation (e.g.

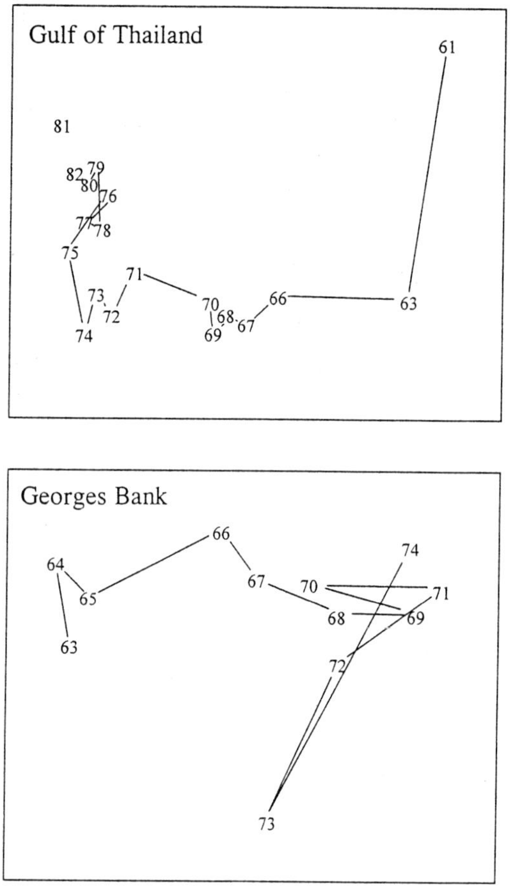

Fig. 7.3 MDS ordination plots of demersal species composition for Gulf of Thailand (1961–1981) and Georges Bank (1963–1974), showing directed temporal trends in community composition. Plots are from Bray–Curtiss dissimilarity matrices calculated on untransformed data. Data for Gulf of Thailand is taken from Pauly (1987); data for Georges Bank is taken from Clark & Brown (1977).

Sheldon *et al.*, 1972). One might choose, therefore, to take an alternative view of community organisation and forget for the moment that systems contain many different species. Instead, we simply ask how individuals are distributed among size classes, irrespective of species identity. This approach has a number of potential advantages. First, we know that fishing is often size selective, so focussing on community size structure could be an especially promising approach for examining fisheries effects – it is easy to envisage fisheries-related mechanisms that might lead to changes in size structure. Second, for fish assemblages, it has been observed that the biomass and numbers of individuals in size classes (pooled across all species) is relatively stable (Sheldon *et al.*, 1972; Pope & Knights, 1982; Pope *et al.*, 1988; Murawski & Idoine, 1992). This stability occurs even though the richness and relative abundances of species in a series of samples is highly variable. Thus, in looking at the community as a whole, rather than its component species (for which data are often rather noisy), we might get a more easily interpreted and integrated picture of how the community is organised.

One of the first efforts to explore the size-structure of a fish community and its possible relation with fishing was undertaken by Pope & Knights (1982). Figure 7.4a shows the results of their analysis, with numbers at length for two areas, one around the Faroes and another in the northern North Sea. These areas are at similar water depths and have a similar species complement but show strikingly different size distributions. The northern North Sea has far fewer large fish and many more small ones. Pope & Knights (1982) speculated that these differences may have resulted from different patterns of exploitation in the two areas – the mesh size used at the Faroes when the samples were taken was 110 mm, but it was 80 mm in the North Sea. Perhaps the greater intensity of fishing in the North Sea led to a reduction in the numbers of large fish in the system and (more speculatively) an increase in the numbers of small fish through a release from predation.

Further work by Pope *et al.* (1988) contrasted data for the North Sea with that of Georges Bank and showed that the trends for the two systems differed markedly. For the North Sea there was a decline in the proportion of larger fish, as indicated by an increase in the slope of the numbers at length curve computed for different years. In contrast, for Georges Bank there was a slight increase in the numbers of larger fish and a consequent decrease in the slope of the size spectrum. The slope of the size spectrum was also much greater for the North Sea indicating a much faster decline in the numbers of individuals as body size increased (Fig. 7.4b). This increase in large fish for Georges Bank is almost wholly accounted for by an increase in the numbers of elasmobranchs in the system, a feature of the areas ecology that was discussed in Chapter 6 (Murawski & Idoine, 1992).

In re-evaluating these data Murawski & Idoine (1992) made the important observation that the difference in the size spectra between the two regions was much greater than any variation over years within regions. This larger difference between areas could be explained in three ways: (1) it is a manifestation of differences in the way the communities are organised (e.g. different rates of predation mortality at length, proportion of total food comprised of fish etc); (2) it reflects differences in the pattern of fisheries exploitation, or (3) there are biases in the way survey data are collected. However, as Murawski & Idoine (1992) noted at the time, without additional data or convincing theoretical analysis, deciding among these alternatives was difficult, if not impossible.

Further work by the ICES Working Group on the Ecosystem Effects of Fishing Activities in 1994 provided such additional data and theory for the North Sea. The work of this group suggested that when log abundance by size class was regressed against size for different years the slope of the spectrum became increasingly negative over time. In contrast, the intercept changed only little (Anon, 1994). The underlying basis for such change was later examined by Rice & Gislason (1996) using MSFOR, the forward projecting version of multispecies virtual population

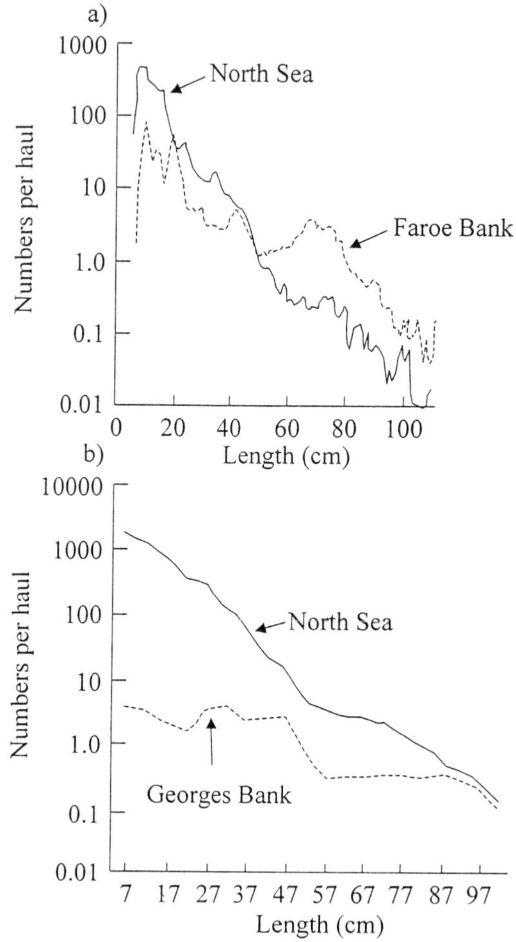

Fig. 7.4 (a) Average numbers at length from Faroe Bank and northern North Sea groundfish surveys. Redrawn from Fig. 2 Pope & Knights, 1982. (b) Average numbers at length from North Sea English groundfish surveys and Georges Bank USA bottom trawl surveys. Redrawn from Fig. 4 Murawski & Idoine (1992).

analysis (MSVPA) described in Chapter 6. This model allowed predatory interactions between species to be accounted and predicted the equilibrium length distribution for 11 North Sea species.

Figure 7.5a shows how the size spectrum is predicted to change at three different levels of exploitation, the current level, a 50% increase and a 50% decrease. Beyond 30 cm these graphs are well described by a straight line. Figure 7.5b summarises the trends in predicted slopes and intercepts of the spectra in this range with changes in fishing effort. The trends agree with our commonsense notion of how the size spectrum ought to respond to fishing, with a loss of large individuals and an increase in small ones.

These results are important because they show that changes in size spectra can

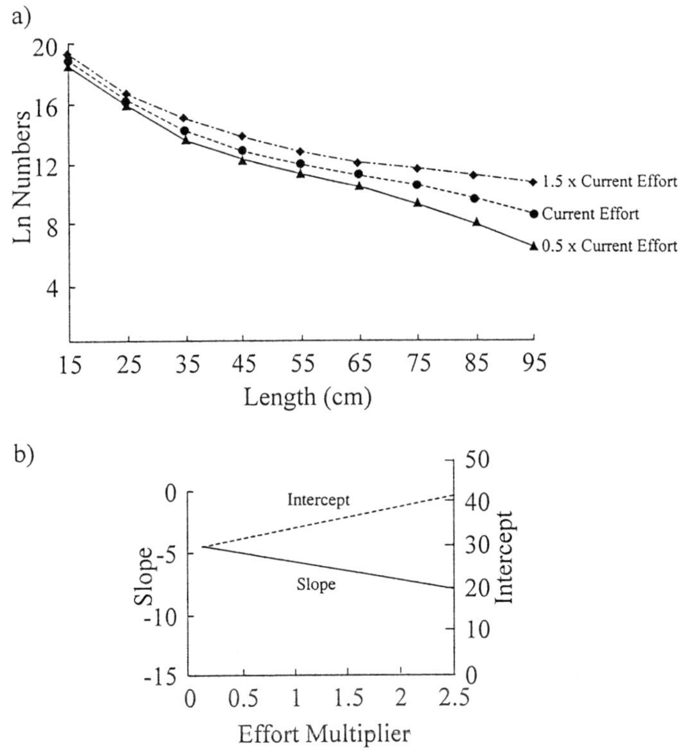

Fig. 7.5 (a) The distribution of numbers at length for a community of 11 fish species under different levels of fishing effort using single species length based VPA. (b) Trends in the predicted slopes and intercepts of the size spectra for fish beyond 30 cm length. Adapted from Figures 3 and 4 of Rice & Gislason (1996). For further explanation see text.

occur as a result of fishing and suggest that changes between years in the North Sea size spectra might be accounted for in this way. But what of the differences between the North Sea and Georges Bank? It is doubtful that the differences in response between these geographic regions can be solely explained by different histories of fishing. They are probably more a reflection of fundamental differences in community structure between the systems. When making these comparisons it is important to bear in mind that the North Sea is a much larger area than Georges Bank (approximately 11 times bigger) and that it is much more heterogeneous, with a greater depth range and a thermocline in summer. From an oceanographic standpoint, therefore, the areas are not strictly comparable, although they are nevertheless both temperate shelf seas, with broadly similar trophic structures.

Although contrasts between areas might be difficult, we ought to expect changes in size spectra within any given area with the development of a fishery. Focussing on these changes might be rather useful for fisheries management in some cases. In tropical fisheries, for example, three problems arise: (1) a wide range of shorter-lived species are caught, (2) individual species assessments are impossible, and (3) time-series of catch and effort data are often lacking. Analysis of size structure from trawl surveys could provide an easy and cost effective way of measuring the development of the impact of the fishery on the fish fauna in these circumstances. This is an area of research that certainly warrants further study.

Extending size-based analyses to include diversity

An alternative to the purely size-based analyses described above is to consider how species, rather than individuals, are distributed among size categories. This approach occupies potentially informative middle ground between a purely size-based analysis and analysis of fully resolved data sets that retain the identity of species, their sizes and their abundances. If number or biomass at size spectra reflect important, organisational principles that are determined by feeding relationships, one might expect regularities in the species diversity–body size relationship that could be explicable in terms of the partitioning of food resources.

In a preliminary effort to explore such issues, members of the ICES Working Group on the Ecosystem Effects of Fishing Activity analysed trends in species diversity (expressed as Shannon's H′) against length class for English Groundfish trawl survey data (Anon, 1994) – work that was subsequently built upon by Rice & Gislason (1996). In both analyses, least-squares linear regressions of diversity per size category versus body size were used to describe the spectrum. Interestingly, the slope of these regressions calculated for data from each year separately showed no significant overall trend over the time period analysed, although there were periods within the time-series where trends were apparent. This result was in contrast to the slope of numbers at length analysis discussed in the last section that showed a progressive increase in the steepness of the decline. Recall that the latter change was interpreted as being a direct consequence of heavy fishing pressure decreasing the numbers of large fish.

In addition to modelling the expected responses of length distributions to fishing (see above), Rice & Gislason (1996) also examined model predictions for how diversity size spectra would change under various fishing pressures. As with the analysis of survey data, they found that calculating the Shannon index by size class provided a rather poor indicator of changes in fishing intensity. This lack of correlation led the authors to conclude that this way of looking at fisheries data does not provide a sufficiently sensitive metric that would be useful in monitoring fisheries impact on North Sea fish assemblages. However, while the above approach is probably not very useful, another way of presenting the same data, inspired by work on insects, perhaps shows more promise.

Lessons from insects?

Of course, analysis of patterns in the distribution of abundance, species diversity and body size is not only conducted by those interested in fish. There is a large literature on the subject, in which the relationships between these properties have been variously examined, either individually or in combination, for other communities, particularly insects. Although many have looked at these kinds of patterns in their data sets, recent work by Siemann *et al.* (1996) is probably the most comprehensive effort at a unified analysis to date. Their work is particularly thought-provoking. These authors studied a grassland insect community and, in common with others, they found a dome-shaped relation between both species richness (S) and number of individuals (I) with body size, when species are classified into $\log_2$ size class intervals. In addition, however, Siemann *et al.* found a previously unreported, but surprisingly simple power law relationship between S and I within size classes where $S = aI^b$. Plotting log S versus log I gives a straight line, the slope of which corresponds to exponent b and the intercept to the coefficient a.

Stimulated by these findings a preliminary analysis of fisheries trawl survey data was undertaken to examine whether these same relationships held for marine fish (Hall & Greenstreet, 1996). Similar dome-shaped curves were observed for a number of individuals and species richness versus body size along with a power law relationship between S and I within size classes. Could such a relationship form the basis for a community level analysis of fishing effects? To investigate this possibility the analysis was extended to examine how the relationship between S and I within size classes changed over time and over geographic region. Figure 7.6 shows the results for changes between the periods 1929–1956 and 1981–1993. The coefficients of the regressions of log S versus log I indicate a significant decline in intercept between survey periods that is not associated with a change in slope. Additional analysis showed that this decline could not be accounted for by any sample size effects and appears to reflect an overall reduction in species richness within body

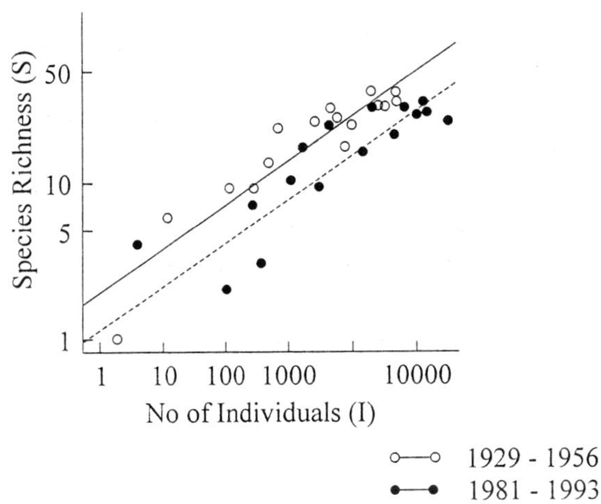

Fig. 7.6 The relationship between S and I within $\log_2$ size classes for fish in the northern North Sea for the periods 1929–1956 (open symbols) and 1981–1993 (solid symbols).

size categories. Looking more carefully at the patterns in these data, the most striking trend over time was a decline in species richness between 1929 and 1950 for both small and large size classes. As discussed earlier, declines in the abundance of large fish are to be expected with fisheries exploitation, and it is perhaps unsurprising that species richness should decline in this category. Previous work, for example, has noted the absence of species such as the large skate *Raja batis* from the Irish Sea (Brander, 1981) and large cod are increasingly rare in the North Sea (Cook *et al.*, 1997). Declines in the abundance of skates and rays in the North Sea has also been documented (see Section 2.2) and it seems likely that other large-bodied species have also been similarly affected.

Although a fisheries effect on large individuals is to be expected, a response in smaller fish is much more difficult to explain. Given the undoubted problems of sampling small fish adequately, it is tempting to discount this result altogether. The trends in species richness, however, run counter to our expectations based on sampling effort because the number of species in samples declined despite the fact that more individuals were sampled during the later time period. It is likely, therefore, that the decline is not a sampling artefact. Changes in size at settlement or increases in growth rate of fish could account for the trend. (The alternative is that the species complement changed over time with smaller species replacing larger ones, but this was not the case). One interesting possibility is that it is changes in the within cohort variability in growth rate that has led to such effects.

Importantly, much of the change occurred in the early part of the time-series suggesting, perhaps, that for established fisheries systems for which only more recent data are available, much of the effect has already occurred. Despite the lack of a firm explanation for the observed trends at this stage, this analysis suggests that the coefficient a of the power law relationship $S = aI^b$ might be a valuable measure of community effect. However, interpreting such trends is undoubtedly hampered by the absence of a theoretical foundation for understanding species richness body size relations. One approach that may ultimately shed light on this issue is to use multispecies models to examine the effects of fishing on model assemblages, as Rice and Gislason (1996) did with MSVPA to examine how biomass and diversity indices within body size classes responded to fishery exploitation. Unfortunately, only 11 species have been included in the MSVPA system so far and so the power to look at diversity effects is limited, perhaps explaining why Rice & Gislason (1996) found no diversity effect under fishing. Given the empirical results described here, and the recognition of biodiversity as a property of systems that needs to be accounted for, further efforts to explicitly incorporate some aspect of species identity into quantitative analyses remains highly desirable. What seems certain, though, is that the theory will need to be further developed and evaluated before the mechanisms underlying observed trends can be explained and the full significance of these findings can be determined.

Another point that needs to be borne in mind when evaluating the approaches described above is the enormous data demands that they make. For example, we can only derive diversity size spectra because some fisheries data is so comprehensive and has been gathered over relatively long time-scales. Routine collection of such data on time-scales appropriate for management would often be impractical. It may well be that the interest and use of such approaches will wane once the few comprehensive data sets that are available have been fully explored and analysed.

7.4 Energy budgets and other approaches

So far this chapter has focussed on fish assemblages and the community level analyses that can be undertaken with fisheries data. But there are, of course, other ways of looking at marine systems which might offer additional insights and which have the advantage of including non-fish components. One such alternative is to consider how energy or matter flows around the system. In discussing Georges Bank, for example, Murawski & Idoine (1992) identify changes in energy flow as a likely consequence of fishing:

> 'One would assume that given the profound changes in biomass, production and species, composition [caused by fishing], that food webs have undergone major restructuring, with energy transfer rates being extensively modified.'

Since the International Biological Program emphasised ecosystem research in the late-1960s there have been many efforts to document energy transfer in both marine and terrestrial systems. These efforts were often a prelude to constructing large ecosystem models which, it was hoped, would ultimately provide tools for management. It is fair to say that the early optimism about the value of the energy budget approach and the value of such large models has waned somewhat (see comments, for example, in Pimm, 1991). Recently, however, there seems to have been something of a resurgence of interest in what such trophic flow models might tell us. Dynamic energy or biomass budgets for whole systems (i.e. ecosystem models) predict how the standing stocks in system compartments and the flows of matter between them change. They are not especially common, are often difficult to construct and are certainly difficult to parameterise with any confidence. Moreover, their focus for marine systems has traditionally been on the dynamics of lower levels (nutrients, phytoplankton and zooplankton), largely because these biological components can be coupled to physical circulation models using standard advection and diffusion approaches. Because fish move independently of the patterns of circulation they cannot be treated in this way and therefore are often ignored. For these reasons, the larger dynamic ecosystem models are probably of limited utility in exploring fishing effects. In contrast, descriptions of stocks and flows at any one point in time are often achievable, sometimes even for different seasons (see, for example, Greenstreet *et al.*, 1997). With the recent development of new software tools, this static approach has become even easier. In particular, a software tool called ECOPATH, originally described in 1983, and subsequently developed and extended by Christensen and Pauly (see Christensen & Pauly, 1992) has encouraged many groups to try and document trophic transfers for their particular system (see contributions in Christensen & Pauly, 1993).

The essence of ECOPATH is an equation which balances inputs and outputs from every compartment or box in the ecosystem description. The equation can be most simply expressed as: Consumption + Imports = Production + Respiration (where Production = Exports + Predation Mortality + Flow to detritus). Thus, in a balanced system where the biomass or energy in each box remains constant, the consumption of a predator compartment generates the predation mortality for its prey. Therefore, the whole system can be expressed as a set of linear equations, one for each system component. Each equation requires four basic parameters: biomass, consumption, production rate and ecotrophic efficiency. The last parameter, ecotrophic efficiency, expresses the proportion of the production for any group that is utilised for either predation, catches or biomass accumulation in the system. The main attribute of this approach is that by linking boxes it becomes possible to estimate one missing parameter for each equation.

One interesting aspect of ECOPATH is that it can provide an estimate of the amount of primary production (and detritus) required to sustain the consumption by each of the compartments represented in the model. Similarly, the amount of primary production required to sustain fish catches can be estimated from the catches and the energy transfer along food chains leading to the catches. In a sense, estimates of the primary production required to sustain fisheries provides us with a system level index of the impact of fishing activity because it gives us an appreciation of the human influence on the lowest trophic levels in the ocean. Of course, ECOPATH itself is not essential for making such calculations, but it does make the task a lot easier.

One early effort to calculate such an index was by Vitousek *et al.* (1986), who assumed that the average fish fed two trophic levels above primary producers. With other assumptions about the efficiency of transfer of matter between trophic levels, Vitousek *et al.* (1986) estimated that 2.2% of the world's aquatic primary production was required to sustain fisheries. Such a low figure led the authors to conclude that the effects of fishing on the lowest trophic levels in the ocean were minimal and were insufficient in themselves to cause any large-scale effects on anything other than target species or those which interact closely with them. More recently, however, Pauly & Christensen (1995) re-examined this issue after breaking global fisheries down into classes of system and fishery type. Independent estimates of the primary production requirements for each were then obtained, with more defensible estimates of the trophic level at which each fishery operated and of the efficiency of transfer between levels. Combining these estimates to arrive at more robust global figures the authors concluded that 8% of global aquatic primary production was necessary to support the world's fish catches in the early 1980s, including a 27 million tonne estimate of discards (see Chapter 2). Although, one might query some of assumptions underlying Pauly and Christensen's calculation, it is unlikely the 8% figure is far wrong.

Against what yardstick should one compare this value? Perhaps the most appropriate comparison is with terrestrial systems, where almost 40% of primary productivity is used directly or indirectly by humans (Vitousek *et al.*, 1986). Although revised upward from the 2.2% estimated by Vitousek *et al.* (1986), 8% may still seem a rather moderate figure in the light of terrestrial demands. Unfortunately, however, it seems unlikely that there is much potential for increase in global fisheries production. This is because, if one looks on a regional basis, the requirements for upwelling and shelf systems are comparable to the terrestrial situation, ranging from 24–35% (Table 7.1). This range of figures is also supported by a rather comprehensive analysis for the North Sea by Christensen (1995) who estimated that 29% of overall primary production and detritus is required to sustain fish catches. Bearing in mind that the coastal seas are rather less accessible to humans than the land, these values for fisheries seem considerable, leading Beddington (1995) to suggest that 'current levels of fishing – and certainly any increases – are likely to result in substantial changes in the ecosystems involved'.

Table 7.1 Global estimates of primary production and the proportion of primary production required to sustain global fish catches in various classes of marine system. (Adapted from Pauly and Christensen, 1995.)

Ecosystem type	Area ($10^6 km^2$)	Primary production ($gCm^{-2}yr^{-1}$)	Catch ($gm^{-2}yr^{-1}$)	Discards ($gm^{-2}yr^{-1}$)	Mean % of primary production	95% CI
Open ocean	332.0	103	0.01	0.002	1.8	1.3–2.7
Upwellings	0.8	973	22.2	3.36	25.1	17.8–47.9
Tropical shelves	8.6	310	2.2	0.671	24.2	16.1–48.8
Non-tropical shelves	18.4	310	1.6	0.706	35.3	19.2–85.5
Coastal reef systems	2.0	890	8.0	2.51	8.3	5.4–19.8

The reason that the overall global figure is so low compared to the land is that, in contrast to the coastal and upwelling systems, the primary production required to support open ocean fisheries was about 2%. As Pauly & Christensen (1995) note, while the bulk of aquatic primary productivity occurs in the open ocean owing to their vast extent, the higher trophic levels (e.g. the tunas), which dominate high seas fisheries must range widely to find scattered food patches. Because these species are searching for food in an aquatic desert the prospect for increasing yields are not good. Energy budget analysis does nothing to contradict the conclusions of conventional fisheries assessments that the majority of the world's fish stocks are fully or over-exploited.

Another perspective on fisheries that the ECOPATH approach has provided comes from work by Pauly *et al.* (1998), who examined changes over the last 45 years in the average trophic level at which landed fish were feeding. To do this they first estimated the fractional trophic level of 220 species or groups of fish or invertebrates, covering all the statistical categories found in the FAO global statistics of fisheries landings. These estimates were obtained from published ECOPATH models from different regions around the world, each of which used local data that included diet composition. Having obtained the trophic level for each fish and invertebrate category, the authors then calculated the mean trophic level of fisheries landings using the FAO's global landings statistics.

Figure 7.7a shows the global estimate from this analysis and indicates that there has been a decline in mean trophic level from about 3.3 in the early-1950s to 3.1 in 1994. Very large landings of Peruvian anchoveta, which feeds at a low trophic level, account for the marked dip in the time-series in the 1960s and early 1970s. When this fishery crashed in 1972–1973 mean trophic level of global landings rose again. Trends for particular regions are somewhat variable, and can be readily explained by the details of fishery development in each area, but where fisheries have been most developed there have been generally consistent declines in trophic level over the last two decades.

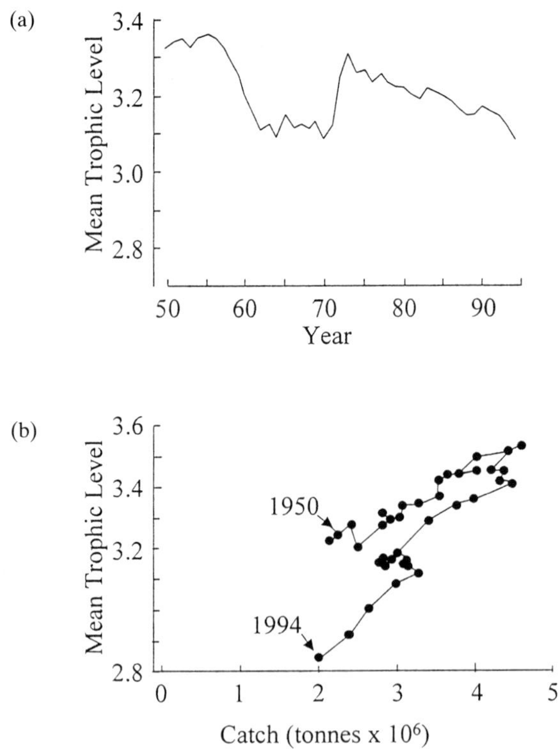

Fig. 7.7 (a) Global trends in mean trophic level of fisheries landings from 1950–1954. (b) Plot of mean trophic level versus catch for the northwest Atlantic. (Redrawn from Pauly *et al.*, 1998.)

Although the trend shown above is interesting, plots of mean trophic level against catches give a more revealing insight into the system level dynamics of fisheries (Fig. 7.7b). This graph shows one surprising feature – contrary to expectations from simple trophic pyramid arguments, highest catches are not associated with the lowest trophic levels. Instead, the time-series tends to bend backward. A number of explanations for this trend are discussed by Pauly *et al.* (1998), among them the influence of discards, reduced catchability of species at lower trophic levels owing to their smaller body size and fishing induced changes in food webs from which the landings were extracted. All explanations are speculative, but one thing seems undeniable – given the consistency of the result across regions the notion that fishing down food webs will increase yields is unsupportable. This is important because it has been suggested in the past that fishing at lower trophic levels will give greater yields because energy losses from transfers up the food chain will be less.

Insights from stable isotopes?

Another way to examine energy flow through a system is by analysing stable isotopes in animal and plant tissues. The basic principle of the approach is that different food sources have characteristic ratios of heavy and light isotopes of different elements and that the tissues of consumers will also reflect these ratios. In particular the ratio of ^{14}N to ^{15}N (usually termed $\delta^{15}N$) in consumer tissues is often used as an indicator of the trophic level at which the animal is feeding. Similarly the ratio of ^{12}C to ^{13}C can be used to identify the primary carbon source in systems where more than one source is available. There are many assumptions associated with these kinds of analyses and care must be taken in interpretation (see, for example, Gannes *et al.*, 1997), but data of this kind can give important insights into how food webs are structured. One rather interesting result comes from work by Wainright *et al.* (1993) who examined trends in the stable isotopic composition of fish scales for samples taken from George's Bank. This study is of particular interest because scale samples were available from as far back as 1929.

Among other things Wainright *et al.* (1993) found that $\delta^{15}N$ for haddock has shown a long-term decline. This result suggests that haddock were feeding at $\frac{2}{3}$ of one trophic level lower in the food web in 1987 than they were in 1929. A change in the feeding ecology and trophic flows could explain such changes in the system, or an isotopic change at a lower trophic level or both. It is important to bear in mind the undoubted difficulties in interpreting stable isotope data, but a decrease in the abundance of fish prey and/or an increase in the abundance of invertebrates that occupy lower trophic levels could certainly explain the result. The authors found that isotopic variation was significantly correlated with a combination of environmental and population variables, including stock size and fishing mortality. It is tempting to speculate, therefore, about the role that fishing might have played, despite the fact that there was not a significant correlation between isotopic composition and fishing effort itself. However, while one might advance mechanisms to implicate fisheries activities in changes in the trophic level at which species feed, any conclusions about their relative importance would be speculative at best. One can but hope that other analyses of this kind using stable isotope approaches will be conducted to tell us whether there are similar long-term decreases in other systems with a history of fishing activity.

7.5 Conclusions

System level approaches have revealed important trends

System level approaches to the analysis of fisheries effects are important. The scale of biomass removals and its spatial extent make fishing activity a strong candidate for effecting large-scale change to marine systems and we need to find ways to examine these wider effects that are practical and informative. At the coarsest scale, the kinds of global analyses performed by Pauly & Christensen (1995) paint a picture of our fish harvesting activities as being comparable to terrestrial agriculture, when expressed as a proportion of the earth's productive capacity. On these grounds alone, a careful examination of ecosystem level effects is warranted. Moreover, the related work by Pauly *et al.* (1998) reveals that worldwide, we are harvesting at progressively lower trophic levels. Yet in the long term, catches do not increase when we do so. By any rational argument a global switch towards lower trophic levels, which generally comprise lower value species, *and lower catches* is indicative of fisheries' regimes that are badly in need of restoration. These analyses are clear warning signs that global fisheries are operating at levels that are certainly inefficient and probably beyond those that are prudent if we wish to prevent continuing change in the trophic structure of marine ecosystems.

The need for theory and mechanisms

With respect to more fine-grained system level analysis of particular areas, it is probably fair to say that there is no clear consensus on the best approach. Nor indeed are we clear on the extent to which fishing is responsible for many of the observed changes in system level properties. Part of this problem lies in the fact that mechanistic theory to explain observed community patterns remains limited. We are developing increasingly sophisticated approaches to data analysis, but with relatively little understanding of how to interpret patterns of constancy or change in, for example, species diversity or patterns of species richness among body sizes. The key to developing the required theory probably lies in the construction and careful analysis of theoretical models.

Perhaps the most tractable problem is to explain patterns in size or biomass distribution. Certainly, it is in this area that most work has been done and most progress made. Beginning with Sheldon *et al.* (1972) who first identified the issue for marine systems, through theoretical efforts such as that of Thiebaux & Dickie (1993), to tests of models with real data (e.g. Sprules & Goyke, 1994; Sprules & Stockwell, 1995; Rice & Gislason, 1996), we are beginning to appreciate how trophic interactions might determine community size distributions and how fishing might modify them. If and when we get a feel for how these underlying mechanisms determine patterns, one might hope to be in a better position to predict the effects of fishing, or perhaps more optimistically, to decide how seriously we should view any structural changes that can be attributed to it. Work such as that by Rice & Gislason (1996) certainly seems to be in the right direction. With the exception of biomass spectra, however, approaches remain to a large extent phenomenological – we describe changes (or lack of them) but underlying causes and consequences are unknown. At this stage, therefore, most analysis of community structure are probably some way from being of practical value for examining fishing effects, or in making decisions about management actions. At the very least, we need to convince ourselves that there is a plausible mechanism by which fishing could effect a particular community change before we can move forward.

There is also a deeper issue that needs to be addressed, however, since without a mechanistic understanding it is difficult to argue that we have a reasonable perspective on what a particular change might mean. Put another way, we have no a priori basis for saying that one location on an ordination plot, or a shift in diversity among body sizes is any more or less desirable or important than any other. For community patterns of the kind described here it is probably only once we understand why changes occur that we can begin to appreciate their importance. The need to understand mechanisms is made doubly important by the results of the global analyses by Pauly and co-workers. The warning signs from such work are clear, but we remain ignorant of the ecological inner workings that might explain some or all of the trends. Without this understanding we have little basis for deciding whether undesirable rapid and large-scale shifts in trophic structure are on the horizon if we continue with current levels of exploitation.

Energy budgets and system models

Tracking the energy (or biomass) flow through a system is a fundamentally different activity from documenting patterns in diversity etc., and we get a complementary perspective on the system when such analyses are performed. At the global level this can give valuable insights into what fishing is doing (e.g. Pauly *et al.*, 1998). Energy budget approaches also require less data than most of the other methods described in this chapter (at least for a first cut), and the ECOPATH software makes their construction relatively simple. These attributes add greatly to the attractiveness of the approach.

But how far can an energy budget take us? At a local level when trying to describe a particular system, their most useful role may be in providing a way to organise disparate data into a coherent whole. This act of data compilation alone, using ones own data and information from colleagues and the literature, can make the relationships between various trophic groups in a system clearer – something that can only be a good thing. Constructing an energy budget can also be invaluable as a vehicle for bringing together disparate groups of scientists to focus on a common goal, and often leads to the identification of information gaps and critical process oriented studies that need to be conducted – again a worthwhile outcome. Once complete, however, it is rather difficult to know what one should actually do with an energy budget for a particular system.

This scepticism about the ultimate use of an energy budget is partly a function of the uncertainties that remain, regarding the input parameters. However, it is also a function of the compromises that have to be made to represent the system structure. One must decide, for example, about how many compartments should be explicitly represented and how the data for individual species and groups should be allocated to them. Clearly these decisions will reflect the particular aspect of the system one is interested in and how much data are available on each component. While the budget is only used to identify gaps that need to be studied this is not a problem. If, however, static energy flow models are used to infer how the system will behave following perturbation the dangers are considerable. Among other things, the choice about which groups to represent might dramatically affect the conclusions that are drawn. Somewhat paradoxically, perhaps the greatest value of the energy budget approach is when many models are compiled for different regions. Such compilations permit the kind of large-scale global analysis performed by Pauly *et al.* (1998). The uncertainties about particular models in a compiled set will still exist, but these local uncertainties are unlikely to alter the broad-scale global conclusions that can be drawn. It is more detailed conclusions about the local situation that may be dangerous to make. There are a number of examples where simplistic energy budget analyses have been used to justify arguments for fisheries management actions in particular areas. Such efforts are fraught with difficulties and extreme caution is required.

The problem of choice of model structure is also one for dynamic models which predict changes in biomass of species or species groups over time. When constructing such models, it is very important to recognise that one might draw completely opposing conclusions about how a system will behave, depending on which one of two equally defensible model representations is chosen. Since the recent development of ECOSIM, a dynamic counterpart to ECOPATH has made the process of model construction much easier for novices in using ecological simulations, this cautionary note is perhaps especially important. While the problem of deciding on an appropriate model structure is a common one (see, for example, Gurney *et al.*, 1996), it may be especially problematic for marine systems where recruitment variability is often so great. To be fair most scientists are fully aware of the limitations of ecological models, and would see them as part of a suite of approaches that can be used to address a problem. Those less familiar with the approach, however, should be aware that all model output should be treated with a healthy degree of scepticism.

Chapter 8
Ecosystem Health, Function and Stress

This chapter moves from asking about the kinds of system level attributes that have been explored to consider the concepts of ecosystem health, function and stress. Examining changes and the evidence and likelihood that fishing might cause such change is only part of the task. We must also consider what changes in structure we are prepared to accept. To do this, it is worth considering how concepts of health and stress have been applied in an ecosystem context and whether they are useful. Discussion of this topic occupies the first part of this chapter.

A key consideration when suffering from ill-health or stress, of course, is whether recovery from the condition is possible. Following a discussion of the general concepts of health and stress, therefore, the question of whether the effects of fishing are generally reversible is considered. Finally, deciding whether a system is healthy or stressed will be partly determined by whether the changes have functional consequences for the ecosystem as a whole, and what those consequences are. Given the scale of the activity, one could easily imagine fisheries having the potential to detrimentally affect the way an ecosystem functions – a powerful additional motivation for changes in our approach to fisheries management. It seems sensible, therefore, to briefly review what we know about ecosystem function and place it in a fisheries context. This is the topic for the last part of the chapter.

8.1 Health, integrity and stress

Ecosystem health

The analogy between the health of human beings and health of ecosystems is one that is finding growing acceptance. This is important because mental images of ill-health for our ecological commons are powerful motivators of public opinion. It is worth considering, therefore, the extent to which an analogy with human health can be applied. Calow (1992) identifies two forms of the analogy, a weak form in which the term health simply signals normality (the converse ill-health signals abnormality) and a strong form in which health defines a condition that is favourable (i.e. optimal) for the functioning of the system. In the latter case, homeostatic processes would actively defend this favourable state. Ideally, a healthy state should be generalisable between systems because only then can objective health criteria be defined.

Can the strong form of the health analogy be applied to an ecosystem? Calow argues that the answer is no because for such a form to be valid the existence of a controlled 'optimum' state is necessary. Control is occurring if a system remains unchanged with perturbation (i.e. change is resisted), or if it has the ability to return to a previous state after perturbation (i.e. it is resilient to it). Resistance and resilience are properties that have been discussed and explored widely in the ecological literature within the context of community stability. Whether marine systems show such features is considered a little later in this chapter. Regardless of the conclusion, however, it is unlikely that component parts of ecosystems are programmed for *active* control in a way similar to the homeostatic processes in the human body. This is because natural selection on individuals and populations will favour those that maximise command of resources even if it is at the expense of the rest of the ecosystem. Thus, the strong sense of the analogy with health in humans is flawed. One cannot, therefore, use actively controlled properties or the control processes themselves to define health for ecosystems. We do not have an ecosystem analogy for body temperature that can provide a barometer for health.

To be fair, most of the proponents of the ecosystem health paradigm agree that the strong form of the health analogy is invalid. They argue, however, that we can use the weak form usefully by adopting the concept of human health as a metaphor and working from the premise that there is a definable 'normal' or baseline state that would constitute a healthy system. One approach to defining such baselines might be to list the properties of putatively pristine systems (i.e. those that have been unaffected by man's activity). This is analogous to what happened in early medicine where physicians sought to correlate body states with conditions of health and ill-health. A problem here is that the structure of the biotic components of a system (i.e. biodiversity in all senses of the word) varies with 'natural' environmental conditions. Thus, to use some single ecosystem state as a baseline from which to judge the effects of our activities would be to take a dangerously static view of a dynamic system. Instead, one is faced with having to characterise the bounds of variability that constitute a normal system. In this respect, it may be that the kind of system level properties discussed in Chapter 7 may be much less affected by environmentally driven or other stochastic variation than the size of individual species populations or the structure of the habitat. Such integrated measures may, therefore, turn out to be more appropriate as indicators of health, although considerable further work needs to be done before they might be useable.

Ecosystem integrity

System integrity is another concept that has been promoted as being useful in discussions of fishing effects, but again what is meant by the term is a little elusive. The notion of ecological integrity takes a rather different perspective from that of health by giving primacy to conservation – not of species, but of ecological organisation. The goal here is the preservation of the full complement of species and interrelationships that would prevail in the absence of human influence (Rapport, 1998). Westra (1996) states, for example, that an ecosystem can be said to possess integrity 'when it is wild, and free as much as possible today from human intervention, that is, when it is an un-managed ecosystem, although clearly not a pristine one'. Westra argues is that it is this absence of influence by man that distinguishes the concept of integrity from that of health. This distinction arises because the notion of health could be taken to imply active support and manipulation to maintain it. An agricultural monoculture might satisfy health criteria, but would not constitute a system with integrity in Westra's book. Rather, maintaining a system's integrity depends on retaining it in a state with the maximum capacity for change and development in response to a changing environment.

In practical terms the ethics of integrity lead, in Westra's view, to the conclusion that we should be maintaining large areas in states that are 'as wild as possible', not only for their intrinsic value, but also because they help support the function of the rest of the system which is managed or used. What this could be taken to signify is a case for the establishment of large protected (closed) areas in which fishing and other activities are prevented. This topic is considered in Chapter 9.

Ecosystem stress

The related concept of stress on ecosystems has also reached some prominence, because an alternative way of viewing the behaviour of an ecosystem under harvesting is to consider the reaction in terms of a stress response. This approach draws strong parallels with the responses of individual organisms – humans become stressed when excessive demands are made upon them, so perhaps effects on ecosystems could be viewed in a similar way. In considering individuals, three stages of reaction to a perturbation from its 'nominal' (equilibrium if you will) state have been postulated; an initial alarm reaction, then a stress resistance stage, followed eventually by exhaustion or passive compliance with the stress. Rapport *et al.* (1985), argue that these three stages also have parallels with the response of an ecosystem to perturbation.

But what diagnostics might one measure to determine the level of stress an ecosystem is experiencing? Rapport *et al.* (1985) suggest the following:

- An increase in the rate of loss of nutrients from the system (i.e. partial breakdown of the dissipation structure of the food web).
- Changes in primary productivity.
- Increase in P/B ratios.
- Decreases in species diversity.
- Retrogression, defined as a shift in species composition to those that are best adapted to new and more difficult environmental conditions.

- Changes in the size composition of the community, often towards smaller mean sizes and shorter life spans – relates to the increase in P/B ratio postulated above.

Other signs of distress could include increased incidence of disease, changes in growth rates and mass mortalities.

If one accepts these diagnostics, it is difficult to argue against fishing as a stressor of ecosystems. The pages of this book provide numerous examples of changes in such properties that seem to be, or theoretically could be, attributed to fishing. Unfortunately, however, recognising that fishing imposes stress does not get us especially far. It is not simply a question of whether fishing is a stress, but rather how much? In the case of humans, moderate levels of stress actually seem to be beneficial to well-being (the so-called eustress that stops us getting bored etc.). It would clearly be foolish to ascribe such positive benefits to the ecosystem from moderate levels of fishing, but it would be equally foolish to argue a priori that the same level of stress was 'bad for it'. We are necessarily faced with the problem of deciding *how much* of a change in the system we are prepared to accept – a decision that cannot be guided by science alone. Some of the ethical and social issues surrounding fisheries effects are considered more fully in Chapter 9, but one fairly clear yardstick is that of reversibility. This reversibility is tied to the ethical notion of 'incompensable harm', a concept that has received most attention with respect to the prospect of loss of human life – compensating the dead is tricky at best. Of more relevance in the present context, however, is irreversible environmental damage (Schraeder-Frechette & McCoy, 1993) – the issue to which we now turn.

8.2 The reversibility of effects

Species extinctions

By definition if a species is extinct, the effect is irreversible, and if fishing activity is causing such irreversible changes, the ethical arguments for curbs on fishing will have considerable public support. We ought then to explicitly consider the likelihood that fishing leads to extinctions.

With respect to the exploited species themselves, a stock collapse is often viewed in the eyes of the popular press and many others as a small step from biological extinction (Canadian cod being a good recent example). This is a falsehood that needs to be put to rest. A clear distinction must be made between a fisheries collapse and biological extinction. With a collapse, animal abundances fall to low levels where fishing becomes uneconomic, or controls on the fishery become imperative. A risk of true biological extinction comes at *much* lower levels of abundance where the species is so rare that it is unlikely to recover regardless of any action we might take to arrest its decline. This distinction is made plain by the frequency with which fish stocks recover after a collapse. In one recent analysis Hilborn (1996) took data for 129 stocks where there was at least a 20-year time-series and calculated 5-year running averages of abundance estimates. He then calculated the frequency with which stocks increased or decreased by 10, 5 and 3-fold for each of the major fish groups (Table 8.1). This approach is certainly rather simplistic and makes no effort to distinguish the relative roles of environmental events and fishing as agents for collapse, but it is nevertheless revealing. In particular, the results show that 10-fold changes in abundance occur with surprising frequency (about 10% overall) and that increases occur almost as often as decreases. As one might expect, large changes occur most often with pelagic fisheries, but interestingly, it is the gadoids that seem to show fewest recoveries.

Table 8.1 Frequencies of declines and increases in spawning stock biomasses. (Taken from Table 1 of Hilborn, 1996.)

Taxonomic group	n	Decline (%)			Increase (%)		
		10-fold	5-fold	3-fold	10-fold	5-fold	3-fold
Clupeiformes	31	23	39	65	16	35	58
Gadiformes	43	12	26	40	0	2	21
Perciformes	4	0	25	50	0	25	25
Pleuronectiformes	15	0	13	33	0	0	0
Salmoniformes	52	8	10	17	10	13	35
Total	**145**	**11**	**21**	**40**	**7**	**14**	**32**

Of course, there are strong caveats to an analysis of this sort, not least the possibility that the data set is biased against any species that collapsed and did not recover before efforts to survey stock abundance were halted. Nevertheless, the analysis lends weight to the contention that biological extinction of exploited species is rarely an issue. Exceptions to this would be the great whales, which are now of course protected. Certainly exploited finfish species have been suggested for listing as endangered, or as candidates for CITES classification (Convention for International Trade in Endangered Species), but with a few exceptions these seem to be as much political manoeuvrings as considered scientific assessment (Hilborn, 1996). One could be forgiven, however, for thinking differently. Even in the scientific literature one can find general reference to species extinctions through fisheries that are simply not supported by data. Discussing the issue of the collapse of Northern cod, for example, Westra (1996) states

> 'Canada and all other countries must learn to curb their overall economic goals, and reduce them from previous years expectations, if they are not prepared to face complete extinctions, not only of specific fish [my underline], but also of their traditional lifestyles.'

The dismay at the collapse of a major fishery is understandable, but such hyperbole should not go unchallenged.

Perhaps because of the open nature of most marine systems and the economics of fishing we are certainly not dealing with the scale of species loss that obtains on land, where the clear felling of rain forest and increases in agricultural acreage continue apace. This is not to say, however, that there are *no* species (target or non-target) that fisheries do not threaten. Chapter 2, for example, describes the fate of a number of bird, mammal, and turtle species that are caught as by-catch or perhaps for the curio trade that we may lose. Clearly, efforts need to made in these cases to mitigate effects. However, given the global extent of fishing activity the actual extinction of species at the hands of fishermen is rare.

Persistence, resilience and variability

Another issue that needs to be considered with respect to fishing is whether the resilience and stability of marine ecosystems is being weakened. Ecological stability can be formally defined mathematically as the return of all component species in a system to their equilibrium densities following a perturbation. Analysis of abstract model communities for their stability properties (more usually their local stability properties, following very small perturbations from equilibrium) has provided a rich vein for theorists and has given a considerable stimulus to the development of ideas about community structure and function. Unfortunately, empirical testing of such ideas based on stability criteria is probably a hopeless task, but Pimm & Hyman (1987) offer a useful alternative framework.

In earlier work Pimm (1984) argued that concepts of persistence, resistance, resilience and variability were more amenable to measurement in the field than stability *per se*. Pimm argues that these other terms offer a much more tractable route for empirical analysis. The terms can be briefly defined in the following way.

- Persistence: The propensity for communities to admit invaders – a more persistent community is one that is more resistant to alien species.
- Resilience: The speed with which populations return to their long-term trend – more resilient populations are those that return more quickly.
- Variability: The degree of variation in populations about their long-term trends in abundance.

For marine communities relevant to fisheries, there seem to be little data on the establishment of foreign invaders, suggesting, perhaps, that marine fish communities are persistent. Pimm & Hyman (1987) cite two examples which suggest that this is indeed the case: the North Sea is given as an example because it 'has not admitted any new species of fish, despite the obvious potential of colonists to reach the area, carried in the ballast water of ships' and the Pacific Ocean off Hawaii where there has been a lack of success with attempted introductions of exotic fish, compared to inland non-marine habitats. The apparent absence of other examples of fish invasions may indeed indicate that such marine communities are persistent (particularly in contrast to freshwater systems), but such a conclusion can only be tentative. One difficulty is that we have no idea of the rate at which marine systems are challenged by potential invaders. Contrasts with freshwater systems may be confounded, for example, by the fact that deliberate introductions of species into lakes and rivers are much more likely than into the sea. Without information on challenge rates, it is impossible to truly judge persistence. Evidence from other taxa suggests that marine systems are by no means immune to invasion. Shellfish pests are a good example, as is the ctenophore *Meniopsis* which invaded the Black Sea and which preys upon fish eggs and larvae. This species may have been a key contributor to the collapse of Black Sea anchovy in the late 1980s (see Chapter 5).

It is not surprising that fisheries biologists have a keen interest in resilience and variability; if a fish stock is depleted it would be nice to know how quickly it will recover. Knowing population variability and its causes is a great help for predicting future catches. When recruitment has shown a major collapse there is usually insufficient known to predict the recovery rate when fishing is reduced or stopped. As noted in Chapter 5, while some fish species recover relatively quickly (e.g. North Sea herring), others seem to stay at low levels for a long time, for example the Californian sardine (McCall, 1980) and the Georges Bank haddock (Anthony & Waring, 1980). For mammal stocks that have been heavily depleted, recoveries are more, but by no means completely, predictable (Laws, 1985).

There have been a number of efforts to explore theoretically the effects of increasing harvesting effort on return times (the inverse of resilience) and population variability. Most of this work has been done with single-species models and Pimm & Hyman (1987) cite the study by May *et al.* (1978) as being representative of these efforts. May *et al.* (1978) show that the response of populations to increased harvesting rates is dependent on the type of population growth function used. For most models return times increase linearly with fishing mortality, but according to the classic Beverton & Holt model, return times fall as fishing effort approaches the level at which the Maximum Sustainable Yield would be obtained, and then increases thereafter. (For one other model variability remained unchanged). One feature that is ignored in these analyses is the possibility that compensatory processes that affect resilience may operate. A number of such mechanisms can be suggested and there is evidence that they do in fact operate when mortality rate increases (i.e. increased pre-recruit survival, decreasing age at sexual maturity, increase in reproductive output). If these effects are important, for the mortality range over which they are effective, resilience of a harvested population may increase.

In considering the way that such changes might translate into variability when random fluctuations are imposed Pimm (1984) considers three possible responses, two depending on whether the stochastic perturbations are density dependent or independent and one relating to the mortality rates and life history characteristics. In the density dependent case, a long return time means that the population is less responsive to changes in the environment. Here, when harvesting decreases resilience (i.e. return times increase), its variability should decrease because the population will less closely track environmental fluctuations. In the density independent case, a population with a long return time is more likely to be at some distance from a mean (equilibrium) value when the next perturbation occurs, so the population will reach more extreme values. Thus, a decrease in resilience leads to increased variability. The third possibility is that variability and resilience will be inversely related for highly fecund species (high resilience) when the mortality rate is very high on the early life stages because small changes in mortality rate between years will produce a large change in recruitment. To date, there seems to be little empirical evidence for a positive correlation between the degree of population fluctuations and abundance.

Work by Steele & Henderson (1984) adds an additional and important perspective to the debate. These authors examined the effects of environmental variability on a simple population model with multiple equilibrium states. A key point about this work is that environmental stochasticity was modelled as a random process, but with a red-noise spectrum so that variance in the environment increased with time (as physical variables do in marine systems). With this kind of environmental variability, populations could be made to switch rapidly between alternative stable conditions at intervals of 50–100 years. These findings coincide with the fluctuations observed in marine pelagic species and provide one possible explanation for the decline of herring in the 1970s in the North Sea. The authors also found that harvesting the population increased the frequency with which such state changes occurred. It is interesting to speculate that, for some systems, fishing may not affect the absolute magnitudes of population densities that occur, but the frequency of the fluctuations between high and low levels.

The picture that emerges from much of the theoretical treatment of the issues of resilience and variability is that harvesting can have effects in any direction, depending on the circumstances. Although, it ought to be possible to use fisheries data to determine which relationships actually operate, there are important scale considerations which need to be taken into account if analyses are to be meaningful. For example, many analyses fail to take account of fisheries behaviour over time and to partition sources of variation resulting from processes operating at different time-scales (Pimm & Hyman, 1987). Without distinguishing the variability due to trends in the data from the variability about such trends, the effects of harvesting on variability will remain unresolved. It should be recognised at the outset, however, that inadequacies in the data may yet prove insurmountable. Similarly with persistence and resilience, despite the importance of the question, we are far from being able to make definitive statements about trends in these properties or the factors which are most important in affecting them.

Alternative stable states

Although global extinctions are not a major issue for fisheries, and fish stocks often recover from collapse, there is hardly room for complacency. Indeed, the fact that fish stocks have recovered in the past is no guarantee that they will do so in future, given the other pressures that we are placing on most marine systems. It is encouraging that stocks such as the Californian sardine, for which fishing stopped in the 1950s has finally started to recover to reasonable numbers in recent years, but the prospect that other currently collapsed stocks may not still looms. In such cases thought inevitably turns to whether one is doing more than simply waiting for especially favourable years for recruitment so that the stock can rebuild. Perhaps instead, other predator and competitor populations have reached levels that prevent the stock of concern from increasing. Without some further perturbation that depletes these other populations, the system might stay in its current configuration.

The idea that systems can occupy two different states and that switching from one to another requires a rare, and/or intense perturbation, is one that has intrigued ecologists. When this occurs, a system has been said to occupy alternative stable states. As a general proposition, could fishing be forcing systems into other states from which they cannot return? Theoretical analysis suggests that fishery-induced alternative stable states are certainly possible. Work by Gilpin & Case (1976), for example, suggests that complicated (species rich) communities, once perturbed by fishing, are unlikely to return to the same state, even if fishing ceases. The theoretical message from multispecies systems seems to be that the probability of alternative stable states is high. It must be admitted, however, that ecologists have little real idea of how likely such a scenario is. There is, however, good evidence from South Africa that such a situation can occur in marine systems.

Barkai & McQuaid (1988) showed that two islands, approximately 15 km apart off South Africa, support markedly different benthic communities. One island was dominated by seaweeds and rock lobsters which prevent the establishment of mussel beds. On the other island mussels dominate and rock lobsters and seaweeds are almost absent. On this second island a whelk species that is usually prey for lobsters is also very abundant. Lobster transplant experiments showed that at these high densities whelks overwhelm lobsters, thereby reversing the usual predator/prey roles. This prevents a lobster population from establishing on the island. Local fishermen report that the two islands were similar some 25 years ago, but the original cause of the disappearance of lobsters is not known. One possibility is a period of low oxygen levels that was known to have occurred near the island in the 1970s, or perhaps there was a disease outbreak at one site. Apparently in this case it seems unlikely that exploitation by fishermen was responsible for the changes, but the example clearly illustrates the potential for fisheries to push communities into another stable domain. The prospect that such a situation has occurred on Georges Bank with the increase in elasmobranchs was discussed in Chapter 6.

Another compelling example where fisheries may well have contributed to a marked switch in system structure is provided by Costanza & Greer (1998) for Chesapeake Bay. These authors consider the role of the oyster fishery in the region. This fishery yielded a harvest of 15 million bushels per year in the early nineteenth century, maintained a steady production of 2 million bushels per year in the decades up to the early 1980s but now provides less that 500 000 bushels. After years of harvesting, oyster bars are now fragmented low-lying structures on the seabed, but in earlier times they are reported to have been substantial reefs reaching almost to the water surface and lining both sides of Chesapeake Bay along the shallow margins. It is highly likely that these reefs provided a habitat for many species of fish and invertebrates and were important consumers of phytoplankton production. It has been estimated, for example, that the reefs were so abundant that the entire volume of water in the Bay would have been filtered by oysters once a week – today that figure has fallen to about a year. It would be difficult to argue that water clarity, phytoplankton population dynamics and the structure of this system had not been altered substantially by the loss of oyster reefs and the earlier state is unlikely to return. Of course, many other factors will also have played their part in changing the Chesapeake Bay system (agriculture and urban development especially), but the case for a fishing effect is compelling. Sadly, our cultural memory is so short that few will appreciate what the system really used to look like.

Habitat destruction

The final and, perhaps, most compelling mechanism leading to potentially irreversible effects is habitat destruction. With the exception of continued unrestrained exploitation, removing habitat that is utilised by fish stocks is perhaps the single most effective way of slowing or preventing stock recoveries. In particular, the destruction or degradation of spawning grounds and nursery areas for juvenile fish is cause for concern. With respect to spawning areas, it is often difficult to pin down precisely why fish spawn where they do, but in some cases the physical characteristics of the benthic environment will be important. North Sea herring, for example, spawn on gravel banks where the eggs coat the sediment surface in a mat, multiple egg layers thick. Bottom trawling over such areas, either in or out of the spawning season could conceivably alter the suitability of the site for spawners. Such a possibility has not to my knowledge been examined, but it is certainly plausible and leases for oil exploration or gravel extraction often take into account proximity to spawning sites.

This same degradation of physical architecture may also be important for nursery areas, where juvenile fish are likely to utilise physical features on the seabed, either for protection from predators or as shelter from currents (see also Section 3.7). Scientific evaluations of the relationship between physical habitat characteristics and the sustainability of stocks is lacking in most cases, but it is certainly true that comparative surveys of areas with marked physical features and those without show distinct differences in the fish and invertebrate assemblages that are supported. More physically complex habitats generally contain much richer faunal communities and there are sound arguments for believing that they are often of greater functional significance than their areal extent suggests, owing to the protection they afford juveniles. The marked differences between seagrass beds and adjacent sediment areas is a case in point where fish abundances can be many-fold greater in the seagrass (e.g. Jenkins *et al.*, 1997). Destruction of such areas by fishermen is a short-sighted strategy that may cost dearly in the long term.

With some biological structure such as seagrass there is of course the potential for recovery (although other pressures from coastal development on seagrass beds often makes re-establishment difficult). For very slow-growing organisms such as corals, or non-biological structures, however, the prospects are less good. In some cases positive action might be taken to rehabilitate habitats, and structures such as artificial reefs have been variously suggested as a mechanism for rehabilitation. Unfortunately, such reefs often seem to concentrate existing fish biomass rather than rehabilitate stocks. Carefully conducted, locally specific studies will probably be required to establish the value of the approach on a case by case basis. Even if successful, however, it is unlikely that such intervention is a tractable proposition other than on local scales.

While removal of large biotic structures such as corals or seagrasses is quite noticeable, in other cases change in habitat may be rather subtle. One could imagine, for example, that juveniles of demersal fish on continental shelves benefit from a high abundance of relatively small physical features (sponges, empty shells and small rocks etc), but that over time trawling will gradually lower the physical relief of the habitat with deleterious consequences. In turn this lowered relief may make the habitat relatively more suitable for flatfish, which would explain their apparent increase in heavily fished areas (see Chapter 6). Although the validity of such an hypothesis has rarely been established, the positive association between fish or other organisms and physical structure is highly suggestive. The work by Sainsbury (1987, 1988), discussed in Section 3.5 is one example that seems to bear out the idea and argues strongly that habitat architecture is important.

There are of course many other threats to habitats associated with coastal development, marine pollution etc., and for many local stocks these threats are probably much more important than habitat destruction by fishing *per se*. Nevertheless, the potential for bottom trawling in particular to degrade the physical architecture of benthic habitats is palpable and rehabilitation of such habitat may be difficult. It is also worth bearing in mind that protecting only the habitat that we think is critical may not be sufficient. Olver *et al.* (1995), for example, note that habitat conservation is often based on the simplistic assumptions that: (a) the level of productivity will be roughly proportional to the supply of suitable habitat, and (b) the loss of non-critical habitat will not noticeably affect sustainable levels of exploitation. Such assumptions may not be warranted in the case of fishes. The 'critical' habitat that is often required for the earlier juvenile phases may facilitate survival of juveniles, but other less suitable regions may be required to accommodate the accumulated biomass of older fish.

8.3 Ecosystem function

The term ecosystem function is being increasingly used in the ecological literature and concerns about loss of function through man's activity have a marked sense of urgency about them. Ecosystem function is usually taken to denote those ecosystem level processes that contribute to the well being of mankind. Such beneficial processes are often termed ecosystem services, and a recent study by Constanza *et al.* (1997) suggests that the oceans account for approximately two-thirds of the value of ecosystem services on a global basis. The oceans play a major role in climate regulation, provide protein for human consumption, and regulate global water, carbon and nutrient recycling. By implication, if fisheries affect how ecosystems work, rather than simply how they look, we ought to be especially worried.

Mindful of the accelerating rate of species loss on earth, perhaps the most commonly asked question with respect to ecosystem function is whether continued species losses are likely to seriously impair ecosystem function. This was an idea first raised to prominence by Ehrlich & Ehrlich (1981) who advanced what has since been termed the 'rivet hypothesis' (Lawton & Brown, 1993). This hypothesis views each species as making a contribution in the same way that rivets collectively act to keep an aeroplane in the air. In contrast, what Lawton & Brown (1993) term the 'redundant species hypothesis' states that species richness is irrelevant and that all that matters is that the biomass of primary producers, consumers and decomposers is maintained. In other words, as far as the functioning of the system is concerned, we can get on very well with just a few species. Although in this context the term redundancy is used in a very specific manner, many people feel uncomfortable with using it, because it might be taken to imply that species deemed to be 'redundant' do not need to be conserved. A conservation policy based purely on the grounds of functionality would be ridiculously narrow. To avoid providing ammunition to those who view nature in such strictly utilitarian terms Roberts (1995) suggests the term 'substitutable' as a less contentious alternative.

As discussed earlier, notions of large-scale species loss are largely unwarranted with respect to fishing effects (see Section 8.2). However, even if species are not actually lost, many are substantially reduced in abundance. One might well ask, therefore, whether marked changes in the relative abundance of taxa affect ecosystem function. To date, most of the work on this topic has been undertaken in terrestrial systems and the results of early experiments were considered by many to show that higher levels of biodiversity benefited ecosystem function (Naeem *et al.*, 1994; Tilman & Downing, 1994). In many respects, this conclusion was precisely what people wanted to hear – a link between species loss and ecosystem services of benefit to man adds an additional and important justification for conservation. Indeed, despite the preliminary nature of the findings, the idea that species diversity was important in this way even crept into a prominent marine ecology text book as a statement of fact (Castro & Huber, 1997, page 215). Such enthusiasm for the theory seems premature, however, because both comparative studies of natural terrestrial systems and subsequent criticisms of the early experimental work have cast doubt on the notion that it is biodiversity *per se* that determines ecosystem function (e.g. MacGillivray and Grime, 1995; Huston, 1997; Grime, 1997). Rather it would seem that variation in ecosystem properties is related primarily to the functional characteristics of the dominant plants in the system, notably their capacities for resource capture and uptake.

The recent swing away from a key role for biodiversity itself as an important determinant of ecosystem services implies that the redundant species hypotheses should be taken more seriously. Indeed, in a discussion of these issues for coral reefs, Roberts (1995) asserts that 'there is ample evidence that reef communities are characterised by substantial substitutability [redundancy] of species', noting that, although species differ in the details of their biology, 'there are only a limited number of ways of making a living'. The highly speciose functional groups on reefs imply that loss of particular species may have little overall effect on, for example, rates of planktivory or herbivory, provided that some other species population with the same functional attributes shows compensatory behaviour, either through increased densities or increased per capita grazing rates. Similarly, it seems likely that the role of piscivore, for example, could be filled by only one or a few species, and we know that reef calcification rates are similar on the Australian GBR and the eastern Pacific, despite an order of magnitude difference in the number of coral genera between the regions (Roberts, 1995). For benthos also, Snelgrove *et al.* (1997) concluded that, there is probably considerable redundancy within groups and that many species could be lost without really changing the system. While the roles of the main groups (bacteria, meiofauna and macrofauna) are fundamentally different from one another, and loss of one entire component could not compensate for loss of another, it is difficult to find evidence that benthic biodiversity *per se* is needed for healthy ecosystem function.

Despite the fact that many species are almost certainly substitutable, it would be rather foolish to embrace the notion of species redundancy too enthusiastically. Clearly a point could be reached at which further loss of key species might impair functioning, with marked (probably undesirable) consequences for the system. Moreover, a species which might be able to substitute for another under one circumstance, may not be able to do so if environmental conditions change. This is part of the 'insurance argument' for conserving biodiversity, which says that species that seems to be unimportant at present may be more important under other circumstances. Given the often non-selective nature of fishing methods the problem then is to determine whether there are situations where ecosystem viability depends on an unusually high species richness and/or to identify particular species which might play a key functional role. The possible existence of key species or groups that might be vulnerable to fishing activity was of course, the topic of Chapters 5 and 6.

8.4 Conclusions

Concepts of health and stress for ecosystems are important because human beings identify strongly with them and arguments that are made using them can have a powerful effect on public opinion. As a means for communicating messages about desires or concerns for the marine environment, they carry considerable weight. The term 'health of an ecosystem' can usefully be used in summary statements, in much the same way that the term 'health of the economy' is used. Similarly, the notion that a particular ecosystem is 'in poor health' or 'good health', with respect to a stated set of criteria about what the system is used for (e.g. healthy fishery production) has a valuable role in communication. Difficulties arise, however, if one tries to define management goals in terms of these concepts. At this stage, there is no general consensus about what characteristics or integrated measures of a system could be used to assess health or the acceptable functioning of a system. This is a research area of great interest, but it is doubtful that integrated measures of the kind described in Chapter 7 will ever be used to provide early warnings of undesirable changes. Indeed, it is not even certain that they are necessary for this purpose since, in most cases, the signals that a system has reached an undesirable condition are all too obvious from more easily gathered data on fisheries yields, water quality or other simple measures. It may be, however, that analysis of such properties provides an additional political incentive to do something about the problem and they could provide goals to aim for in restoration attempts. Yet, even when diagnostics of health or stress are agreed upon, the target that should be adopted will be difficult to arrive at. This does not mean that the concepts of health and stress are of no use, but there are severe limits on the degree to which they can be made operational in a way that would allow them to be included in a management process.

With respect to the notion of integrity in particular, there seem to be problems with an approach that equates 'normality' or 'health' with the absence of man's influence. This is because it implies that systems affected by man are inherently abnormal or 'unhealthy'. This is not to say that the changes mankind has made to systems are desirable or morally defensible – clearly many of them are not. It is simply that there is no a priori reasons why a system we have affected should be viewed as functioning any more or less well than one in which our influence is minimal. Indeed I would be more comfortable if one were to substitute this kind of analogy for one about appearance or looks. We are happy with the notion that there are degrees of attractiveness in people or art and that our judgements are subjective. It seems to me that we should be prepared to acknowledge and accept that we must make the same judgements about the 'state' of ecosystems, rather than investing ostensibly pristine environments with some form of moral superiority.

To make the difficulties with the integrity concept more concrete. Consider a reef that has been surveyed and shown to have rich kelp beds and a diverse fish and invertebrate fauna. In others words the reef is in a state that most people would be happy to describe as healthy and showing an acceptable degree of integrity. Now imagine that before a second survey a year later (and unbeknown to the surveyor) a storm removes most of the kelp from the reef and there was a very successful recruitment of sea urchins. These circumstances conspired such that most of the algae cover was removed and the reef was a depauperate urchin barren. Would the surveyor be correct to call the second state less healthy than the first? It is easy to see that it is less desirable from a human perspective but if the criterion for integrety is a state engendered by the activities of man it clearly does not qualify. Although it is powerfully emotive to call undesirable system states unhealthy or lacking integrity, it is probably more reasonable and intellectually honest to consider systems as being in desirable or undesirable states. This does not of course remove the difficulty of determining the controls on those states and the degree to which our actions can effect changes, but it does remove some of the hyperbole surrounding the issues.

It is also worth bearing in mind that many people's perspective on what a 'normal' system should look like are probably more a reflection of recollections from childhood, or those of parents, than of a 'pristine' system. For terrestrial systems there is certainly a compelling case for widespread structural change by primitive man (Budiansky, 1995). Although it seems highly unlikely that on a global scale pre-industrial effects were anywhere near as marked in marine systems, it would be foolish to imagine that they are only restricted to recent decades.

In summary, there is a case to be made for more attention being paid to how we want systems to look, rather than trying to argue for ways they are 'supposed to be' (i.e. healthy, or unstressed). Arguing that one particular system state has moral primacy over, or is more 'normal' than, another seems rather pointless. Steele (1998), for example, argues that sustainability and biodiversity are difficult to apply in marine systems where decadal scale regime shifts are to be expected. He introduces the term fungibility (cf. substitutability) from economics, which means replacing one entity with another of equal utility, and suggests that it is this fungibility that characterises systems dominated by regime shifts and prevents notions of sustainability or biodiversity from being usefully applied. The latter because conventional notions of biodiversity (e.g. diversity indices) do not change despite large changes in species composition. Steele (1998) also notes that there is no evidence that major shifts in species composition result in a breakdown within the ecosystem comparable to, for example, forest clear cutting. In his words 'Generally, the new marine ecosystem is as diverse and ecologically acceptable as the previous one, even if it is economically much less valuable'. This is of course a personal value judgement about what is ecologically acceptable, but it is difficult to disagree with this viewpoint. Better to focus on what we want from a system (in the broadest sense) and then decide how to go about getting or keeping it.

Part 4
Fisheries Management

Introduction

Precaution in principle

Even governments seem to now accept (in theory at least) that we need to be more prudent with our resources and our relationship with the natural environment. Indeed, a key outcome of the Rio Declaration of the United Nations Conference on Environment and Development in 1992 was the following principle:

> 'In order to protect the environment, the precautionary approach shall be widely applied by States according to their capabilities. Where there are threats of serious or irreversible damage, lack of full scientific certainty shall not be used as a reason for postponing cost-effective measures to prevent environmental degradation.'

At about the same time the FAO started developing its Code of Conduct for Responsible Fisheries and drew its own conclusions about what a precautionary approach would mean for capture fisheries (FAO, 1996). The outcome of their deliberations bears repeating:

The precautionary approach involves the application of prudent foresight. Taking account of the uncertainties in fisheries systems and the need to take action with incomplete knowledge. It requires *inter alia*:

- consideration of the needs of future generations and avoidance of changes that are not potentially reversible;
- prior identification of undesirable outcomes and of measures that will avoid them or correct them promptly;
- that any necessary corrective measures are initiated without delay and that they should achieve their purpose promptly, on a time-scale not exceeding two or three decades;
- that where the likely impact of resource use is uncertain, priority should be given to conserving the productive capacity of the resource;
- that harvesting and processing capacity should be commensurate with estimated

sustainable levels of resource, and that increases in capacity should be contained when resource productivity is highly uncertain;

- all fishing activities must have prior management authorization and be subject to periodic review;
- an established legal and institutional framework for fishery management within which management plans that implement the above points are instituted for each fishery;
- appropriate placement of the burden of proof by adhering to the requirements above.

Often the precautionary approach has been taken to mean that human actions are assumed to be harmful unless proven otherwise. This is an extreme position, but it must be recognised that all fishing activities have some environmental impact, and one cannot *assume* that they are negligible. However, while one might prudently require that fishing activities cease if they have potentially serious environmental consequences, this should not imply that no fishing can take place until all potential impacts have been assessed and found to be negligible. What the precautionary approach should require, however, is that all fishing activities be subject to prior review and authorisation and that a management plan be in place that clearly specifies: (a) the management objectives, and (b) how the impacts of fishing are to be assessed, monitored and addressed. It should also require that interim measures be put in place while a management procedure is agreed. Finally, the FAO argue that the standard of proof to be used in decisions about authorisation of fishing should be commensurate with the potential risk to the resource, while also taking account of the potential benefits of the activity.

There is little doubt that if these principles were fully implemented, not only would we obtain more fish from the sea, but that many of the wider effects of fisheries activity would be mitigated. But stating the principles is one thing, it is quite another to decide on the actual risk from a particular action (such as stock collapse) or the level of risk one is prepared to accept. Unfortunately, the way science is traditionally done has tended to weight debates towards accepting more risk-prone strategies and reducing the level of precaution.

Type I and Type II error

Most scientists are trained in the Popperian hypothetico-deductive (H-D) method of science, where hypotheses are tested through explicit formulation of null models. A null hypothesis is constructed stating that a particular factor has no effect and then tests are devised to falsify it. For non-scientists this might seem a back to front way of doing things, but it stems from the idea that one can never say definitely that a hypothesis is true, only that it has not yet been falsified. This is the most common theory for how science is supposed to operate, but many argue that it is not always the practice. For ecology in particular, the H-D emphasis was a welcome departure from the earlier phenomenological descriptive phase, but the discipline has had difficulty fitting comfortably into the classical H-D mould. This is largely because it has proved difficult to arrive at unambiguous null models and many ecological concepts are rather fuzzy and difficult to operationally define. Despite this difficulty, there remains considerable emphasis on what Shraeder-Frechette & McCoy (1993) term scientific rationality, which typically involves minimising Type I rather than Type II statistical error.

Type I errors occur when one rejects a null hypothesis that is true – saying there is an effect when there isn't one. In contrast, a Type II error occurs when one fails to reject the null hypothesis that is false – saying there is no effect when there is one. One important value judgement that scientists must make concerns which type of error is preferable. Ecologists traditionally attempt to minimise Type I error to limit the chances of claiming false positives. In other words, they attribute greater to loss to accepting a falsehood than not accepting a truth – getting it wrong is worse than not getting it right. When we minimise Type I error we minimise the error of rejecting a harmless development; this has been termed 'producer' or 'developer' risk (Shraeder-Frechette & McCoy, 1993; Churchman, 1947). When we minimise Type II error, we minimise the error of accepting a harmful development. This has been termed 'consumer' or 'public risk'. For environmental issues many have argued that efforts to limit Type II errors should be given more emphasis, often citing precautionary principles or approaches as justification (e.g. Dayton, 1998). Also, because of the predominantly practical and applied nature of ecology there is a case to be made that it is practitioners, perhaps more than most other disciplines, who must consider both ethical and scientific perspectives when giving advice. If one accepts this, then one is faced with the prima facie ethical case that one should avoid harm before enhancing welfare – limiting Type II errors (i.e. adopting a more precautionary approach to scientific judgements) then becomes inescapable.

Unfortunately, however, while we might have a formal statistical framework for determining the probabilities of particular outcomes, agreement on what those probabilities actually are will often be difficult to find. If expert estimates for the risks of something as apparently straightforward as living near liquefied natural gas terminals vary by a factor 100 million (Otway & Peltu, 1985), what level of disagreement can we expect about the likelihood of a stock collapse, or more difficult still, of flipping an ecosystem into an alternative stable state? Moreover, even if we are confident that the risks of something occurring are high, there are some classes of fisheries effects for which the response in some quarters will still be 'so what?' Few would question that foregone resources through over-exploitation, global subsidies far in excess of revenue, or diminution of food security should be avoided, but not all would agree that altering the species composition of benthic communities, or the structure of habitats is also *in itself* a harmful act. Many would consider such changes in structure rather irrelevant, unless they can be shown to matter in some other more tangible sense. As a basic principle, while one might not share this view, it is difficult to argue that all change is bad – as with most things it is a matter of degree.

Precaution in practice?

Appeals and actions to widen the scope of fisheries management and adopt a more precautionary approach have often been described as an ecosystem management approach. The Ecological Society of America defines Ecosystem Management as:

> '. . . Management driven by explicit goals, executed by policies, protocols, and practices, and made adaptable by monitoring and research based on our best understanding of the ecological interactions and processes necessary to sustain ecosystem composition, structure and function.' (Fogarty & Murawski, 1998)

As with all such definitions, precise operational criteria for sustaining 'ecosystem composition, structure and function' are elusive at best, but the spirit is clear – we should not be taking a blinkered one-dimensional approach to natural resource management. Within this general framework then, it is worth considering those strategies that might be adopted to help limit the community and ecosystem effects described earlier and minimise the risk that other more catastrophic changes might occur.

Chapter 9
Mitigating Effects

The cross-disciplinary nature of the fisheries management problem, with its clear ecological, economic and social dimensions is likely to make fisheries management solutions a continuing topic of debate in the coming years. One could not hope to cover all aspect effectively in this book. This chapter is, therefore, somewhat selective, dealing with only a sub-set of the issues that deserve attention. It is, however, a sub-set that is being debated at present and it includes the issues that are perhaps most pertinent to a book on the wider effects of fishing. Each topic covered deserves consideration in any search for improved fisheries management, although some may hold more promise than others. Importantly, however, the order in which they are dealt with here holds no particular significance.

9.1 Dealing with uncertainty – biological reference points

Fish populations fluctuate, whether because of deterministic chaos, environmental stochasticity, demographic stochasticity or a combination of factors. Fisheries management must deal with the uncertainty that results from this high level of variability – a fact well recognised by the fisheries scientist. How can this uncertainty be coped with?

The current strategy is quite straightforward. Every effort is made to obtain the best estimate possible for a stock's current abundance and future trends and catch quotas are then set with this current information. There is a substantial literature on how harvest rates should be adjusted over time in response to natural variability, particularly in recruitment rates (e.g. Mangel, 1985). In general, however, the best policy to give acceptable yields and levels of variability involves some feedback mechanism between harvest rate and stock size so that catches are regulated up or down in response to stock size. A fairly stable escapement strategy, or some constant catch rate as a function of stock size (which generally leads to lower variance in catches) can generally be found to do the job (see Hilborn & Walters, 1992, Chapter 15). If short-term variation can be forecast the best policy will be to increase effort on a rising trend and decrease on a falling one. The problem of course is the uncertainty in the forecasts and actually making the required adjustments in the fishery.

Everyone recognises that the information available to help make management decisions is uncertain, but taking this uncertainty into account is another matter. The most common approach is probably to accept the best scientific estimate and proceed on that basis. As Lauck *et al.* (1998) note, one of the strongest reasons for doing this is that any admission of doubt on the part of the fisheries managers opens a chink in the armour, allowing the fishing industry to argue for quotas at the upper limit of the confidence bound. The industry can argue that there is 'no proof' that a lower level is necessary. The problem with this approach is that fisheries are usually not sufficiently self correcting, so high levels one year would not lead to much lower estimates of stock abundance and hence much lower quotas in subsequent years. The structure of most fisheries is such that annual fine-tuning of this kind just does not seem to work very well, usually because lowering high quotas is difficult to achieve politically when fishermen have loans to pay on expensive boats. In discussing the political realities, Larkin (1977) puts it well: 'Unfortunately fishermen vote; once a person has become a fisherman, he can be almost counted on to vote against anyone who doesn't help him continue to be a fisherman and ensure a decent standard of living'. A good example of the destructive compromises that can be made can be found in the management of the cod fishery in Atlantic Canada. Here a target of $F_{0.1}$ was considered too stringent when the stock was assessed to be low because catches would be insufficient for the industry. As a result a 50% approach rule was adopted where the industry would go half-way towards what was required to 'minimise the impact of drastic reductions in catches between consecutive years' (Rivard & Maguire, 1993).

One resolution to this difficulty of effort adjustment is, in addition to setting annual catch quotas, to agree in advance on minimum biologically acceptable levels for populations as management reference points. Within ICES, for example, such a reference point defines (in principle at least) a stock biomass level below which a strong conservation strategy would be required. In essence, this is an effort to define a cut-off point, below which recruitment overfishing and risk of stock collapse is unacceptably high and drastic management action needs to be taken. The situation ought to be improved by setting up such management procedures with industry *in advance of a problem.* The details of this procedure should include a clear and agreed specification of the data that should be collected, how it should be analysed, and the action that should be taken, particularly should a reference point be reached. Such an approach has a great deal to commend it.

9.2 By-catch and discards

As noted in Chapter 2, with respect to environmental effects, by-catch and discarding is probably one of the most important problems facing the global fishing industry today. The threat to species populations, the wastefulness of the activity and the problems that undocumented discarding pose to stock assessment are all major issues. Discarding is also a source of conflict between fishers themselves as one sector of the fishery can often catch commercial species which are the target of another and for which it does not have a quota. On a practical level, handling sorting through unwanted by-catch is also something many fishers would like to avoid.

What can be done to mitigate the problem? Perhaps the first point to make is that there is no universally applicable solution. Each fishery has to be examined separately and the relative merits of alternative approaches assessed. One activity that must be undertaken in all cases, however, is a clear definition of what the problem actually is. It is difficult to see how this can be achieved without the implementation of observer programmes. Someone has to go out with fishermen and document what is caught and what is thrown back. This is an expensive undertaking, but there are many examples of such programmes around the world – they are an essential prerequisite for arriving at a successful by-catch mitigation strategy.

One obvious route to reducing unwanted catch is to increase the selectivity of the fishing method in some way. In trawl fisheries, in particular, technical advances, combined with a greater understanding of the behaviour of fish in nets has led to the development of new methods to increase selectivity. These methods adopt one of two strategies. The first is to exploit behavioural differences between the various fished species, using devices such as separator trawls, modified ground gear (i.e. the parts of the net that touch the seabed) or modifications to the sweep ropes and bridles that are attached to the trawl doors. For example, separator trawls in the Barents Sea have been shown to successfully segregate cod and plaice into a lower net compartment from haddock, which are caught in an upper compartment (Valdermarsen *et al.*, 1985). A similar approach has been used in the North Sea to separate nephrops from haddock (Galbraith & Main, 1989), while in Alaska the approach has been used to allow 40% of bottom-associated halibut to escape while retaining 94% of cod, the target species (Stone & Bublitz, 1995). Other approaches that are now receiving attention include the use of modified ground ropes. In the northwestern Atlantic silver hake fishery, for example, such modifications appear to allow juvenile flatfish to escape, while still preserving hake catches (DeAlteris *et al.*, 1996). In this region, juvenile flatfish are thought to be an important food source for the adjacent northwest Atlantic groundfish fishery, so minimising their incidental mortality is of significance for both fishery and conservation reasons.

The second approach is to exploit the different sizes of species. In many fisheries it is the capture of undersized fish that is the main problem and regulation of minimum permissible mesh size is, of course, a cornerstone of most fisheries management regimes. Such a measure can often, however, be improved upon. For example, the inclusion of square mesh panels in front of the codend can often allow a greater number of escapees, because the meshes do not close up when the codend becomes full. In addition, recent work which alters the visual stimulus that the net provides by using different coloured netting in different parts has been shown to improve the efficiency of such panels considerably (Glass *et al.*, 1995). At the other end of the scale excluding large sharks, rays or turtles from the catch can be achieved by fitting solid grids of various kinds. In some fisheries such devices are now mandatory (e.g. turtle exclusion devices in some prawn fisheries), but there is often resistance from fishermen because they can be difficult to handle and catches of target species can fall. For non-trawl fisheries examples of technical solutions can also often be found. The research into how best to lay long-lines to avoid incidental bird capture was discussed in Chapter 2 and is a good example, as are the dolphin escapement procedures that are now used in high seas purse seine fleets.

Many technical approaches have met with considerable success in different parts of the world and such measures are certainly worth exploring to mitigate by-catch problems. It is important to recognise, however, that the greatest chance of success is when it can be demonstrated to fishermen that the modified gear works in a real world setting without compromising catches. If this can be done gear modifications are likely to be readily adopted. The behaviour of one prawn fishery in South Australia is a testament to this. Here, following a simple demonstration project, a by-catch reductions device was immediately and voluntarily adopted by the whole fleet once the fishers were convinced of the benefits. On the other hand, if fishers are not convinced, even if a device is legally required, individuals can and sometimes will find ways to minimize their effectiveness.

It is also important to recognise that technical fixes are only part of the solution – the system in which they have to operate must also be considered. The regulations that govern fisheries and the vagaries of the market place often create a complex web of incentives and disincentives that drive the discarding practices of fishermen. The situation can be especially complicated in multispecies fisheries, where it is very important to decide what the unit of management actually is. For example, the notion of setting catch quotas for each species separately from each other and from by-catch species is highly questionable. In these circumstances a system level focus is more sensible and alternative quota management arrangements might be preferred. Potential solutions might include seasonal and/or area closures or continuous monitoring of the fishery and closure once a given by-catch quota has been reached. It is also important to establish incentives that encourage compliance with management controls. For example, flexible quota arrangements might be offered as a reward for compliance. By offering fishing firms rebates on management levies in return for achieving agreed by-catch reduction objectives a greater level of compliance might be achieved.

There is clearly no magic bullet for reducing unwanted catches, and the problem is likely to remain central to any debate about the ecological impact of fishing for some years to come. Fishers and fisheries managers are fully aware of this and it would be wrong to suggest that efforts to reduce levels have not been made. However, while the problem will never be solved completely, there is no doubt that we can do better. Perhaps the difference now is that the general public is also more aware of the issue. The kind of pressure that can be imposed by public opinion and consumer preferences for products that are perceived to be environmentally friendly is likely to ensure that even more effort will be put into finding solutions in the future. It is essential that it is.

9.3 Broader-based harvesting?

Fogarty & Murawski (1998) suggest that much of the problem of community change on Georges Bank can be attributed to the selective nature of the fisheries harvest, where particular species were the primary focus for attention at any particular time. The authors suggest that broader-based harvesting strategies coupled with lower overall levels of effort would help prevent the dramatic shifts in community structure that have been observed in that region. In effect this is an argument for harvesting each species at levels that are closer to their proportional abundance than we have achieved to date – a strategy which has, in principle, some merit. In practice, however, even the authors note that species with low reproductive rates will be very vulnerable to over-exploitation and that the system would have to be managed with great care. This fact casts doubt on the true usefulness of the approach. In the multispecies trawl fisheries of tropical regions, for example, one would imagine that fishing is already as close as one could realistically get to harvesting in proportion to abundance. Yet as we have seen in the Gulf of Thailand marked changes in community structure occurred since the fishery began. Catching a wider variety of fish also requires that there be a market for them all – a situation that will often not hold, particularly in temperate regions. Arguing for the harvest of species that no-one wants to buy as a management strategy to protect trophic structure would be difficult to sustain on any level.

9.4 Environmental research

Hofmann & Powell (1998) give four excellent examples of how environmental variability can affect fisheries recruitment. In each case large-scale oceanographic features have been linked in process-oriented studies to the survivorship of larvae; there are many other studies that indicate similar influences. The authors argue that the changing nature of marine fisheries requires management that recognises such environmental effects, suggesting that having identified favourable and unfavourable environmental conditions fishing levels should be adjusted downwards during the unfavourable periods. This approach argues for development of integrated sampling programmes, which measure both biological and physical properties of the system and for advances in circulation, ecosystem and fisheries models. There are certainly some signs of it happening. For example, stock assessments for walleye pollock (*Theragra chalcogramma*) now include a qualitative index of advective losses of larvae by unfavourable currents (Megrey *et al.*, 1996).

Despite this trend, however, and the optimism about the incorporation of environmental measures in some quarters, cautionary notes by Walters & Collie (1988) are worth bearing in mind. Among other things, these authors note that recruitment tends to be log-normally distributed with many average or below average years, punctuated by a few very good ones. This distribution is consistent with the idea that recruitment is the product of survival rates of several independent life stages, all of which must be high to produce a strong year class in a given year. If this is so, the search for a single environmental index to predict recruitment may be of only moderate utility, since it is likely to only represent one of the factors. Using it may, therefore, improve prediction of average, or below average years, but it will miss the really good ones. Of course, this may not be considered a major flaw in the approach, since failure to predict especially good years is the least of the problems fisheries managers face! More importantly, however, the question of where the assessment dollar should best be spent has to be addressed. In many cases it may well be that investing in pre-recruit surveys or some other fishery independent index of year class strength will provide a more valuable index to work with.

Incorporating environmental data into assessments will occur increasingly as understanding improves, but the time base on which the information will be useful needs to be considered. Predicting recruitment success in a given year from larval data, gives at best a few years advance warning of likely recruitment to the fishery. However, regime shifts where overfishing can convert declines into economic collapse, occur on decadal time-scales and make the concepts of sustainability difficult to apply (Steele, 1996). Identifying early signs of regime shifts may assist with the longer-term planning of fisheries activity, perhaps with negotiated regulatory changes for the fishing sector affected. Knowing that a certain stock or a fish assemblage is likely to be at low abundance for the next decade or so, and that some other assemblage will be higher over the same period, is likely to be is something that fisheries can act upon. Such action would be similar to the way that farmers prepare (or brace themselves) for El Niño periods when the ENSO index suggests they should.

9.5 Accounting for species interactions

Notwithstanding the development of multispecies fisheries methods, the dominant tools and paradigms of fisheries management evolved in the Northern hemisphere and the roots of this development lay in the single species population dynamics theory of the early 1950s. The legacy of a population-based rather than a community-based view is still with us today, and for many fisheries it remains a valid approach for assessment purposes. Hilborn & Walters (1992) offer a number of reasons why, from a fisheries management viewpoint, one might ignore the interactions between species. First, the interactions may not matter very much. If the dynamics of fish populations are tightly coupled to one another, altering the densities of one species through fishing will have large effects for the other species on which it preys, which prey upon it and with which it competes for resources. But it doesn't *have* to be that way. A predator species might, for example, simply crop the surplus production of a prey population that would have died anyway before reproducing. In these circumstances, removing the predator would have no effect on the prey population. Even if effects are measurable, from an assessment point of view, it may not be necessary to worry about them if the behaviour of the stock can be adequately assessed with data on the stock size and structure, the harvest and perhaps the physical environment. A second reason for ignoring species interactions is that the data requirements for incorporating them may be too onerous to make it a practical proposition. Third, even if it is possible to estimate the parameters of the interaction, it may not be possible to predict the dynamics of the other species in the system or control them. In this case it may be better to treat the effects as unexplained noise in the system.

The above are undoubtedly valid arguments but, even if successful management strategies can be formulated from a single-species view, there are good reasons why biological interactions should not be ignored. Perhaps the most compelling reason is fishing nations are often not very good at implementing the advice of fisheries biologists. We have a formidable track record of exploiting populations at levels well beyond those that would optimise yields. While advice on fisheries management might be sound when the system is relatively close to its unexploited state, it is by no means certain that it will continue to be so when the system is pressed to extremes. It is at these extremes, perhaps typified by the collapse of a particular fish resource, that biological interactions might play a critical role in determining how the system behaves. In other words, biological interactions may not be that important for well-managed fisheries, but understanding them may be vital when management fails.

There are also good practical reasons why one might wish to include species interactions in well-managed fisheries. First, there is sometimes sufficient information to include interaction effects that can improve predictive capability and suggest strategies for improving the harvestable productivity of the community as a whole. The fact remains that the principal agent for mortality in fish is probably predation by other fish (Bax, 1991). Second, even if biological interactions are unimportant, it is common for several species to be caught with the same fishing gear and it is often impossible to adopt independent harvesting strategies for all the species of interest. This is especially true for tropical systems where the mixed species nature of the system is a dominant feature of the biology.

There can be little doubt that understanding the implications of ecological interactions for fisheries is becoming increasingly important. Many scientists are now advocating the replacement of single-species based management with multispecies analysis. Although analytical tools such as multispecies virtual population analysis (MSVPA) are primarily directed at exploring the effects of harvesting one target species on the abundances of other target populations, mainly by taking account of between species predation, there is no reason why such efforts should not be extended to non-target taxa. This is perhaps one tangible step towards the ultimate goal that we should adopt measures which manage the target species with an ecosystem level perspective which ensures that the species' environment and that of the remaining biota is also taken into account (Kessler *et al.*, 1992). With this perspective, management of the fishery is regarded as a sub-set of managing the whole ecosystem. The state of the fishery depends on the state of the ecosystem in which it is prosecuted and the state of the ecosystem in turn depends on careful management of the fishery.

This holistic perspective is enshrined in the term 'ecologically sustainable development,' the stated goals of which are:

- Sustainable use of both the species and the ecosystem.
- Maintenance of essential ecological processes.
- Preservation of biological diversity at all levels from ecosystem to gene.

Achieving these goals requires a strategic approach to protecting marine systems from abuse and a clear understanding of the role and consequences of fishing activity. Interactions between species are an important component of this problem.

Despite such strong arguments, however, large uncertainties are likely to remain even with the best assessment methods, and there is now a growing interest in more explicitly incorporating uncertainty into management advice. Lauck *et al.* (1998) make a compelling case for abandoning the idea that all will be well if we simply try harder to incorporate an ecosystem perspective into management by doing more and better marine research and using this to develop multispecies and ecosystem models to manage fisheries. The examples in Chapters 5 and 6 are a testament to how difficult it is to identify key ecological interactions between species, let alone manage for them. Instead, the authors argue that we should be seeking better ways to deal with the fact that the world will always be uncertain and that there are limits to what we can know. Efforts to understand the mechanisms underlying recruitment processes and changes in fish distributions and abundance are important and may well help fisheries managers achieve their goals, as Walters & Collie (1988) note. However, one's enthusiasm for the approach should perhaps be a qualified one. Our scientific culture is firmly wedded to the idea that we can reduce uncertainty substantially by applying the scientific method. It would certainly be foolish to suggest that we cannot improve our predictive capacity, but one has to recognise that there will always be limits to what we can know. These limits may be closer than we care to admit. There is also, of course, a financial price tag that goes with these efforts and the costs of improvement, which may be very substantial, will need to be weighed against the benefits.

9.6 Adaptive management and the Bayesian perspective

If we are going to widen the information base upon which management is founded a number of challenges need to be met. In particular, we need to find ways to include new information in the management process and to adapt our management actions in the light of it. The inclusion of new information, however, should be done in such a way that we can judge its value and the extent to which it should change our views about how the system works. One problem that I have repeatedly stressed is that the information we obtain about complex communities is uncertain and our hypotheses about key controls are largely untested. Moreover, the prospect of conducting critical tests of hypotheses through independent process-oriented research programmes seems rather remote, albeit highly desirable. Rather, it seems likely that learning how communities respond to the pressures we impose on them will continue to be largely empirical – we do something (with a greater or lesser degree of premeditation) and then see what happens. It clearly makes sense to consider how best to obtain and use such information on how the systems respond while still maintaining revenue.

One approach to this problem has been termed active adaptive management because it includes the prospect of taking management action that intentionally increases the contrast between key variables. By doing so important processes can be identified and the underlying management model can be improved (Walters & Hilborn, 1978; Walters, 1986). The basic principle of adaptive management is that fishery management itself can implement experiments by adopting sets of regulations to control fishing activity. By observing the responses of the system to these 'experiments' more can be learnt about how the biological processes operate and management can be adapted in the light of the new information. More formally, the approach consists of a short-term management regime W that is applied during a 'learning period' of duration (t). During this period certain revenues are obtained, monitoring costs are incurred, and observations are made. At the end of the learning period, the observations are used to guide a long-term policy choice. Walters (1986) provides a complete account of the principles.

To adopt such an approach, however, a framework for evaluating the information is required. The following conditions are desirable (Scandol & Walters, 1997):

(1) It should be possible to revert to the results that would have been obtained if the new information had not been included.
(2) There should be a clear and fixed objective for the management of the system so that the performance of new approaches can be evaluated.
(3) It should be possible for managers to alter the weighting or influence they give to new information.
(4) There should be a mechanism for identifying when new information contradicts the existing ideas (models) about how the system works so that ideas can be discarded or modified.

Achieving these conditions seems most tractable through use of a branch of statistics known as Bayesian. Bayesian statistics depart from the more familiar frequentist approach to statistical inference through a different interpretation of probability. The frequentist interpretation of probability is that it is the long-term expected frequency that a proposition is true. In other words, the frequentist approach gives us the probability of obtaining one's data given the truth of a particular hypothesis. In contrast, Bayesian statisticians take a more subjective perspective and interpret probability as an individual's degree of belief in a proposition, i.e. the probability that a particular hypothesis is true given the data we have available. Many argue that this is a more natural form of inference that reflects more closely how we think about scientific and management problems. For an excellent account of the underlying philosophy and examples of the power of this approach for unravelling how biological systems work, Hilborn and Mangel (1997) is highly recommended.

There is a long running and heated debate about the relative merits of Bayesian and frequentist approaches and there are reliable tales of statisticians coming to blows over their religion. However, use of the Bayesian approach to fisheries is not new and Hilborn & Walters (1992) provide a good account of how it can be applied. There is also at least one good example where adaptive management using Bayesian statistics to evaluate alternative management regimes has been used to examine fisheries issues from an ecosystem perspective. This example bears closer examination.

The northwest shelf

Section 3.5 described the history of the northwest shelf fishery and how four alternative hypotheses were developed to explain observed changes in the fish community (Sainsbury, 1988). Briefly, the community composition altered over time, with higher value Lethrinids and Lutjanids declining and Saurids and Nemipterids increasing. The four hypotheses were:

(1) Intra-specific mechanisms: the observed changes result from independent responses of each species.
(2) Competitive release due to fishing: there was a negative influence of *Lethrinus* and *Lutjanus* on the population growth rate of *Saurida* and *Nemipterus* so that when the Lethrinids and Lutjanids were removed by fishing the latter experienced a release from competition and increased in abundance.
(3) Competitive depression: *Saurida* and *Nemipterus* have a negative influence on the population growth rate of Lethrinids and Lutjanids and the abundance of these species declined because the former increased for reasons independent of the fishery.
(4) Habitat modification: habitat characteristics determine the carrying capacity of each genus separately so that trawl-induced modification of the abundance of the habitat type alters the carrying capacity of the different genera.

A formal evaluation procedure was adopted to examine these alternatives by stating the hypotheses as explicit mathematical models which, given appropriate data, could be ranked probabilistically. The key data for this evaluation of alternative models came from the management experiment conducted during the 'learning period' in which the area was subdivided into three zones. One part was left open to trawlers, a second part was closed to trawlers in 1985 and a third was closed in 1987. Since the management experiment itself could affect the fishery, a key issue was which combination of experimental regime and duration would give the best overall expected revenue. A good experimental regime would result in a high probability being assigned to whichever of the alternative models is true, so that the appropriate long-term policy is selected. A second consideration was that the experiment would achieve this goal at low net cost. The costs of making the observations and foregone revenue for the fishery during the learning period needed to be balanced against the potential gains in information from the experiment.

A Bayesian analytic framework provided the basis for choosing between long-term management strategies, by up-dating the degree of belief in each of the four hypotheses using the data from the management experiment. A choice of strategy could then be made that depended in part on the probability placed on each of the models. On completion of the experiment, the probability placed on the habitat modification mechanism increased substantially. Indeed, the value for this hypothesis rose to about twice that of the closest contender and indicated that a high value *Lethrinus* and *Lutjanus* fishery could be established on the northwest shelf if the habitat could be protected. Changes in fish community structure could probably be attributed in large part to habitat modification by trawling.

Basing management decisions on a formal evaluation of observed effects in real systems is a real strength and the northwest shelf provides a good example of where an ecosystem level analysis suggested a strategy that both conserves benthic communities and should enhance yields. One cannot help feeling that considerable benefits would follow from more widespread adoption of this approach. However, it does demand that management authorities and fishermen, be prepared to do the experiments. There will often be major political obstacles to doing so, but a clear analysis of the problem, an explicit description of alternative management solutions and a temporary fishery regime that would provide the experimental contrasts from which to make a decision are goals that should be aimed for.

Unfortunately, as noted in Section 3.5, in the case of the northwest shelf, the timescales for recovery of epifaunal benthos are rather slow and there is a high probability that large benthos will be removed by a trawling. This means that enforcement measures to protect the habitat would need to be very effective to maintain the habitat structure required. The issue of monitoring and enforcement are likely to be critical to the success of many management experiments and the regimes that are adopted.

9.7 Marine protected areas

One management approach that is being increasingly promoted as a tractable device for precautionary fisheries management is the establishment of marine protected areas (MPAs). Of course, various levels of protection can be afforded to a geographic region, ranging from strict exclusion of all fishing activity (so-called 'no-take' areas) to more relaxed approaches which restrict fishing to certain gear types or particular seasons. For a clear taxonomy of marine protected areas, from small closed areas to multi-zoned biosphere reserves, the interested reader should consult Agardy (1997). Each type of area is likely to have its place as a management or conservation tool, and Agardy (1997) also outlines the array of legal instruments that support their establishment and the steps that need to be taken to create them. The following discussion focusses on the value of the strictest form of MPA, the no-take reserve, where fishing is entirely prevented. It should be borne in mind, however, that these other classes of reserve are also used and that effects can differ dramatically between full and partial protection. Watson & Ormond (1994), for example, studied two reefs, one in a marine park that allowed no fishing and another in a reserve which only allowed artisanal fishing. The abundance of commercially important species was ten times greater in the fully protected area.

MPAs for nature conservation

The MPA approach has come under particular scrutiny in coral reef systems. A number of authorities have imposed temporary or permanent closure to fishing of portions of, or in some cases whole, coral reefs in the hope that it will prevent fish stocks being depleted and maintain or even enhance yields in adjacent areas. Additional benefits might accrue if tourist activity is to be promoted in an area. Unspoilt coral and an abundant fish life are a prerequisite for attracting visitors. Of course, given the destructive nature of some fishing practices on coral reefs, the imposition of closed areas serves a conservation purpose that does not require positive benefits to the fishery. Simple protection is often justification enough in the eyes of many conservationists and no additional benefit to tourists or fisheries may be required for them. Happily affording such protection, even to relatively small areas does lead to positive effects on the resident fauna. When reserves are established, the densities and biomasses of target species generally increase within the reserve area. Perhaps the best demonstration of this effect is from the Philippines, where Russ & Alcala (1996a) used visual census techniques to compare fish populations in two small areas (Sumilon and Apo) where protection from fishing was variously established and then relaxed over a period of 10 years. Figure 9.1 summarises the results of this study. For Sumilon densities of large predatory fish decreased significantly when it was opened to fishing in 1985 and 1993, and increased significantly three times following periods of protection. In contrast, at

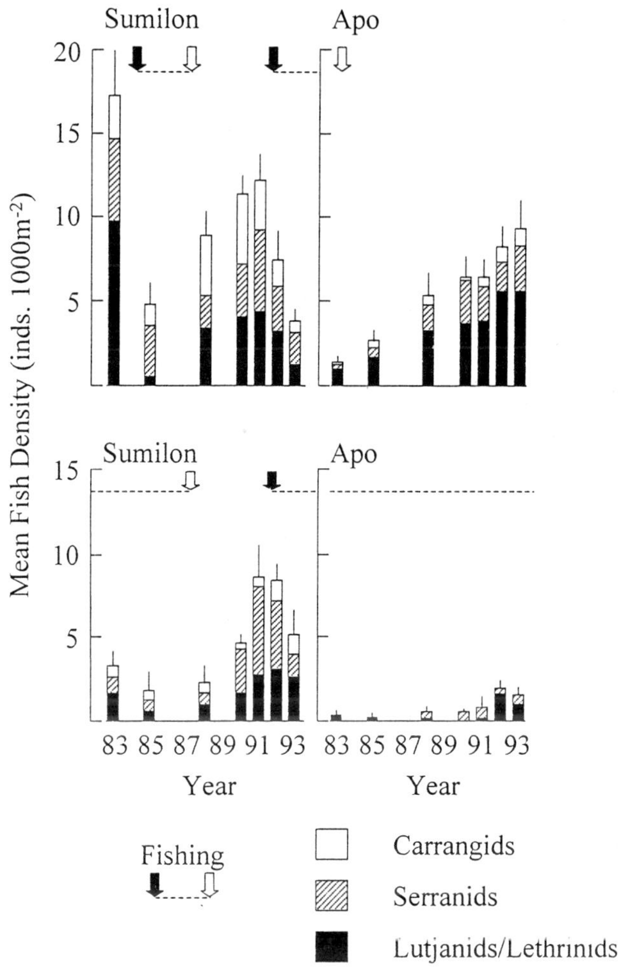

Fig. 9.1 Densities of large predatory fish at four sites at Sumilon and Apo, Philippines, with contrasting fishing histories. Periods of fishing at each site are indicated. (Adapted from Fig. 3 of Russ & Alcala, 1996b.)

Apo there was a steady increase in densities over an 11-year period of protection while comparable non-protected areas showed little change.

The above results and similar results from other areas show that fish populations in both temperate and tropical regions respond, even in relatively small areas, if one protects them from fishing. One rather unsettling observation from Russ & Alcala's (1996a) study, however, is that it took only 1.5 and 2 years of unregulated access to an area to eliminate density and biomass gains accumulated over 5 and 9 years of marine reserve protection.

Growth in numbers is the most commonly documented effect of marine reserve protection, with a number of studies in both tropical temperate regions showing the effect when abundances in fished and protected areas are compared (Dugan &

Davis, 1993). Another response that is observed is an increase in species richness in protected areas. For example, McClanahan (1994) found that 52 of 110 species that were found on protected Kenyan coral reefs were not found on unprotected reefs. Of these, 44 species were unique to the protected coral reefs. Similar responses have been observed by others (e.g. Jennings *et al.*, 1996).

Although increases in abundance (and perhaps of species richness) are a general rule, the establishment of no-take reserves must usually be justified on the grounds that it will offer other benefits. This is especially true if the reserve is to be of any great size, since few authorities in either the industrialised or developing countries are willing to close off large areas to fishermen to keep conservationists happy. Setting up a few small conscience salving marine parks is one thing, but however desirable one might find them from a simple nature conservation perspective, establishing large-scale MPAs is quite another. Realistically, if this is to be achieved, the utilitarian arguments will probably need to be stronger. In particular, the possibility that MPAs might enhance fisheries themselves needs to be carefully addressed, because fishers are those who stand to be most disadvantaged by closure.

MPAs for fish stock management

The idea that MPAs might be a cost-effective means for managing coral reef fisheries is becoming particularly prominent (Alcala, 1988; Bohnsack, 1993; Man *et al.*, 1995), but it is also now being advocated for other areas (e.g. Dugan & Davis, 1993; Allison *et al.*, 1998; Lauck *et al.*, 1998; Pauly *et al.*, 1998). The major objectives for such reserves is to protect critical spawning stock biomass to ensure recruitment supply to fished areas via larval dispersal, and possibly to maintain or enhance yields in areas adjacent to reserves by spill-over of adults. Closed areas may be especially attractive for coral reef fisheries which are almost always multispecies, with many artisanal or subsistence fishers using a wide variety of gears and landing their catch at many sites over a wide area. These features make it difficult to collect even the most basic information such as catch and effort that are required for conventional management. With respect to enforcement there are also clear advantages to a reserve system, especially one in which the local fishing community supports the initiative and polices it themselves. For example, control of fishing effort by closed areas or seasons seems to be one of the few options open to managers of marine fisheries in the Philippines. Even for industrialised nations, however, enforcement is a major problem and simple no-take areas could be particularly effective, especially if they are combined with satellite surveillance techniques.

But will marine reserves protect stocks? In the limit of course, where most of an area is closed to fishing, the answer has to be 'yes' (unless the small piece that was left open happened to be the only spawning ground for the resource). A more appropriate question, however, is under what range of circumstances are marine reserves likely to succeed in this objective and by what mechanisms?

One mechanism that has been examined to a limited degree is whether reserves on coral reefs maintain or enhance yields in areas adjacent to reserves by adult (post-settlement) movements from protected to fished areas. This has been referred to as the 'spill-over' effect. An opportunity to examine this question occurred at the Sumilon reserve in the Philippines (see above), which after 10 years protection was re-opened to fishing for 18 months. With data on catch per unit effort from the fishery for a period before and after this event, Alcala & Russ (1990) showed that there was a decline in CPUE following the loss of the reserve and a fall in total yield from 36.9 to 19.87 tonnes km^{-2} – this despite a 33% increase in fishing grounds. The authors argued that these changes occurred because the reserve was no longer supplying adults to the fishable area. These data are suggestive of an effect, but they are by no means conclusive.

Russ & Alcala (1996b) have also been able to examine this question to a limited degree for Apo reserve. They show positive correlations between mean density and species richness with duration of closure for large predatory fish in both the protected and the adjacent unprotected area (Fig. 9.2). This is, of course, also consistent with a 'spill-over' effect as density and species richness built up in the reserve over time. The same result could also be obtained if there was a general increase in fish abundance over the duration of the study. However, the authors also showed that the build up in the non-reserve areas was greater nearest to the reserve, thereby supporting the spill-over hypothesis. Furthermore, the results could not be accounted for by differences in habitat in areas at different distances from the reserve.

Unfortunately, the establishment of the Apo reserve for fisheries management was somewhat opportunistic and there are no systematic data on fish catches which

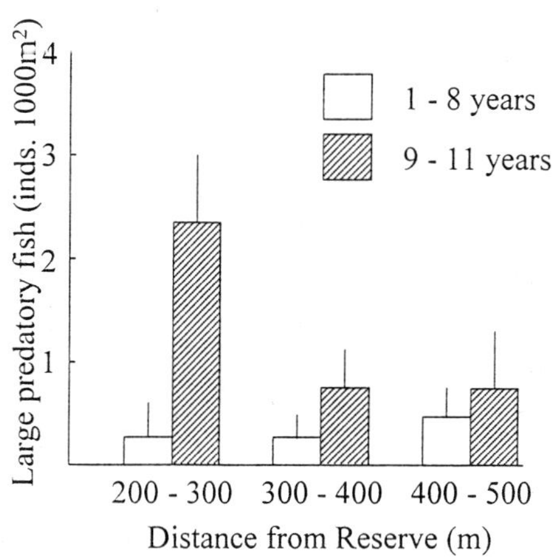

Fig. 9.2 The densities of large predatory fish at different distances from the Apo reserve boundary during the first 8 years of reserve protection and from years 9 to 11. Data are means ±1 sem. (Redrawn from Fig. 3 of Russ & Alcala, 1996a.)

could be used to more critically examine the value of the reserve approach. Despite the lack of hard data, however, and even though the reserve reduced available fishing grounds by 10%, interviews with local fishermen indicate that they perceive the loss of fishing grounds to have been more than compensated for by increased yields (Russ & Alcala, 1996b). While one cannot be certain, taking Russ & Alcala's results and those of others together, it seems fair to conclude that reserves on reefs may export adult fish biomass to adjacent areas. In doing so, however, it should be borne in mind that effects may be slow to develop and may perhaps be important only on scales of a few hundred metres (Russ & Alcala, 1996b) – a feature that probably reflects relatively small-scale movements of fish. Other studies that have examined the question of 'spill-over' are relatively few in number. Indeed, Russ & Alcala (1996b) could only find one (Yamasaki & Kuwahara, 1990), which showed increased catch rates in a fished area surrounding a reserve after five years of protection.

What about closed areas to protect stocks in temperate regions? It has to be accepted that the data are scant, but for Georges Bank perhaps we will find out over the next few years whether the idea works. This is because long-term areal closures were set up in 1994 to try and improve the fisheries of the region (Fogarty & Murawski, 1998; Fig. 9.3). The history of fishing on Georges Bank has been fairly typical of many temperate demersal systems, with overall biomass conservation failing because there was no direct control on fishing effort. Qualitative controls (e.g. on mesh sizes, minimum fish sizes, and seasonal areal closures) simply failed to meet the stated objectives of conserving biomass and retaining the broad character

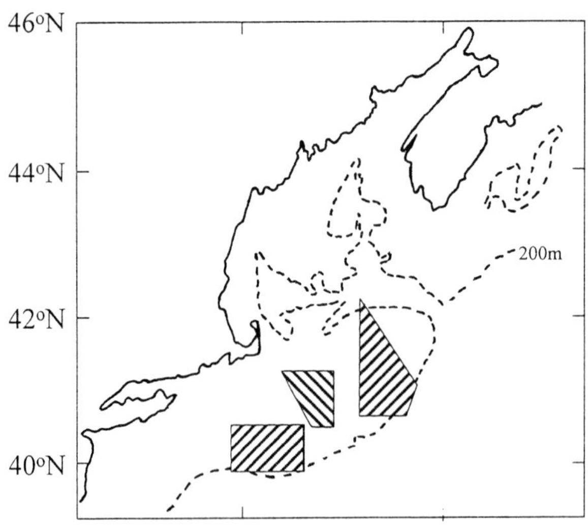

Fig. 9.3 The location of the closed areas established in 1994 on Georges Bank. (Adapted from Fig. 9 of Fogarty & Murawski, 1998.)

of the fish community. If there is one lesson that seems universal it is that indirect controls are ineffective. Changes in community structure occurred largely as a consequence of highly species-specific harvesting patterns driven by market considerations.

The closed areas on Georges Bank encompass areas of traditionally high fishing effort (i.e. good fishing grounds), including part of the scallop grounds of the region and important spawning grounds for cod, haddock and yellowtail flounder. These areas are now protected along with some of the sand/gravel areas that may be important for, among other things, juvenile survivorship. Unfortunately, however, large gravel areas also occur in the northeast of the region, where no protection is afforded. Herein, lies a dilemma for all such management experiments, if it does not work *ad hoc* explanations or excuses can always be made, based on what you did not protect. It remains to be seen how successful the approach will be.

Some theoretical underpinnings

As described earlier, uncertainty about the status of stocks and about true catch levels is a major problem for both fishery assessment and management. In other uncertain situations the normal response would be to hedge one's bets in some way and adopt a range of activities to reduce risk of catastrophe. Could closing areas to fishing be a suitable bet-hedging tactic? The argument and analysis put forward by Lauck *et al.* (1998) with respect to demersal fish stocks that inhabit large areas of ocean floor suggests that it can.

A reasonable management goal for the demersal fish resources of most nations might be that the stock should remain at > 60% of the carrying capacity over a given time horizon, say 20 years. Maintaining such levels would put the stock in the region of optimal sustainable yield that many fisheries aspire to. Lauck *et al.* (1998) consider how the establishment of an MPA might help achieve this goal using a mathematical model that incorporates uncertainty in harvest rates. The model considers a single harvested stock that grows according to the familiar discrete logistic (Ricker) equation. In the absence of harvest the stock in year $N(t)$ and year $(t + 1)$, $N(t+1)$ are related by:

$$N(t+1) = N(t)\exp\left[r\left(1 - \frac{N(t)}{K}\right)\right]$$

where K is the carrying capacity, assuming a stable steady state, and e^r is the maximum per capita growth rate of the population. To model the effects of an MPA, it is assumed that only a fraction A of the area in which the stock exists is available for harvesting. Thus, if $N(t)$ is the total number in the area, the number in the reserve $N_r(t) = (1-A)N(t)$ and the stock available for fishing $N_f(t) = AN(t)$.

It is further assumed that fish are harvested at a target rate u, so the total stock remaining after fishing is:

$$N_r(t) + (1-u)N_f(t) = (1-A)N(t) + (1-u)AN(t)$$
$$= (1-uA)N(t)$$

Because controlling fishing effort is notoriously difficult to achieve, the target rate u is assumed to be uncertain. This uncertainty was modelled by allowing the actual harvest rate to follow a beta probability function with a mean value set to the target rate u. Selecting a criterion for successful management to be that the stock should remain at > 60% of the carrying capacity over a given time horizon, the probability of achieving this goal was calculated.

The model results are striking (Fig. 9.4), indicating that even when the levels of uncertainty in harvest are moderate (CV < 50%), the chances of achieving the management goal drop dramatically once the fraction of the total area available for harvesting becomes greater than approx 30%. Perhaps more significantly, when the levels of uncertainty are larger, even allowing fishers access to only very small proportions of the area still gives a very high probability of failure. This analysis suggests, for example, that the MPAs on Georges Bank may need to be substantially larger to be effective.

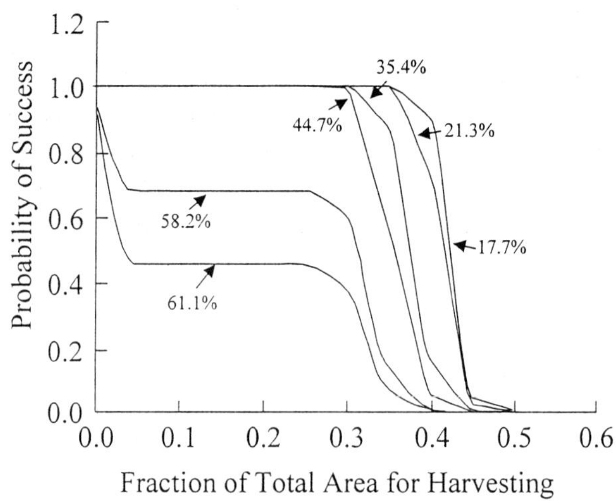

Fig. 9.4 The probability that the stock $N(t)$ remains $> 0.6\,K$ for a 20 yr time horizon, as a function of the fraction A of area available for harvesting, for six different coefficients of variation in the harvest distribution. The model assumes $N(1) = K$ and uses beta distributions for the harvest with the assumption that the harvest rate is 50% in the fished region. Redrawn from Fig. 1 of Lauck *et al.* (1998).

Among other things, these results show that high levels of uncertainty require a considerable degree of management control to assure a sustainable future for stocks, regardless of whether it is through MPAs or simply controlling effort. Interestingly, however, the imposition of an MPA may not lead to reduced catches. On the contrary, when Lauck *et al.* (1998) compared the effect of simply lowering the allowable catch level to achieve the management goal, rather than impose an MPA they found that up to 50% lower catches were obtained for the same level of protection. It would appear, therefore, that MPAs could make sense for fishermen and conservationists.

Of course, the model that was used to draw these conclusions is extremely simple and can be challenged on many grounds. It is important to recognise, however, that while such strategic models do not reflect reality they alert us to important underlying principles. In this case they show us that MPAs can be an important bet-hedging strategy in an uncertain world and could act as an effective insurance policy that could protect both the long-term future of stocks, protect habitat from trawlers *and* yield higher average catches. On this basis alone they deserve careful consideration by the fisheries management establishment, particularly since, in many respects, this model is no simpler than those that have under-pinned fisheries for the last 30 years.

Practical considerations

It would be quite foolish to suggest that the above analysis is sufficient in itself to justify the establishment of MPAs to protect fisheries. Indeed, there can be no substitute for careful ecological research to ensure that key criteria for reserve establishment are met. Importantly, however, the theoretical foundations that justify the field research effort that will be required are now being laid.

But what are the key criteria that should be adopted for establishing an MPA? Lauck *et al.* (1998) note the following as 'desirable' features of an MPA for fisheries management:

(1) It should be large enough to protect the resource in the event of overfishing in the unprotected area.
(2) The reserve should serve as a source capable of replenishing the exploited stock in the event of its depletion. In particular, reserves should protect spawning grounds and any other areas critical to the viability of the population.
(3) The reserve must be completely protected, since the almost certain build up of biomass inside the reserve will be very attractive to poachers.

It could be argued that these features are not just desirable, but essential if an MPA is going to work to protect stocks. Indeed, if these criteria are met, there would appear to be a prima facie case for using MPAs, given the failure of current management practices.

With respect to point (3), there should be relatively little difficulty working out what is required – it is simply a matter of establishing the political will to meet the costs. In contrast, points (1) and (2) are a major challenge for ecologists. Identifying appropriate sizes and locations for closed areas requires consideration of the relative proportions of the populations and communities of interest within the protected region, their potential to serve as source populations for unprotected areas, and the location of any sensitive habitat types, which need to be included in the protected area to maximise the benefits of the approach. Clearly, to satisfy these requirements a considerable amount of biological and ecological information is required before informed decisions can be made. Even then the issue is likely to be contentious – the SLOSS debate for terrestrial reserves (Single Large Or Several Small) still seems to rage despite years of scientific endeavour. There seems to be no prospect of the question being easier to answer in marine systems. In fact, in many ways these questions may be harder to answer in the marine environment given the practical difficulties of making observations and the open nature of the system. Interestingly, McClanahan & Kaunda-Arara (1996) suggest that it may be that many small reserves are to be preferred from a fisheries perspective. These authors found that small reserves increased the total catch in adjacent areas, but a larger park did not. This effect may be due to the lower ratio of edge to park area in a large reserve. In contrast, to effectively conserve highly mobile organisms, large areas are probably required.

In many cases, the information for basing a decision about a closed area will probably be poor or unavailable. It is worth asking, therefore, how much information we really need before we have a go? The answer is probably not much – but if we proceed without a sound baseline of data our efforts are likely to be sub-optimal. Regardless of the prior information base, however, if an MPA is set up, it would be wise to take steps to evaluate its performance.

One should be sceptical of some of the bolder claims that are made for the success of reserves. It is not axiomatic that they must enhance fisheries. This is not to suggest, however, that uncertainty about their value as a management tool should be used as a reason not to establish them. On the contrary, the case for trying the reserve approach is rather compelling. However, we ought to ensure that we learn from the experiments, by making efforts to understand the underlying mechanisms that determine success or failure. A programme of research that is closely allied with the implementation of the management measure is a pre-requisite for doing this. There are also perhaps more politically compelling reasons for mounting research programmes in association with MPAs. Consider the following scenario.

A marine reserve is set up using arguments that benthic habitats will be conserved and fish stocks will be enhanced. This reserve was established in the face of great resistance by fishermen, who perceived the measure as an unnecessary constraint on their trade. In the end the fishing industry accepted the measure, albeit grudgingly. Imagine now that after five years, there were no detectable effects on either the benthic communities or the stock status. One could imagine at least four reasons why this might be:

(1) the reserve was not big enough
(2) the area had not been closed for long enough

(3) the reserve was in the wrong place
(4) reserves do not work in this system.

Reasons (1) and (2) argue for even stricter constraints, and acting on reason (3) is likely to be politically very difficult. Fishermen of course are likely to argue for reason (4) and for a re-opening of the region. The point is that in the absence of information about the mechanisms that operate in the region, there is no basis for saying which of these explanations is the most likely. Thus, one is unable to decide whether to make the reserve bigger, move it, or abandon reserves altogether. Of course, even with a directed research programme the information will not be perfect, but adopting the kind of Bayesian approaches described earlier seems to me to be a sensible route towards a sensible decision. Without efforts to monitor the effectiveness of MPAs and understand why they succeed or fail, the goal of protecting fish stocks and the marine system in general may well be compromised.

Some cautionary thoughts

It is too early to tell whether closed areas really can be a panacea for the fisheries crisis. Certainly if the pattern from small-scale closures in the tropics is anything to go by, we should expect larger fish and greater biomass in the closed area. Interestingly, Fogarty & Murawski (1998) note that there is already an indication of such a difference for scallops on Georges Bank. For some, simply seeing such differences between protected and unprotected areas may, in itself, vindicate the approach. However, if this turns out to be the only benefit, we should recognise that we have set up a marine park that no-one (or very few) can visit, rather than a management regime that improves the sustainability of fisheries. It might make us all feel better, but unrestrained fishing outside may still eventually lead to stock collapses unless reserves are set up appropriately. Even if one accepts that implementation of a fishery reserve will provide higher production levels in adjacent fished areas, the potential benefits may often be in danger of being largely dissipated. If, for example, the fishery remains open access, the increased production is likely to attract new entrants into the fishery, thereby driving it back towards bioeconomic equilibrium. In this as in other cases initiatives to optimise fisheries production will usually need to be coupled with efforts to limit entry to the fishery. Indeed, without other actions, establishing a closed area may simply displace effort elsewhere, with consequent increases in conflict between users of different gear types and continued political pressure to maintain access. Coupling area closures with reductions in fishing capacity seems to be by far the better strategy. One option that has been considered is rotational closures – akin to fallowing in agricultural systems. However, given the apparently slow build up of biomass in reserve areas and the very rapid depletion that can occur once an area is opened, it seems unlikely that rotational closures will be worthwhile. The vote is out on whether benefits to fisheries will be seen from

MPAs, although it is encouraging that in at least one case (Apo in the Philippines) the fishermen themselves seem to think they do.

9.8 Understanding and involving the fishers

It would appear that the general model for the Tragedy of the Commons is usually taken for granted, including the inherent assumptions about human behaviour. McGoodwin (1990), however, argues that rather than take the Tragedy of the Commons as an inevitability more effort should be put into understanding what motivates fishermen to 'fish as if there were no tomorrow'. While one might be sceptical about the success of systems where fishermen manage themselves, there is no doubt that – at the very least – understanding fishermen and what motivates their behaviour will pay dividends.

Lessons from traditional cultures?

It has been argued that there are lessons to be learnt from the fisheries management approaches adopted in non-industrialised societies. Unfortunately, however, while traditional systems have been loosely documented for many areas around the world, good accounts are rare. Old management systems are also decaying or adapting rapidly as western influences become more pervasive (Ruddle, 1996). Yet, while the details are difficult to obtain, most traditional systems have some common features. In particular, they are often common property fisheries, but with ownership restricted to a very localised community. Within a localised area, specific fishing sites might be owned by individuals, families, clans or other small groups, but the local community is often the sole owner that controls the marine habitats in its immediate area (Ruddle, 1996). Thus, the social milieu in which the fishery is embedded can help serve to regulate, control and police the activities of fishers. Some authors suggest traditional systems should be viewed as being explicit marine conservation practices (e.g. Johannes, 1982), while others take the view that, although conservation may be an outcome, it is essentially an unintended consequence of community law based on resolving conflict and community concerns (e.g. Polunin, 1989). There is evidence, for example, that management systems in some Pacific islands evolved to control access and the distribution of benefits in order that the political system be maintained, rather than to conserve stocks (Ruddle, 1996). The notion that traditional fishers are in harmony with nature in some way does not seem sustainable.

Traditional fisheries systems do, however, seem to have two advantageous features. First, almost all traditional systems are governed by use rights to a property. A property right is a claim to a resource and/or the benefits that derive from it, protected by customary law and practice. These rights operate at two levels. At an upper level access for a local community against outsiders is protected, which includes the right to enforce and apprehend those who do not respect this exclusivity. Within this community domain, a lower level set of rights defines the rules for individual fishers. These rights commonly include exclusive use by the owner, which can serve to act as an incentive to exploit the fishery in a sustainable manner. Second, management is locally based with a focus on resolving gear usage and allocation problems using defined areas and controlled access. Importantly the fishery is self monitored by local fishers and enforced by local moral and political authority. It is, perhaps, by drawing on these latter features that one might improve modern management systems.

The argument that is generally advanced in favour of the above approach is that the more responsibility for control of local resources that can be left to local traditional users, the fewer the social, political, legal, conservation and management problems that must be addressed by governments (Ruddle, 1996). In a similar vein, Agardy (1997) argues that the best coastal management will be one that is fine-tuned to local conditions, implying that management of human activity may be most effective when it is done on local scales. Before one rushes to embrace such an approach, however, the cautionary notes of Ruddle (1996) bear repeating. First, it should be recognised that most of the positive benefits of adopting these traditional models for modern management have yet to be verified. Second, transferring the concepts is likely to be hazardous, because they arise in specific social and cultural contexts. The difficulty for Western societies in drawing lessons from indigenous cultures is not in assessing efficacy of management techniques but rather in translating this experience to a very different contextual setting (Fogarty, 1995). There are likely to be marked differences in the value system and motivation, and in the potential to affect the resource in small-scale or artisanal fisheries relative to industrial fisheries. In particular, the relationship between subsistence and revenue-generating production need to be understood.

One should also not get too romantic about fisheries management in non-industrial societies. It would be wrong to assume, for example, that overfishing is restricted to western style harvesting strategies and management. Even local small-scale fishers can over-exploit a resource if there is no restriction on the number of local participants and there are numerous examples of local stocks being totally destroyed by artisanal fisheries (Kesteven, 1976). This seems to become more likely where local fishing communities begin to adopt more effective fishing techniques than those traditionally available, particularly when this is coupled with a switch from a subsistence fishery to one which serves a local market to generate cash.

Fishers in fisheries management

Most nations, developed or otherwise, face the problem of determining the rights and responsibilities in marine resource management. What institutions should manage and enforce regulations? What role should central government play in resource management? What is the feasibility of centralised management plans over local decision making? How should the consultative process be structured to ensure that fishermen, local authorities and national governments are adequately represented? These questions are complex and largely outside the scope of this book. What seems certain, however, is that whether it be community based decision making, or representation by industry on fisheries management bodies, fishers must be engaged in the process of setting management objectives and the procedures for achieving them. This, coupled with controls on access can go a long way to securing the future of fish stocks.

Given the difficulties of drawing on systems from very different cultural settings, Fogarty (1995) suggests that one should concentrate on examples of self-regulated fisheries in industrialised sectors as a guide to action in fisheries in the developed world. He cites two examples in a western setting using different forms of direct control. First is the silver hake trawl fishery prosecuted in the mid-Atlantic region of the USA, which is controlled by limited entry and voluntary quotas established through a fisheries co-operative. The primary motivation for the quota constraints was marketing considerations but the benefits for conservation are palpable. Direct controls on effort and entry are also a dominant component of the US lobster fishery around offshore islands off the coast of Maine. Here there is a system of territories, trap limits and a *de facto* limited entry system that is more stringent than the state legislation and has measurable benefits in terms of resource status and profitability (Acheson, 1987). The key to the success of both these fisheries, and indeed to the success of traditional management schemes is self-regulation at local scales. When fishers see themselves engaged in the process of fisheries management, rather than as recipients of edicts from a centralised authority, they are more likely to identify with the management goals and police the activity themselves. When direct controls fail, it seems to be because fishers disregard regulations and there is inadequate enforcement. This argues for an acceptance of regulations by fishers. There is a strong case to be made that decentralised fisheries management regimes that involve fishing communities in the management process will be most effective in many cases. This is particularly true of poor nations and/or those with extensive coastlines where the costs of centralised policing of fisheries management are prohibitive. Clearly, such a model would be difficult to apply and probably inappropriate for regulation of high seas industrial fleets because the concept of a local community breaks down. For many fisheries, however, the model can work and even where it cannot, the principle of involving fishermen themselves in the management process is still sound.

There is growing recognition that effective fisheries management, requires active participation by those for whom regulations must apply. Without, the assent of fishers and the acceptance that the necessary regulations to limit or direct fishing activity in some way make sense, there is little hope of sustained success. Put simply, there are just too many ways to buck the system without detection. A key emphasis then is on modifying attitudes and, where practicable, involving entire communities in the management process should be encouraged. As MacManus

(1996) puts it, 'community based management facilitates the development of attitudes such that undesirable practices become not merely illegal, but socially unacceptable as well.'

The mechanisms for achieving a community-based approach are likely to differ for each situation, but there appear to be five important elements:

(1) information on existing resource use and management systems
(2) definition and provision of legal instruments
(3) use of a participatory planning approach
(4) definition of clear management agreements
(5) building and development of community institutions.

While such bottom-up approaches have considerable merit, they are unlikely to succeed in isolation. Neither the social, economic, nor ecological systems in which a particular fishery occurs will be immune from outside influence. National and international support and reinforcement through scientific resource assessment provision and legislative support will often be required.

Much of the literature dealing with community-based management in fisheries is for developing countries, often for coral reef fisheries. MacManus (1996) and references therein, give a good overview and examples for these areas. It seems self-evident, however, that the principles of consultation and active involvement apply universally. Fisheries management plans involving strong industry participation are being used increasingly in developed countries and while one could not pretend that the development of such plans are without dispute they are preferable to simple imposition of legislation from above. It seems trite to say it, but consultation and dialogue are the most likely routes to success. This dialogue, however, should also include other interest groups. Representatives of organisations with a legitimate and informed interest in marine conservation should also be heard. Engaging such an organisation in a dialogue about how to reconcile conflicting objectives imposes a level of responsibility that is absent when shouting from the sidelines.

There are promising signs. In a number of countries, for example, the philosophy that the industry, rather than government, should pay for the stock assessment and research work that benefits their industry is taking hold. These costs are often recovered through industry levies and licence fees, and in many cases the industry itself commissions the research provider. Because research costs are tied clearly to a particular industry sector it helps to ensure fishermen's participation in the management process, and give a sense of ownership of the results. This approach does not of course remove the role of government as overseer or watchdog, but it does help to involve all sectors in both the problem and the solution. In Australia, for example, the Australian Fisheries Management Authority acts to coordinate fisheries management decision making through Management Advisory Committees (MACs). These MACs involve both industry and government, but government has an advisory role, with decision making being led by industry. An important aspect of this shift will be a move away from the adversarial relationship between industry and government towards a situation which Caddy (1995) describes as 'enlightened self-interest by an industry that is convinced that short-term interests should not predominate over long-term objectives'.

9.9 Concluding thoughts

Most would agree that fisheries management needs to change, but the best means to achieve it is a topic for hot debate and has many political and social as well as scientific facets. I believe that the widening of the focus for fisheries management will ultimately be to the benefit of fishers and, as noted above, engaging them in the process of management is essential. In particular, it should be recognised that in many respects the objectives of fishers and conservationists coincide. Both sides often find it convenient to misunderstand or misrepresent the other, but in reality their objectives are largely complementary. One good example concerns the introduction of escapement measures for dolphin, trapped in seine-nets during fishing operations for yellowfin tuna. Dramatic reductions in by-catch mortalities have been achieved (from 500 000 to 5000 *per annum*). Such reductions are to the benefit of fishers who would otherwise have to waste time and effort removing dolphins from the catch and would suffer the negative public reactions to the practice, such as boycotts of tuna products. Similarly, in prawn trawl fisheries, the adoption of by-catch reduction devices can be of substantial benefit to fishers. In South Australia for example, recent trials have so impressed fishermen that uptake in one fishing fleet has been 100%. Clean catches and an environmentally friendly image have economic benefits.

Of course, these perceptions of mutual benefit are not universally held and the successful implementation of mitigation measures will often be a complicated matter. In the case of dolphin escapement, for example, improvements can be attributed to a combination of factors.

(1) methods were developed by fishing gear technologists and the fishermen themselves which would allow animals to escape after they have been encircled by the net;
(2) the US Marine Mammal Protection Act was passed;
(3) declining limits on catch quotas were authorised;
(4) tuna import embargoes were established and enforced against nations whose fleets did not meet the by-catch reduction targets adopted by the US fleet;
(5) the Inter American Tropical Tuna Commission established performance standards and dolphin mortality limits for individual vessels and coordinated an observer program to monitor progress.

Greater efforts to learn from these successes are needed.

The role of science – the cook and waiter

What of the role of biologists in the decision-making process? One might argue that because we have failed to manage fisheries appropriately thus far, the resultant over-exploitation in itself gives sufficient grounds for concern about ecosystem level impacts and that the wider potential risk should be emphasised. Perhaps, if this wider concern generates the political will to manage fisheries better it is all to the good, regardless of how much agreement there is on ecosystem effects. On ethical grounds alone, some feel that ecologists should stress (or even overemphasise) the potential risks of current practices for the system. But here one must be careful, because grossly over-stating the case could backfire. This might be termed the 'cry wolf' argument after the fairy tale about the boy who kept doing just that to gain attention. Simberlof (1987), recounted in Shraeder-Frechette & McCoy (1993, pp. 185–87) makes the same point. In debates about how big nature reserves should be and whether there should be corridors between them, Simberlof argued that erring in favour of Type II error and being wrong about the effect that it was claimed would occur, could lead to a loss of credibility. This loss could then jeopardise the opportunity to contribute to future environmental policy decisions. Shraeder-Frechette & McCoy (1993) point out this argument may to some extent be fallacious, since experts that have been proved wrong often do not lose credibility – government economists and city stockbrokers are surely cases in point! Retaining credibility may be easier, however, for experts that represent the establishment than those that do not, and we should, perhaps remain uncomfortable about scientists adopting an over-zealous emphasis on the negative impact of fisheries on ecosystems to push along the political process. A more measured assessment is preferable, where scientists are perceived to help the process in the right direction rather than force it with threats of imminent doom.

The preferred model for scientists in this debate could be considered as something of a cross between a cook and a waiter. In one sense we ought to be cooks who decide what the outcomes of particular actions are likely to be for marine populations, communities and ecosystems and prepare a menu of alternative actions and outcomes for the customer (society, fisheries managers, elected politicians) to choose from. In an ideal world each of the alternatives on the menu would have a cost associated with it, but unlike in a restaurant there would also be a probability of success. Scientists should also have a role as a waiter, however, to explain the alternatives and perhaps recommend the dish of the day. In the end the decision lies with the customer, but the waiter can influence the choice through weak advocacy of particular options. Clearly ecologists cannot take sole responsibility for setting out the alternatives and for fish stock management and Caddy (1995) suggests that their role should be primarily at the limit, where the stock becomes endangered. In other words, ecologists and fisheries biologists should not be given much say in setting exploitation objectives as long as the resource is kept in a sustainable condition. In situations where the stock is in good shape, questions of equity and allocation should predominate over biology, with a dominant role in decision making going to fishermen. At present, however, most fisheries biologists are not in the luxurious position of being on the sidelines while well-managed stocks are being harvested prudently.

Reconciling conflicting aspirations for the marine environment – the ecosystem approach?

This book has naturally focussed on the effects of fishing, but of course, fishing activity has to be viewed as one of a number of activities and uses we have for the marine environment. It is increasingly recognised that local fisheries management in particular, has to be approached as part of a wider framework of coastal zone management. Constructing this framework, however, requires political will, a wide range of expertise from many disciplines and, most importantly, active consultation with the entire constituency of users. Given these components, the benefits can be considerable. For example, MacManus (1996) cites a study where cost-benefit analysis for a coastal region in the Philippines showed that logging activity on a hillside was in direct conflict with a nearby reef fishery. Logging activity was shown to contribute to increased sediment loads in the water column which would affect the ecology of the reef, fishing patterns, fishery productivity and the tourist industry (Fig. 9.5). Careful analysis of present status, the ecological trajectory the system was likely to follow and the economic benefits that would obtain convinced the authorities that the optimal strategy was to discontinue logging. In this case, the outcome was the cessation of one activity, but under most circumstances the various activities will need to be accommodated in ways that minimise conflict and allow all parties to operate.

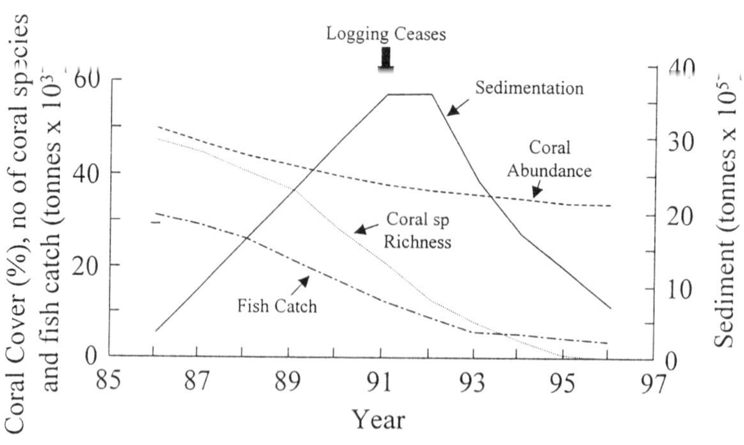

Fig. 9.5 Predicted changes in the levels of sedimentation, the number of coral species richness, the abundance of corals and fish catches if logging were to continue for a 10-year period at El Nido, Palawan, Philippines. (Redrawn from Fig. 10.3 of MacManus, 1996.)

Another example of where alternative claims on the commons must be reconciled concerns tourism. When viewed in isolation it might make sense (from a commercial viewpoint) to harvest reef fish, but the loss from tourist revenue may more than offset the benefits. Tourism is a major source of income in many developing nations, generating US$ 1–2 billion annually in some countries of South East Asia (MacManus, 1996). Coral reefs in particular can be of significant economic benefit. In Australia, for example, the Great Barrier Reef is believed to generate directly and indirectly, tourist incomes in excess US$1 billion annually. Part of this attraction is the presence of a diverse and abundant assemblage of fish species, particularly larger-bodied species, which are often not present on reefs that are heavily fished. Such effects and the general incompatibility between fishing and tourist activities in the same area mean that a balance has to be struck.

Analyses, even in terms solely of ecological impacts are difficult. I recall, for example, a meeting at which it was claimed that shark fishing was affecting, not only the ecology of a region, but also the tourist revenues from operators who took divers out to sit in shark cages and observe the animals in the wild. It was estimated that a single shark viewed regularly could generate of the order of US$1 million annually for the local economy. It was argued, therefore, that the tourist income that would follow from the assured presence of sharks, and the removal of any other fishing effect through a ban on shark fishing was by far the most sensible strategy both ecologically and economically. This was a compelling argument, but in a wider context, if one considers the ecological impacts associated with coastal development for increased tourist activity, the equation is much less easy to solve. Increased tourism is likely to be a major factor in the economic growth of many developing countries and it will essential for the well-being of people in many regions. Along with such growth, however, are threats to coastal marine ecosystems that may greatly exceed those posed by fishing. Given that tourism too can have destructive effects on marine systems, careful analyses of cost and benefits of the two activities, explicit statements of ecological and environmental objectives, and the development of ecologically sensitive tourism strategies are required. The issue of reconciling the various activities and uses to which the marine environment is put is an entire area of research in itself. Suffice it to say that if, in considering the wider implications of fishing activity for the biota, we are also stimulated to look beyond to the place of fishing within the social milieu that can be no bad thing.

Reduce effort, reduce effort, reduce effort

There is little doubt that the record of managing fisheries is poor, but for the most part, the successes there have been have not been achieved without direct controls on effort of some sort. In the northern US, for example, between 1982 and 1994 an attempt was made to manage some fisheries solely on the basis of mesh-size restrictions, closed areas and closed seasons. This approach led to sharp declines in the abundance and yield of principal groundfish because fishing effort was unconstrained and dramatically increased (Parsons, 1993). This contrasts with the success of direct controls for regulating fisheries such as the Alaskan pollock and others. In general it would appear that fisheries under direct effort control are in better condition that those under parametric regulation (Parsons, 1993). We cannot afford to *substitute* an alternative 'ecosystem' management perspective for direct controls on fisheries. Moreover, even if we were to do so, there seems to be no prospect of reducing the information requirements (and hence costs) of managing the system with this approach (Fogarty, 1995). Instead we have to adopt a two-pronged approach by making direct controls and traditional management measures more effective, while at the same time seeking to better incorporating the ecosystem perspective into the management process. As with many things, it is the middle ground rather than one of the poles of opinion that probably makes most sense.

What then of fishing and the environment? This book has tried to illustrate the effects of fishing on target stocks, the direct consequences of these changes in species densities for other components of the ecosystem and the incidental effects of the fishing process. Based on the available data, one must inevitably conclude that the effects of fishing have been profound. For the most part, however, these effects are probably reversible, provided fishing effort is reduced. It is here that the objectives of fisheries managers and environmental movements coincide most clearly. Simply stated, a reduction in fishing effort to levels that would lead to high yields and healthy fish stocks would also be one that markedly reduced the detrimental effects that fishing imposes on the system as a whole. Indeed, one could argue that achieving such reductions would be the single most effective means for mitigating the environmental effects of fishing on a global scale. This may be especially effective if combined with the establishment of MPAs.

That the nub of the problem is fishing effort is graphically illustrated by FAO statistics which show that, from 1970–1989, global industrial fleet size increased at a relative rate that was double that shown by global landings. There are simply too many fishermen, fishing too hard. The equally startling observation that annual operating costs of the global marine fishing fleet in 1989 was around $22 000 million *greater* than total revenues, even before including capital costs, brings the scale of the problem into even sharper focus. These figures are a testament to the massive overcapacity that exists for exploiting the majority of fishery resources. Mace (1996) for example, suggests that global capacity probably needs to be reduced by 50% to ensure sustainable stocks. In view of this overcapacity, it could be argued that, on a global scale, worrying about the relative merits of one biological reference point or another, or how good the fisheries models are, is largely irrelevant at present. In the final analysis, deciding what to do is fairly easy. Our biological understanding is probably 'good enough' and reducing effort must be the priority. Deciding how to make the reductions, however, is much more difficult. The social and political dimensions to the problem are much less tractable.

While overall effort reduction is by far the most relevant issue for both the environment and human welfare, the legitimate demand to incorporate into fisheries management the wider effects of fishing on the ecosystem, must be recognised and acted upon. There are encouraging signs. For example, in Europe, within the ICES community, a number of study groups are now examining these wider effects and are considering ways to mitigate them. This reflects genuine recognition, by much of the fisheries research and management community of the legitimacy of the wider environmental concerns over the effects of fishing. In other areas also, fishing activities are being altered in line with changing perceptions of what is environmentally acceptable. For example, the Japanese mothership fleet of gill netters was excluded from fishing in the US Exclusive Economic Zone in 1988 because it was unable to obtain a permit to take marine mammals in the course of fishing operations. We can also expect fishers increasingly to adopt more environmentally friendly fishing gears in response to consumer demand and that a process of environmental certification of fishing practices will be adopted in future. This is not to say that the problems are solved, simply that there is some movement in the right directions.

In the final analysis, however, it is in resolving the social and political difficulties associated with harvesting fisheries at sustainable levels that the answer lies. Although understanding biology and ecology is important, solutions to this problem lie ultimately in understanding human behaviour.

References

Acheson, J.M. (1987) The lobster fiefs revisited: economic and ecological effects of territoriality in the Maine lobster fishery. In *The Question of the Commons: the Culture and Ecology of Human Resources* (B.M. McCay & J.M. Acheson, eds), pp. 37–65. University of Arizona Press, Tucson.

Aebischer, N.J., Coulson, J.C. & Colebrook, J.M. (1990) Parallel long-term trends across four marine trophic levels and weather. *Nature*, **347**, 753–55.

Agardy, T.S. (1997). *Marine Protected Areas and Ocean Conservation.* R.G. Landes Company, Austin, Texas.

Alcala, A.C. (1988) Effects of marine reserves on coral fish abundances and yields of Philippines coral reefs. *AMBIO* **17**, 194–99.

Alcala, A.C. & Russ, G.R. (1990) A direct test of the effects of protective management on abundance and yield of tropical marine resources. *Journal du Conseil. Conseil International pour l'exploration de la mer* **46**, 40–47.

Allison, G.W., Lubchenco, J. & Carr, M.H. (1998) Marine reserves are necessary but not sufficient for marine conservation. *Ecological Applications* **8** Supplement, S79–S92.

Alverson, D.L., Freeberg, M.H., Murawski, S.A. & Pope, J.G. (1994) A global assessment of bycatch and discards. *FAO Fisheries Technical Paper* **339**, 233 pp.

Anderson, K.P. & Ursin, E. (1977) A multispecies extension to the Beverton and Holt theory of fishing, with accounts of phosphorous circulation and primary production. *Meddelelser fra Danmarks Fiskeri -og Havundersogelser* **7**, 319–45.

Andrew, N.L. & Pepperell, J.G. (1992) The by-catch of shrimp trawl fisheries. *Oceanography and Marine Biology. An Annual Review*, **30**, 537–65.

Anon. (1988) Report of the multispecies assessment working group. *International Council for the Exploration of the Sea C.M. 1988/Assess:23.*

Anon. (1990) Decline of the sea turtles: Causes and prevention. Committee on Sea Turtle Conservation, National Research Council, National Academy Press, Washington D.C.

Anon. (1994) Report of the Working Group on the Ecosystem Effects of Fishing Activities. ICES C.M. 1994/Assess/Env:1.

Anon. (1995) Report of the Study Group on Ecosystem Effects of Fishing Activities. ICES Cooperative Research Report No 200, Copenhagen, Denmark.

Anon. (1996) Report of the Working Group on the Ecosystem Effects of Fishing Activities. ICES CM 1996/Assess/Env:1.

Anthony, V.A. & Waring, G. (1980) The assessment and management of the Georges Bank herring fishery. Rapport et proces-verbaux des reunions. *Conseil permanent international pour l'exploration de la mer* **177**, 72–111.

Aronson, R.B. (1989) Brittlestar beds: low-predation anachronisms in the British Isles. *Ecology* **70**, 856–65.

Ascioti, A., Beltrami, E., Carrol, T.O. & Wirick, C. (1993) Is there chaos in plankton dynamics? *Journal of Plankton Research* **214**, 603–617.

Auster, P.J., Malatesta, R.J., Langton, R.W., Watling, L., Valentine, P.C., Donaldson, C.L.S., Langton, E.W., Shepard, A.N. & Babb, I.G. (1996) The impacts of mobile fishing gear on seafloor habitats in the Gulf of Maine (Northwest Atlantic): Implications for conservation of fish populations. *Reviews in Fisheries Science* **4**, 185–202.

Avery, M. & Green, R. (1989) Not enough fish in the sea. *New Scientist* **1674**, 28–9.

Barkai, A. & McQuaid, C. (1988) Predator-prey role reversal in a marine benthic ecosystem. *Science* **242**, 62–4.

Bax, N.J. (1991) A comparison on the fish biomass flow to fish, fisheries, and mammals in six marine ecosystems. *ICES Marine Science Symposium* **193**, 217–24.

Beauchamp, K.A. & Gowing, M.M. (1982) A quantitative assessment of human trampling effects on a rocky intertidal community. *Marine Environmental Research* **7**, 279–83.

Beddington, J. (1995) The primary requirements. *Nature* **374**, 213–214.

Beddington, J.R. & May, R.M. (1982) The harvesting of interacting species in a natural ecosystem. *Scientific American* **247**, 66–9.

Beddington, J.R. & Rettig, R.B. (1984) Approaches to the regulation of fishing effort. FAO Fisheries Technical Paper **243**.

BEON (1990) Effects of beamtrawl fishery on the bottom fauna in the North Sea II. Beleidsgericht Ecologisch Onderzoek Noordzee en Waddenzee Report No 13.

BEON (1991) Effects of beamtrawl fishery on the bottom fauna in the North Sea III. Beleidsgericht Ecologisch Onderzoek Noordzee en Waddenzee Report No 16.

Bergman, M.J. & Hup, M. (1992) Direct effects of beamtrawling on macrofauna in a sandy sediment in the southern North Sea. *ICES Marine Science Symposium* **49**, 5–11.

Bernstein, B.B. & Zalinski, J. (1983) An optimum sampling design and power tests for environmental biologists. *Journal of Environmental Management* **16**, 35–43.

Beukema, J.J. (1995) Long term effects of mechanical harvesting of lugworms *Arenicola marina* on the zoobenthic community of a tidal flay in the Wadden Sea. *Netherlands Journal of Sea Research* **33(2)**, 219–27.

Beverton, R.J.H. (1984) Dynamics of single species. In *Exploitation of Marine Communities* (R.M. May, ed.), pp. 13–58. Springer-Verlag, Berlin.

Bohnsack, J.A. (1993) Marine reserves: they enhance fisheries, reduce conflicts, and protect resources. *Oceanus* **36**, 63–71.

Bonfil, R. (1994) Overview of world elasmobranch fisheries. *FAO Fisheries Technical Paper* **341**, 119 pp.

Bowmer, T. & Keegan, B.F. (1983) Field survey of the occurence and significance of regeneration in *Amphiura filiformis* (Echinodermata: Ophiuroidea) from Galway Bay, west coast of Ireland. *Marine Biology* **74**, 65–71.

Bradstock, M. & Gordon, D.P. (1983) Coral-like bryozoan growths in Tasman Bay, and their protection to conserve local fish stocks. *New Zealand Journal of Marine and Freshwater Research* **17**, 159–63.

Brander, K. (1981) Disappearance of the common skate *Raja batis*, from the Irish Sea. *Nature* **290**, 48–50.

Britton, J.C. & Moreton, B. (1994) Marine carrion and scavengers. *Oceanography and Marine Biology. An Annual Review* **32**, 369–434.

Brosnan, D.M. & Crumrine, L.L. (1994) Effects of human trampling on marine rocky shore communities. *Journal of Experimental Marine Biology and Ecology* **177**, 79–97.

Brothers, N. (1991) Albatross mortality and associated bait loss in the Japanese longline fishery in the Southern Ocean. *Biological Conservation* **55**, 255–68.

Brown, B.E. (1987) The fishery resources. In *Georges Bank* (R.H. Backus & D.W. Bouvre, eds), p. 480. MIT Press, Cambridge, Mass.

Brown, B.E., Brennan, J.A., Grosslein, M.D., Heyerdahl, E.G. & Hennemuth, R.C. (1976) The effects of fishing on the marine finfish biomass in the northwest Atlantic from the Gulf of Maine to Cape Hatteras. *ICNAF Research Bulletin* **12**, 49–68.

Brownell, R.L., Ralls, K. & Perrin, W.F. (1989) The plight of the forgotten whales: it's mainly the small cetaceans that are now in peril. *Oceanus* **32**, 5–11.

Buchanan, J.B. (1993) Evidence of benthic pelagic coupling at a station off the Northumberland coast. *Journal of Experimental Marine Biology and Ecology* **72**, 1–10.

Buchanan, J.B. & Moore, J.J. (1986) A broad review of variability and persistence in Northumberland benthic fauna, 1971–1985. *Journal of Manne Biology Ass. UK* **66**, 641–57.

Budiansky, S. (1995). Nature's Keepers. Weidenfield & Nicolson, London.

Butman, C.A., Carlton, J.T. & Palumbi, S.R. (1995) Whaling effects on deep sea biodiversity. *Conservation Biology* **9**, 462–64.

Caddy, J.F. (1973) Underwater observations on tracks of dredges and trawls and some effects of dredging on a scallop ground. *Journal of the Fisheries Research Board of Canada* **30**, 173–80.

Caddy, J.F. (1983) The cephalopods: factors relevant to their population dynamics and to the assessment and management of stocks. *FAO Fisheries Technical Paper* **231**, 416–52.

Caddy, J.F. (1995) Comment – fisheries management science: a plea for conceptual change. *Canadian Journal of Fisheries and Aquatic Science* **52**, 2057–2058.

Calow, P. (1992) Can ecosystems be healthy? Critical consideration of concepts. *Journal of Aquatic Ecosystem Health* **1**, 1–6.

Carcasson, R.H. (1977) *A Field Guide to the Coral Reef Fishes of the Indian and West Pacific Oceans.* Collins, London.

Carpenter, S.R. & Kitchell, J.F. (1993) *The Trophic Cascade in Lakes.* Cambridge University Press, Cambridge.

Carpenter, S.R., Kitchell, J.F. & Hodgson, J.R. (1985) Cascading trophic interactions and lake productivity. *BioScience* **35**, 634–39.

Castilla, J.C. & Duran, L.R. (1985) Human exclusion from the rocky intertidal zone of central Chile: the effects on *Concholepas concholepas* (Gastropoda). *Oikos* **45**, 391–99.

Castilla, J.C. & Paine, R.T. (1987) Predation and community organization on eastern pacific, temperate zone, rocky intertidal shores. *Revista Chilena de Historia Natural* **60**, 131–51.

Castro, P. & Huber, M.E. (1997) *Marine Biology*. Wm. C. Brown, Dubuque.

Christensen, V. (1995) A model of trophic interactions in the North Sea in 1981, the Year of the Stomach. *Dana* **11**, 1–28.

Christensen, V. & Pauly, D. (1992) ECOPATH II – A software for balancing steady state models and calculating network characteristics. *Ecological Modelling* **61**, 169–85.

Christensen, V. & Pauly, D. (1993) Trophic models of aquatic ecosystems. ICLARM Conference Proceedings No 26, 336 pp.

Churchill, J.H. (1989) The effect of commercial trawling on sediment resuspension and transport over the Middle Atlantic Bight continental shelf. *Continental Shelf Research* **9**, 841–64.

Churchman, C.W. (1947) *Theory of Experimental Inference.* MacMillan, New York.

Clapham, P.J. & Brownell, R.L. (1996) The potential for interspecific competition in baleen whales. *Report of the International Whaling Commission* **46**, 361–67.

Clark, C.W. (1973) The economics of overexploitation. *Science* **181**, 630–34.

Clark, C.W. (1976) *Mathematical Bioeconomics: The Optimal Management of Renewable Resources.* John Wiley, New York.

Clark, S.H. & Brown, B.E. (1977) Changes in biomass of finfish and squids from the Gulf of Maine to Cape Hatteras, 1963–74, as determined from research vessel survey data. *Fishery Bulletin* **75**, 1–21.

Collie, J.S. (1991) Adaptive strategies for management of fisheries resources in large marine ecosystems. In *Food Chains, Yields, Models, and Mananagement of Large Marine Ecosystems* (K. Sherman, L.M. Alexander, & B.D. Gold, eds), pp. 225–42. Westview Press, Oxford.

Collie, J.S., Escanero, G.A. & Valentine, P.C. (1997) Effects of bottom fishing on the benthic megafauna of Georges Bank. *Marine Ecology Progress Series* **155**, 159–72.

Cook, R.M., Sinclair, A. & Stefansson, G. (1997) Potential collapse of North Sea cod stocks. *Nature* **385**, 521–22.

Corten, A. (1986) On the causes of the recruitment failure of herring in the central and northern North Sea in the years 1972–78. *Journal du Conseil. Conseil International pour l'exploration de la mer* **42**, 281–91.

Corten, A. (1990) Long-term trends in pelagic fish stocks of the North Sea and adjacent waters and their possible connection to hydrographic changes. *Netherlands Journal of Sea Research* **25**, 227–35.

Costanza, R. & Greer, J. (1998) The Chesapeake Bay and its watershed: A model for sustainable ecosystem management. In *Ecosystem Health* (D. Rapport, R. Costanza, P.R. Epstein, C. Gaudet, & R. Levins, eds), pp. 261–312. Blackwell Science, Oxford.

Constanza, R., d'Arge, R., de Groot, R., Farber, S., Grasso, M.; Hannon, B., Limburg, K., Naeem, S., O'Neil, R.V., Paruelo, J., Raskin, R.G., Sutton, S. & van den Belt, M. (1997) The value of the world's ecosystem services and natural capital. *Nature* **387**, 253–60.

Cramp, S. & Simmons, K.L. (1985) *The Birds of Western Paleartic.* Oxford University Press, Oxford.

Crawford, R.J.M., Underhill, L.G., Shannon, L.V., Lluch-Belda, D., Siegfried, W.R. & Villacastin-Herrero, C.A. (1991) An empirical investigation of trans-oceanic linkages between areas of high abundance of sardine. In *Long-term Variability of Pelagic Fish Populations and Their Environment* (T. Kawasaki, S. Tanaka, Y. Toba, & A. Taniguchi, eds), pp. 319–32. Pergamon Press, New York.

Creutzberg, F., Duineveld, G.C.A. & van Noort, G.J. (1987) The effects of different numbers of tickler chains on beam-trawl catches. *Journal du Conseil. Conseil permanent international pour l'exploration de la mer* **43**, 159–68.

Croxall, J.P. (1987) The status and conservation of Antarctic seals and seabirds: a review. *Environment International* **13**, 55–70.

Croxall, J.P., McCann, T.S., Prince, P.A. & Rothery, P. (1988) Reproductive performance of seabirds and seals at South Georgia and Signy Island, South Orkney Islands, 1976–1987: Implications for Southern Ocean monitoring studies. In *Antarctic Ocean Resources Variability* (D. Sahrhage, ed.), pp. 261–85. Springer-Verlag, Berlin.

Currie, D.R. & Parry, G.D. (1996) Effects of scallop dredging on a soft sediment community – a large-scale experimental study. *Marine Ecology Progress Series* **134**, 131–50.

Cushing, D.H. (1980) The decline of the herring stocks and the gadoid outburst. *Journal du Conseil. Conseil International pour l'exploration de la mer* **39**, 70–81.

Cushing, D.H. (1982) *Climate and Fisheries.* Academic Press, London.

Daan, N. (1978) Changes in cod stocks and cod fisheries in the North Sea. Rapport et proces-verbaux des reunions. *Conseil permanent international pour l'exploration de la mer* **172**, 39–57.

Daan, N. (1980) A review of replacement of depleted stocks by other species and the mechanisms underlying such replacement. Rapport et proces-verbaux des reunions. *Conseil permanent international pour l'exploration de la mer* **177**, 405–421.

Daan, N. (1989) Database report of the stomach sampling project 1981. *Cooperative Research Report Council Internationale Exploration de la Mer* **164**, 177 pp.

Daan, N., Bromley, P.J., Hislop, J.R.G. & Nielsen, N.A. (1990) Ecology of North Sea fish. *Netherlands Journal of Sea Research* **26**, 343–86.

Dawson, S.M. & Slooten, E. (1993) Conservation of Hector's Dolphins: the case and process which led to the establishment of Bands Peninsula Marine Sanctuary. *Aquatic Conservation* **3**, 207–221.

Dayton, P.K. (1998) Reversal of the burden of proof in fisheries management. *Science* **279**, 821–22.

Dayton, P.K., Thrush, S.F., Agardy, M.T. & Hofman, R.J. (1996) Environmental effects of marine fishing. *Aquatic Conservation* **5**, 205–232.

DeAlteris, J., Millikin, H. & Morse, D. (1996) Bycatch reduction in the northwest Atlantic small-mesh bottom trawl fishery for silver hake (*Merliuccius bilinearis*). In *Developing and Sustaining World Fisheries Resources: The State of Science and Management: 2nd World Fisheries Congress proceedings* (D.A. Hancock, D.C. Smith, A. Grant, & J.P. Beumer, eds), pp. 568–73. CSIRO Publishing, Collingwood, VIC.

De Gange, A.R. & Day, R.H. (1991) Mortality of seabirds in the Japanese land-based gill net fishery for salmon. *Condor* **93**, 251–58.

De Gange, A.R., Day, R.H., Takekawa, J.E. & Mendenhall, V.M. (1993) Losses of seabirds in gill nets in the North Pacific. In *The status, ecology and conservation of marine birds of the North Pacific* (K. Vermeer, K.T. Briggs, K.H. Morgan, & D. Siegel-Causey, eds), Canadian Wildlife Special Publication, National Research Press, Ottawa.

De Groot, S.J. (1984) The impact of bottom trawling on benthic fauna of the North Sea. *Ocean Management* **9**, 177–90.

De Groot, S.J. & Apeldoorn, J. (1971) Some experiments on the influence of the beam trawl on the bottom fauna. ICES C.M. 1971/B:2

Dickson, R.R., Kelly, P.M., Colebrook, J.M., Wooster, W.S. & Cushing, D.H. (1988a) North winds and production in the eastern North Atlantic. *Journal of Plankton Research* **10**, 151–69.

Dickson, R.R., Meincke, J., Malmberg, S.A. & Lee, A.J. (1988b) The 'Great Salinity Anomaly' in the northern North Atlantic, 1968–1982. *Progress in Oceanography* **20**, 103–53.

Doidge, D.W. & Croxall, J.P. (1985) Diet and energy budget of the Antarctic Fur Seal, *Arctocephalus gazella*, at South Georgia. In *Antarctic Nutrient Cycles and Food Webs* (W.R. Siegfried, P.R. Condy, & R.M. Laws, eds), pp. 543–50. Springer-Verlag, Berlin.

Donovan, G.P. (1994) Developments on issues relating to the incidental catches of cetaceans since 1992 and the UNCED Conference. Report of the International Whaling Commission, Special Issue 15, 609–613.

Dugan, J.E. & Davis, G.E. (1993) Applications of marine refugia to coastal fisheries management. *Canadian Journal of Fisheries and Aquatic Science* **50**, 2029–42.

Duggins, D.O. (1980) Kelp beds and sea otters: an experimental approach. *Ecology* **61**, 447–53.

Duineveld, G.C.A. & van Noort, G.J. (1986) Observations on the population dynamics of *Amphiura filiformis* (Ophiuroidea: Echinodermata) in the southern North Sea and its exploitation by the dab, *Limanda limanda*. *Netherlands Journal of Sea Research* **20**, 85–94.

Dunnet, G.M., Furness, R.W., Tasker, M.L. & Becker, P.H. (1990) Seabird ecology in the North Sea. *Netherlands Journal of Sea Research* **26**, 387–425.

Duran, L.R. & Castilla, J.C. (1989) Variation and persistence of the middle rocky intertidal community of central Chile, with and without human harvesting. *Marine Biology* **103**, 555–62.

Ehrlich, P.R. & Ehrlich, A.H. (1981) *Extinction: the Causes and Consequences of the Disappearance of Species.* Random House, New York.

Eleftheriou, A. & Robertson, M.R. (1992) The effects of experimental scallop dredging on the fauna and physical environment of a shallow sandy community. *Netherlands Journal of Sea Research* **30**, 289–99.

Elner, R.W. & Vadas, R.L. (1990) Inference in ecology: the sea urchin phenomenon in the northwestern atlantic. *American Naturalist* **136**, 108–125.

Emerson, C.W. & Grant, J. (1992) The control of soft-shell clam (*Mya arenaria*) recruitment on intertidal sandflats by bedload sediment transport. *Limnology and Oceanography* **36**, 1288–1300.

Erickson, A.W. & Hanson, M.B. (1990) Continental estimates and population trends of Antarctic ice seals. In *Antarctic Ecosystems. Ecological Change and Conservation* (K.R. Kerry & G. Hempel, eds), pp. 253–64. Springer-Verlag, Berlin.

Estes, J.A. (1996) The influence of large, mobile predators in aquatic food webs: examples from sea otters and kelp forests. In *Aquatic Predators and their Prey* (S.P.R. Greenstreet & M.L. Tasker, eds), pp. 65–72. Fishing News Books, Oxford.

Estes, J.A. & Palmisano, J.F. (1974) Sea otters: their role in structuring nearshore communities. *Science* **185**, 1058–60.

Estes, J.A. & van Blaricom, G.R. (1985) Sea-otters and shellfisheries. In *Marine Mammals and Fisheries* (J.R. Beddington, R.J.H. Beverton, & D.M. Lavigne, eds), pp. 187–235. Allen & Unwin, London.

Evans, P.G.H. (1984) The seabirds of Greenland: their status and conservation. In *Status and Conservation of the World's Seabirds* (J.P. Croxall, P.G.H. Evans, & R.W. Schreiber, eds), pp. 49–84. International Council for Bird Preservation, Cambridge.

Fanelli, G., Piraino, S., Belmonte, G., Geraci, S. & Boero, F. (1994) Human predation along Apulian rocky coasts (SE Italy): desertification caused by *Lithophaga lithophaga* (Mollusca) fisheries. *Marine Ecology Progress Series* **110**, 1–8.

FAO (1993) Marine fisheries and the law of the sea: a decade of change. Special chapter (revised) of The State of Food and Agriculture 1992. *FAO Fisheries Circular* **853**, 66.

FAO (1994) Review of the state of world marine fishery resources. FAO Fisheries Technical Paper **335**, 1–136.

FAO (1996) Precautionary approach to fisheries. Part 1: Guidelines on the precautionary approach to capture fisheries and species introductions. FAO Fisheries Technical Paper 350/1.

Finlayson, A.C. (1994) *Fishing for Truth (Social and Economic Studies No 52).* Institute of Economic and Social Research, Memorial University of Newfoundland., St Johns, Newfoundland.

Fisher, J. (1952) A history of the fulmar and its population problems. *Ibis* **94**, 334–45.

Fogarty, M.J. (1995) Chaos, complexity and community management of fisheries: an appraisal. *Marine Policy* **19**, 437–44.

Fogarty, M.J. & Murawski, S.A. (1998) Large-scale disturbance and the structure of marine ecosystems: fishery impacts on Georges Bank. *Ecological Applications* **8**, S6–S22.

Fogarty, M.J., Cohen, E.B., Michaels, W.J. & Morse, W.W. (1991) Predation and the regulation of sand lance populations: an exploratory analysis. *ICES Marine Science Symposium* **193**, 120–24.

Fraser, W.R., Trivelpiece, W.Z., Ainley, D.G. & Trivelpiece, S.G. (1992) Increases in Antarctic penguin populations: Reduced competition with whales or a loss of sea ice due to environmental warming? *Polar Biology* **11**, 525–31.

Frid, C.L.J., Buchanan, J.B. & Garwood, P.R. (1996) Variability and stability in the benthos: 22 years of monitoring off Northumberland. *ICES Journal of Marine Science* **53**, 978–80.

Furness, R.W. (1990) A preliminary assessment of the quantities of Shetland sandeels taken by seabirds, seals, predatory fish and the industrial fishery in 1981–83. *Ibis* **132**, 205–217.

Furness, R.W., Hudson, A.V. & Ensor, K. (1988) Interactions between scavenging seabirds and commercial fisheries around the British Isles. In *Seabirds and other Marine Vertebrates: Competition, Predation and other Interactions* (J. Burger, ed.), pp. 232–60. Columbia University Press, New York.

Gabriel, W. (1992) Persistence of demersal fish assemblages between Cape Hatteras and Nova Scotia. *Journal of the Northwest Atlantic Fisheries Science* **14**, 29–46.

Galbraith, R.D. & Main, J. (1989) Separator panels for dual purpose fish/prawn trawls. Scottish Fisheries Information Pamphlet No 16.

Gannes, L.Z., O'Brien, D.M. & Martinez del Rio, C. (1997) Stable isotopes in animal ecology: assumptions, caveats, and a call for more laboratory experiments. *Ecology* **78**, 1271–76.

Garcia, S.M. & Newton, C.H. (1994) Responsible fisheries: an overview of FAO policy developments (1945–1994). *Marine Pollution Bulletin* **29**, 6–12.

Garthe, S., Camphuysen, C.J. & Furness, R.W. (1996) Amounts discarded by commercial fisheries and their significance as food for seabirds in the North Sea. *Marine Ecology Progress Series* **136**, 1–11.

Gilkinson, K., Paulin, M., Hurley, S. & Schwinghamer, P. (1998) Impacts of trawl door scouring on infaunal bivalves: results of a physical trawl door model/dense sand interaction. *Journal of Experimental Marine Biology and Ecology* **224**, 291–312.

Gilpin, M.E. & Case, T.J. (1976) Multiple domains of attraction in model competition communities. *Nature* **261**, 40–42.

Glass, C.W., Wardle, C.S., Gosden, S.J. & Racey, D. (1995) Studies on the visual stimuli to control fish escape from cod-ends. I. Laboratory studies on the effect of a black tunnel on mesh penetration. *Fisheries Research* **23**, 157–64.

Godfray, H.C.J. & Blythe, S.P. (1990) Complex dynamics in multispecies communities. *Proceeding of the Royal Society of London B* **330**, 221–33.

Graham, M. (1955) Effect of trawling on animals of the sea bed. Papers in *Marine Biology and Oceanography*. Deep Sea Res. Suppl. 3, 1–6.

Greenstreet, S.P.R. & Hall, S.J. (1996) Fishing and groundfish assemblage structure in the northwestern North Sea: an analysis of long-term and spatial trends. *Journal of Animal Ecology* **65**, 577–98.

Greenstreet, S.P.R., Bryant, A.D., Broekhuizen, N., Hall, S.J. & Heath, M.R. (1997) Seasonal variation in the consumption of food by fish in the North Sea and implications for food web dynamics. *ICES Journal of Marine Science* **54**, 243–66.

Grime, J.P. (1997) Biodiversity and ecosystem function: the debate deepens. *Science* **277**, 1260–61.

Grosslein, M.D., Langton, R.W. & Sissenwine, M.P. (1980) Recent fluctuations in pelagic fish stocks of the Northwest Atlantic, Georges Bank region, in relation to species interactions. Rapport et proces-verbaux des reunions. *Conseil permanent international pour l'exploration de la mer* **177**, 374–404.

Gulland, J.A. & Garcia, S. (1984) Observed patterns in multispecies fisheries. In *Exploitation of Marine Communities* (R.M. May, ed.), pp. 155–90. Springer-Verlag, Berlin.

Gurney, W.S.C., Middleton, D.A.J., Ross, A.H., Nisbet, R.M., McCauley, E., Murdoch, W.W. & DeRoos, A. (1996) Modelling techniques and data requirements in aquatic systems management. In *Aquatic Predators and their Prey* (S.P.R. Greenstreet & M.L. Tasker, eds), pp. 86–97. Fishing News Books, Oxford.

Hairston, N., Smith, F.E. & Slobodkin, L. (1960) Community structure, population control, and competition. *American Naturalist* **94**, 421–25.

Hall, S.J. (1994) Physical disturbance and marine benthic communities: life in unconsolidated sediments. *Oceanography and Marine Biology. An Annual Review* **32**, 179–239.

Hall, S.J. & Greenstreet, S.P.R. (1996) Diversity, abundance and body size: relationships in the North Sea fish fauna. *Nature* **383**, 133.

Hall, S.J. & Greenstreet, S.P.R. (1998) Taxonomic distinctness and diversity measures: responses in marine fish communities. *Marine Ecology Progress Series* **166**, 227–29.

Hall, S.J. & Harding, M.J.C. (1997) Physical disturbance and marine benthic communities: the effects of mechanical harvesting of cockles on non-target benthic infauna. *Journal of Applied Ecology* **34**, 497–517.

Hall, S.J., Basford, D.J. & Robertson, M.R. (1990) The impact of hydraulic dredging for razor clams *Ensis* sp. on an infaunal community. *Netherlands Journal of Sea Research* **27**, 119–25.

Hall, S.J., Raffaelli, D. & Thrush, S.F. (1994) Patchiness and disturbance in shallow water benthic assemblages. In *Aquatic Ecology: Scale, Pattern and Process* (A.G. Hildrew, D.G. Raffaelli, & P.S. Giller, eds), pp. 333–75. Blackwell Scientific Publications, Oxford.

Hall, S.J., Robertson, M.R., Basford, D.J. & Heaney, S.D. (1993) The possible effects of fishing disturbance in the northern North Sea: an analysis of spatial patterns in community structure around a wreck. *Netherlands Journal of Sea Research* **31**, 201–208.

Hammond, P.S. (1986) Line transect sampling of dolphin populations. In *Research on Dolphins* (M.M. Bryden & R. Harrison, eds), pp. 251–79. Clarendon Press, Oxford.

Hardin, G. (1968) The tragedy of the commons. *Science* **162**, 1243–48.

Harwood, J. & Croxall, J.P. (1988) The assessment of competition between seals and commercial fisheries in the North Sea and the Antarctic. *Marine Mammal Science* **4**, 13–33.

Hay, M.E. (1984) Patterns of fish and urchin grazing on Caribbean coral reefs: are previous results typical? *Ecology* **65**, 446–54.

Heessen, H.J.L. & Daan, N. (1996) Long-term trends in ten non-target North Sea fish species. *ICES Journal of Marine Science* **53**, 1063–78.

Hempel, G. (1978) North Sea fisheries and fish stocks – a review of recent changes. Rapport et proces-verbaux des reunions. *Conseil permanent international pour l'exploration de la mer* **173**, 145–67.

Hilborn, R. (1996) The frequency and severity of fish stock declines and increases. In *Developing and Sustaining World Fisheries Resources: The State of Science and Management: 2nd World Fisheries Congress proceedings* (D.A. Hancock, D.C. Smith, A. Grant, & J.P. Beumer, eds), pp. 36–38. CSIRO Publishing, Collingwood, VIC.

Hilborn, R. & Mangel, M. (1997) *The Ecological Detective: Confronting Models with Data.* Princeton University Press, Princeton, New Jersey.

Hilborn, R. & Walters, C.J. (1992) *Quantitative Fisheries Stock Assessment: Choice, Dynamics and Uncertainty.* Chapman & Hall, New York.

Hill, B.J. & Wassenberg, T.J. (1990) Fate of discards from prawn trawlers in Torres Strait. *Australian Journal of Marine and Freshwater Reseach* **41**, 53–64.

Hislop, J.R.G. (1996) Changes in North Sea gadoid stocks. *ICES Journal of Marine Science* **53**, 1146–56.

Hofmann, E.E. & Powell, T.M. (1998) Environmental variability effects on marine fisheries: four case histories. *Ecological Applications* **8**, S23–S32.

Holden, M. (1994) *The Common Fisheries Policy, Origin, Evaluation and Future.* Blackwell Scientific Publications, Oxford.

Holme, N.A. (1983) Fluctuations in the benthos of the Western Channel. *Fluctuation and Succession in Marine Ecosystems.* Proceedings of the 17th European Symposium on Marine Biology (L. Cabioch, M. Glemarec, & J.F. Samain, eds), pp. 121–24. Oceanol. Acta, Brest.

Holt, R.D. (1997) Community modules. In *Multi-trophic Interactions in Terrestrial Systems* (A.C. Gange & V.K. Brown, eds), pp. 333–50. Blackwell Science, Oxford.

Horn, M.H. (1972) The amount of space available for marine and freshwater fishes. *Fishery Bulletin* **70**, 1295–98.

Hudson, A.V. & Furness, R.W. (1988) Utilization of discarded fish by scavenging seabirds behind whitefish trawlers in Shetland. *Journal of Zoology* **215**, 151–66.

Hughes, T.P., Reed, D.C. & Boyle, M.J. (1987) Herbivory on coral reefs: community structure following mass mortalities of sea urchins. *Journal of Experimental Marine Biology and Ecology* **133**, 39–59.

Hurlbert, S.H. (1984) Pseudo replication and the design of ecological field experiments. *Ecological Monographs* **54**, 187–211.

Huston, M.A. (1997) Hidden treatments in ecological experiments: re-evaluating the ecosystem function of biodiversity. *Oecologia (Berlin)*, **110**, 449.

Idyll, C.P. (1978) *The Sea against Hunger.* Thomas Y. Crowell, New York.

Jamieson, G.S. & Campbell, A. (1985) Sea scallop fishing impact on American Lobsters in the Gulf of St Lawrence. *Fishery Bulletin* **83**, 575–86.

Jehl, J.R. (1984) Conservation problems of seabirds in Baja California and the Pacific Northwest. In *Status and Conservation of the World's Seabirds* (J.P. Croxall, P.G.H. Evans & R.W. Schreiber, eds), pp. 41–48. International Council for Bird Preservation, Cambridge.

Jenkins, G.P., May, H.M.A., Wheatley, M.J. & Holloway, M.G. (1997) Comparison of fish assemblages associated with seagrass and adjacent unvegetated habitats of Port Phillip Bay and Corner Inlet, Victoria, Australia, with emphasis on commercial species. *Estuarine and Coastal Shelf Science* **44**, 569–88.

Jennings, S. & Kaiser, M.J. (1998) The effects of fishing on marine ecosystems. *Advances in Marine Biology* **34**, 201–352.

Jennings, S., Marshall, S.S. & Polunin, N.V.C. (1996) Seychelles' marine protected areas: comparative structure and status of reef fish communities. *Biological Conservation*, **75**, 201–209.

Johannes, R.E. (1982) Traditional conservation methods and protected marine areas in Oceania. *AMBIO* **11**, 258–61.

Jones, L.L. & DeGange, A.R. (1988) Interactions between seabirds and fisheries in the North Pacific Ocean. In *Seabirds and Other Marine Vertebrates: Competition, Predation, and other Interactions* (J. Burger, ed.), pp. 261–83. Columbia University Press, New York.

Jones, R. (1983). The decline in herring and mackerel and the associated increase in other species in the North Sea. Proceedings of the Expert Consultation to Examine Changes in Abundance and Species Composition of Neritic Fish Resources (G.D. Sharp & J. Csirke, eds), pp. 507–520. FAO, Rome.

Jones, R. & Hislop, J.R.G. (1978) Changes in North Sea haddock and whiting. Rapport et proces-verbaux des reunions. *Conseil permanent international pour l'exploration de la mer* **172**, 58–71.

Jones, R. & Richards, J. (1976) Some observations on the inter-relationships between the major fish species in the North Sea. ICES CM 1976/F:35 14 pp.

Judd, W. (1989) Deepwater fishing. *New Zealand Geographic* **4**, 77–99.

Kaiser, M.J. & Spencer, B.E. (1994) Fish scavenging behaviour in recently trawled areas. *Marine Ecology Progress Series* **112**, 41–49.

Kaiser, M.J. & Spencer, B.E. (1995) Survival of by-catch from a beam trawl. *Marine Ecology Progress Series* **126**, 31–38.

Kaiser, M.J. & Spencer, B.E. (1996) The effects of beam trawl disturbance on infaunal communities in different habitats. *Journal of Animal Ecology* **65**, 348–58.

Kaiser, M.J., Edwards, D.B. & Spencer, B.E. (1996) A study of the effects of commercial clam cultivation and harvesting on benthic infauna. *Aquatic Living Resources* **9**, 57–63.

Kawamura, A. (1994) A review on the feeding condition of southern baleen whales. *Report of the International Whaling Commission* **44**, 411–20.

Keough, M.J., Quinn, G.P. & King, A. (1993) Correlations between human collecting and intertidal mollusc populations on rocky shores. *Conservation Biology* **7**, 378–90.

Kessler, W.B., Salwasser, H., Cartwright Jr., C.W. & Caplan, J.A. (1992) New perspectives for sustainable natural resources management. *Ecological Applications* **2**(3), 221–25.

Kesteven, G.L. (1976) Resources availability related to artesanal fisheries. Proceedings of the Seminar Workshop on *Artesanal Fisheries Development and Aquaculture in Central America and Panama* (T.S. Estes, ed.), pp. 130–42. International Centre for Marine Resource Development, University of Rhode Island, Kingston.

King, W.B. (1984) Incidental mortality of seabirds in gillnets in the North Pacific. In *Status and Conservation of the World's Seabirds* (J.P. Croxall, P.G.H. Evans, & R.W. Schreiber, eds), pp. 709–716. International Council for Bird Preservation, Cambridge.

Klaer, N. & Polacheck, T. (1996) By-catch of albatrosses and other seabirds by Japanese longline fishing vessels in the Australian Fishing Zone from April 1992 to March 1995. *Emu* **97**, 150–67.

Kock, K.H. & Shimadzu, Y. (1994) Trophic relationships and trends in population size and reproduction parameters in Antarctic high-level predators. In *Southern Ocean Ecology: the BIOMASS Perspective* (S.Z. El-Sayed, ed.), pp. 287–312. Cambridge University Press, Cambridge.

Koslow, J.A. (1984) Recruitment patterns in northwest Atlantic fish stocks. *Canadian Journal of Fisheries and Aquatic Science* **41**, 1722–29.

Koslow, J.A., Marley, F. & Wicklund, R. (1988) Effects of fishing on reef fish communities at Pedro Bank and Port Royal Cays, Jamaica. *Marine Ecology Progress Series* **43**, 201–12.

Krebs, C.J. (1978) *Ecology: The Experimental Analysis of Distribution and Abundance.* Harper & Row, New York.

Kröncke, I. (1990) Macrofauna standing stock of the Dogger Bank. A comparison: II. 1951–1952 versus 1985–1987. Are changes in the community of the northeastern part of the Dogger Bank due to environmental changes? *Netherlands Journal of Sea Research* **25**, 189–98.

Krost, P., Bernhard, M., Werner, F. & Hukriede, W. (1990) Otter trawl tracks in Kiel Bay (Western Baltic) mapped by side-scan sonar. *Meeresforschung* **32**, 344–53.

Kunzlik, P.A. (1989) Small fish around Shetland. *Seabirds and Sandeels:* Proceedings of a Seminar held in Lerwick, Shetland, 15–16 October 1988 (M Heubeck, ed.) pp. 38–49. Shetland Bird Club, Lerwick.

Larkin, P.A. (1977) An epitaph for the concept of maximum sustained yield. *Transactions of the American Fisheries Society* **106**, 1–11.

Lauck, T., Clark, C.W., Mangel, M. & Munro, G.R. (1998) Implementing the precautionary principle in fisheries management through marine reserves. *Ecological Applications* **8**, S72–S78.

Laws, R.M. (1977) Seals and whales in the Southern Ocean. *Philosophical Transactions of the Royal Society of London Series B* **279**, 81–96.

Laws, R.M. (1985) The ecology of the Southern Ocean. *American Scientist* **73**, 26–40.

Lawton, J.H. & Brown, V.K. (1993) Redundancy in ecosystems. In *Biodiversity and Ecosystem Function* (E.-D. Schulze & H.A. Mooney, eds.), pp. 255–70. Springer-Verlag, Berlin.

Levitan, D.R. (1992) Community structure in times past: influence of human fishing pressure on algal-urchin interactions. *Ecology* **73**, 1597–1605.

Liddell, W.D. & Ohlhorst, S.L. (1993) Ten years of disturbance and change on a Jamaican fringing reef. Proceedings of the 7th International Coral Reef Symposium, **1**, pp 144–50.

Lindley, J.A., Gamble, J.C. & Hunt, H.G. (1995) A change in the zooplankton of the central North Sea 55° to 58°N: a possible consequence of changes in the benthos. *Marine Ecology Progress Series* **119**, 299–303.

Lluch-Belda, D., Crawford, R.J.M., Kawasaki, T., McCall, A.D., Parrish, R.H., Schwartzlose, R.A. & Smith, P.E. (1989) Worldwide fluctuations of sardine and anchovy stocks: the regime problem. *South African Journal of Marine Science* **8**, 195–205.

Lluch-Belda, D., Schwartzlose, R.A., Serra, R., Parrish, R., Kawasaki, T., Hedgecock, D. & Crawford, R.J.M. (1992) Sardine and anchovy regime fluctuations of abundance in four regions of the world oceans: a workshop report. *Fisheries Oceanography* **1**, 339–47.

MacDonald, D.S., Pope, J.G., Daan, N. & Reynolds, J.R. (1994) Impact of fishing on non-target species. Report to the European Commission from MAFF Directorate of Fisheries, University of East Anglia, 85 pp.

Mace, P.M. (1996) Developing and sustaining world fisheries resources: the state of the science and management. In *Developing and Sustaining World Fisheries Resources: the*

State of the Science and Management (D.A. Hancock, D.C. Smith, A. Grant, & J.P. Beumer, eds), pp. 1–20. CSIRO, Collingwood, VIC.

MacGillivray, C.W. & Grime, J.P. (1995) Testing predictions of the resistance and resilience of vegetation subject to extreme events. *Functional Ecology* **9**, 640–49.

MacKenzie, C.L. (1982) Compatibility of invertebrate populations and commercial fishing for Ocean Quahogs. *North American Journal of Fisheries Management* **2**, 270–75.

MacManus, J.W. (1996) Social and economic aspects of reef fisheries and their management. In *Reef Fisheries* (N.V.C. Polunin & C.M. Roberts, eds), pp. 249–82. Chapman & Hall, London.

MacPherson, E. & Duarte, C.M. (1991) Bathymetric trends in demersal fish size: Is there a general relationship? *Marine Ecology Progress Series* **71**, 103–112.

Magnusson, K.G. (1995) An overview of the multispecies vpa – theory and applications. *Reviews in Fish Biology and Fisheries* **5**, 195–212.

Magurran, A.E. (1988) *Ecological Diversity and its Measurement.* Croom Helm, London.

Man, A., Law, R. & Polunin, N.V.C. (1995) Role of marine reserves in recruitment to marine fisheries: a metapopulation model. *Biological Conservation* **71**, 197–204.

Mangel, M. (1985) *Decision and Control in Uncertain Resource Systems.* Academic Press, New York.

Mann, K.H. & Breen, P.A. (1972) The relation between lobster abundance, sea urchins, and kelp beds. *Journal of the Fisheries Research Board of Canada* **29**, 603–605.

May, R.M., Beddington, R.H., Horwood, J.W. & Shepherd, J.G. (1978) Exploiting natural populations in an uncertain world. *Mathematical Biosciences* **42**, 219–52.

McCall, A.D. (1980) Population models for the northern anchovy *Engraulis mordax*. Rapport et proces-verbaux des reunions. *Conseil permanent international pour l'exploration de la mer* **177**, 292.

McCann, T.S. & Rothery, P. (1988) Population size and status of the Southern Elephant Seal (*Mirounga leonina*) at South Georgia, 1951–1985. *Polar Biology* **8**, 305–309.

McCaughran, D.A. (1992) Standardized nomenclature and methods of defining bycatch levels and implications. Proceedings of the National Industry Bycatch Workshop (R.W. Schoning, R.W. Jacobsen, D.L. Alverson, T.G. Gentle, & J. Auyong, eds), pp. 200–201. Natural Resources Consultants, Seattle, Washington.

McClanahan, T.R. (1994) Kenyan coral reef lagoon fish: effects of fishing, substrate complexity and sea urchins. *Coral Reefs* **13**, 231–41.

McClanahan, T.R. & Kaunda-Arara, B. (1996) Fishery recovery in a coral reef marine park and its effect on the adjacent fishery. *Conservation Biology* **10**, 1187–99.

McClanahan, T.R. & Muthiga, N.A. (1988) Changes in Kenyan coral reef community structure and function due to exploitation. *Hydrobiologia*, **166**, 269–76.

McGoodwin, J.R. (1990). Crisis in the Worlds's Fisheries: People, Problems, and Policies. Stanford University Press, Stanford, Ca.

McKay, D.W. (1984) Sprat larvae off the east coast of Scotland. ICES C.M. 1984/H:56, 16 pp.

McLusky, D.S., Anderson, F.E. & Wolfe-Murphy, S. (1983) Distribution and population recovery of *Arenicola marina* and other benthic fauna after bait digging. *Marine Ecology Progress Series* **11**, 173–79.

McQueen, D.J. (1990) Manipulating lake community structure: where do we go from here? *Freshwater Biology* **23**, 613–20.

McQueen, D.J., Mills, E.L., Forney, J.L., Johannes, M.R.S. & Post, J.R. (1992) Trophic level relationships in pelagic food webs – comparisons derived from long-term data sets for

Oneida Lake, New York, (USA), and Lake George, Ontario (Canada). *Canadian Journal of Fisheries and Aquatic Science* **49**, 1588–96.

McQueen, D.J., Post, J.R. & Mills, E.L. (1986) Trophic relationships in freshwater ecosystems. *Canadian Journal of Fisheries and Aquatic Science* **43**, 1571–81.

Megrey, B.A., Hollowed, A.B., Hare, S.R., Macklin, S.A. & Stabeno, P.J. (1996) Contributions of FOCI research to forecasts of year class strength of walleye pollock in Shelikof Strait, Alaska. *Fisheries Oceanography* **5**(Supplement), 189–203.

Menge, B.A. (1995) Indirect effects in marine rocky intertidal interaction webs: patterns and importance. *Ecological Monographs* **65**(1), 21–74.

Monaghan, P. (1992) Seabirds and sandeels: the conflict between exploitation and conservation in the northern North Sea. *Biodiversity and Conservation* **1**, 98–111.

Monaghan, P., Uttley, J.D. & Okill, J.D. (1989) Terns and sandeels: seabirds as indicators of changes in marine fish populations. *Journal of Fish Biology* **35** Supplement A, 339–40.

Montevecchi, W.A. (1993) Avian indicators of marine fish conditions. In *Birds as Monitors of Environmental Change* (R.W. Furness & J.J.D. Greenwood, eds), Chapman & Hall, London.

Montevecchi, W.A. & Myers, R.A. (1995) Prey harvests of seabirds reflect pelagic fish and squid abundance on multiple spatial and temporal scales. *Marine Ecology Progress Series* **117**, 1–9.

Moreno, C.A., Lunecke, K.M. & Lepez, M.I. (1986) The response of an intertidal *Concholepas concholepas* (Gastropoda) population to protection from Man in southern Chile and the effects on benthic sessile assemblages. *Oikos* **46**, 359–64.

Moreno, C.A., Sutherland, J.P. & Jara, F. (1984) Man as a predator in the intertidal zone of southern Chile. *Oikos* **42**, 155–60.

Mountain, D.G. & Murawski, S.A. (1992) Variation in the distribution of fish stocks on the northeast continental shelf in relation to their environment, 1980–1989. *ICES Marine Science Symposium* **195**, 424–32.

Murawski, S.A. (1994) Opportunities in bycatch mitigation. In *Conserving US Fisheries: A National Symposium on the Magnusson Act* (R. Stroud, ed.), National Coalition for Marine Conservation, Savannah, Georgia.

Murawski, S. & Idoine, J. (1992) Multispecies size composition: a conservative property of ezploted fishery systems? *Journal of the Northwest Atlantic Fisheries Science* **14**, 79–85.

Murphy, E.J., Morris, D.J., Watkins, J.L. & Priddle, J. (1988) Scales of interaction between Antarctic krill and the environment. In *Antactic Ocean and Resources Variability* (D. Sahrhage, ed.), pp. 120–30. Springer-Verlag, Berlin.

Murray, T.E., Bartle, J.A., Kalish, S.R. & Taylor, P.R. (1993) Incidental capture of seabirds by Japanese southern bluefin tuna longline vessels in New Zealand waters, 1988–1992. *Bird Conservation International* **3**, 181–210.

Myers, R.A., Bridson, J. & Barrowman, N.J. (1995) Summary of worldwide stock and recruitment data. Canadian Technical Report of Fisheries and Aquatic Sciences.

Myers, R.A., Hutchings, J.A. & Barrowman, N.J. (1997) Why do fish stocks collapse? The example of cod in Atlantic Canada. *Ecological Applications* **7**, 91–106.

Nachtigall, P.E., Lien, J., Whitlow, W.L. & Read, A.J. (1995) *Harbour Porpoises: Laboratory Studies to Reduce Bycatch.* De Spil, Woerden, The Netherlands.

Naeem, S., Thompson, L.J., Lawler, S.P., Lawton, J.H. & Woodfin, R.M. (1994) Declining biodiversity can alter the performance of ecosystems. *Nature* **368**, 734–36.

Nakano, H. & Watanabe, Y. (1992). Effect of high seas driftnet fisheries on blue shark stock in the North Pacific. *Compendium of documents submitted to the scientific review of North Pacific highseas driftnet fisheries, Sidney, B.C., Canada, June 11-14, 1991, Vol. 1.*

Nixon, S.W. (1982) Nutrient dynamics, primary production and fisheries yields in lagoons. *Oceanologica Acta* **1982**, 357–71.

Northridge, S.P. (1991) An updated world review of interactions between marine mammals and fisheries. *FAO Fisheries Technical Paper*, 251, Supplement 1.

Olver, C.H., Shuter, B.J. & Minns, C.K. (1995) Toward a definition of conservation principles for fisheries management. *Canadian Journal of Fisheries and Aquatic Science* **52**, 1584–94.

Oro, D., Jover, L. & Ruiz, X. (1996) Influence of trawling activity on the breeding ecology of a threatened seabird, Audouin's gull *Larus audounii*. *Marine Ecology Progress Series* **139**, 19–29.

Otway, H. & Peltu, M. (1985) *Regulating Industrial Risks.* Butterworths, London.

Overholtz, W.J. & Tyler, A.V. (1985) Long-term responses of the demersal fish assemblages of Georges Bank. *Fishery Bulletin* **83**, 507–20.

Overholtz, W.J. & Tyler, A.V. (1986) An exploratory simulation model of competition and predation in a demersal fish assemblage on Georges Bank. *Transactions of the American Fisheries Society* **115**, 805–817.

Paine, R.T. (1980) Food webs: linkage, interaction strength and community infrastructure. *Journal of Animal Ecology* **49**, 667–85.

Paine, R.T. (1994) *Marine Rocky Shores and Community Ecology: an Experimentalist's Perspective.* Ecology Institute, Oldendorf, Germany.

Parsons, L.S. (1993) Management of marine fisheries in Canada. *Canadian Bulletin of Fisheries and Aquatic Sciences* **225**, 1–763.

Parsons, T.R. (1995) The impact of industrial fisheries on the trophic structure of marine ecosystems. In *Food Webs: Integration of Patterns and Dynamics* (G.A. Polis & K.O. Winemiller, eds), pp. 352–57. Chapman & Hall, New York.

Pauly, D. (1979) Theory and management of tropical multispecies stocks: a review, with emphasis on the Southeast Asian demersal fisheries. ICLARM Studies & Review No 1, pp. 1–35.

Pauly, D. (1987) Theory and practice of overfishing: a southeast Asian perspective. Symposium on the *Exploitation and Management of Marine Fishery Resources in Southeast Asia.* Indo-Pacific Fishery Commission, RAPA Report: 1987/10.

Pauly, D. & Christensen, V. (1995) Primary production required to sustain global fisheries. *Nature* **374**, 255–57.

Pauly, D., Christensen, V., Dalsgaard, J., Forese, R. & Torres, F. (1998) Fishing down marine food webs. *Science* **279**, 860–63.

Pauly, D., Silvestre, G. & Smith, I.R. (1989) On development, fisheries and dynamite: a brief review of tropical fisheries management. *Natural Resource Modeling* **3**, 307–29.

Perez, M.A. & Loughlin, T.R. (1991) Incidental capture of marine mammals by foreign and joint venture trawl vessels in the US EEZ of the North Pacific, 1973–1988. NOAA Technical Report, **104**, 57 pp.

Peterson, C.H. (1993) Improvement of environmental impact analysis by application of principles derived from manipulative ecology: Lessons from coastal marine case-histories. *Australian Journal of Ecology* **18**, 21–52.

Peterson, C.H., Summerson, H.C. & Fegley, S.R. (1987) Ecological consequences of mechanical harvesting of clams. *Fishery Bulletin* **85**, 281–98.

Philippart, C.J.M. (1997) Long-term impact of bottom fisheries on several by-catch species of demersal fish and benthic invertebrates in the south-eastern North Sea. *ICES Journal of Marine Science* **55**, 342–52.

Piatt, J.F. & Nettleship, D.N. (1987) Incidental catch of marine birds and mammals in fishing nets off Newfoundland, Canada. *Marine Pollution Bulletin* **186**, 344–49.

Pielou, E.C. (1969) *An Introduction to Mathematical Ecology*. Wiley, New York.

Pimm, S.L. (1984) The complexity and stability of ecosystems. *Nature* **307**, 321–26.

Pimm, S.L. (1991) *The Balance of Nature? Ecological Issues in the Conservation of Species and Communities.* University of Chicago Press, Chicago.

Pimm, S.L. & Hyman, J.B. (1987) Ecological stability in the context of multispecies fisheries. *Canadian Journal of Fisheries and Aquatic Science* **44**(2), 84–94.

Pitcher, C.R., Burridge, C.Y., Wassenberg, T.J. & Poiner, I.R. (1996) The effects of prawn trawl fisheries on GBR seabed habitats. *The Great Barrier Reef, Science, Use and Management.* National Conference Proceedings, Great Barrier Reef Marine Park Authority, pp. 107–123.

Pitcher, T.J. & Hart, P.J.B. (1982) *Fisheries Ecology*. Croom Helm, London.

Poiner, I.R., Buckworth, R.C. & Harris, A.N.M. (1990) Incidental capture and mortality of sea turtles in Australia's northern prawn fishery. *Australian Journal of Marine and Freshwater Science* **41**, 97–110.

Polunin, N.V.C. (1989) Marine fishes of the Seychelles. In *Biogeography and Ecology of the Seychelle Islands* (D.R. Stoddart, ed.), pp. 171–91. W. Junk, The Hague.

Pope, J.G. & Knights, B.J. (1982) Comparison of length distributions of combined catches of all demersal fishes in surveys in the North Sea and at Faroe Bank. In *Multispecies Approaches to Fisheries Management Advice* (M.C. Mercer, ed.), pp. 116–18. Canadian Special Publication in Fisheries and Aquatic Sciences.

Pope, J.G., Stokes, T.K., Murawski, S.A. & Idoine, S.I. (1988) A comparison of fish size composition in the North Sea and on Georges Bank. In *Ecodynamics: Contributions to Theoretical Ecology* (W. Wolff, C.-J. Soeder, & F.R. Drepper, eds), pp. 146–52. Springer-Verlag, Berlin.

Povey, A. & Keough, M.J. (1991) Effects of trampling on plant and animal populations on rocky shores. *Oikos* **61**, 355–68.

Ramsay, K., Kaiser, M.J. & Hughes, R.N. (1996) Changes in hermit crab feeding patterns in response to trawling disturbance. *Marine Ecology Progress Series* **144**, 63–72.

Rapport, D. (1998) Answering the critics. In *Ecosystem Health* (D. Rapport, R. Costanza, P.R. Epstein, C. Gaudet, & R. Levins, eds), pp. 41–50. Blackwell Science, Oxford.

Rapport, D.J., Regier, H.A. & Hutchinson, T.C. (1985) Ecosystem behaviour under stress. *American Naturalist* **125**, 617–40.

Rasmussen, E. (1977) The wasting disease of eelgrass (*Zostera marina*) and its effects on environmental factors and fauna. In *Seagrass Ecosystems, a Scientific Perspective* (C.P. McRoy & C. Helfferich, eds), Dekker, New York.

Rauck, G. (1985) Wie schadlich ist die Seezungenbaumkurre fur Bodentiere? *Informationen für die fischwirtschaft* **35**, 104–106.

Redant, F. (1991) An updated bibliography on the effects of bottom fishing gear and harvesting techniques on sea bed and benthic biota. Working document to the Study Group on Ecosystem Effects of Fishing Activities ICES.

Reimann, B. & Hoffmann, E. (1991) Ecological consequences of dredging and bottom trawling in the Limfjord, Denmark. *Marine Ecology Progress Series* **69**, 171–78.

Reise, K. (1982) Long-term changes in the macrobenthic invertebrate fauna of the Wadden Sea: are polychaetes about to take over? *Netherlands Journal of Sea Research* **16**, 29–36.

Reisen, W. & Reise, K. (1982) Macrobenthos of the subtidal Wadden Sea: revisited after 55 years. *Helgolanders Wissenschaft Meeresuntersuchungen* **35**, 409–23.

Rice, J. & Gislason, H. (1996) Patterns of change in the size spectra of numbers and diversity of the North Sea fish assemblage, as reflected in surveys and models. *ICES Journal of Marine Science* **53**, 1214–25.

Richards, J., Armstrong, D.W., Hislop, J.R.G., Jermyn, A.S. & Nicholson, M.D. (1978) Trends in Scottish research vessel catches of various fish species in the North Sea, 1922–1971. Rapport et proces-verbaux des reunions. *Conseil permanent international pour l'exploration de la mer* **172**, 211–24.

Riddle, M.J. (1988) Cyclone and bioturbation effects on sediments from coral reef lagoons. *Estuarine and Coastal Shelf Science* **27**, 687–95.

Rijnsdorp, A.D., Buijs, A.M., Storbeck, F. & Visser, E. (1996) Micro-scale distribution of beam trawl effort in the southern North Sea between 1993 and 1996 in relation to the trawling frequency of the sea bed and the impact on benthic organisms. ICES CM 1996/Mini 11.

Rijnsdorp, A.D., Groot, P.J. & van Beek, F.A. (1991) The microdistribution of beam trawl effort in the southern North Sea. ICES CM 1991/G:49.

Rivard, D. & Maguire, J.-J. (1993) Reference points for fisheries management: the eastern Canadian experience. In *Risk Evaluation and Biological Reference Points for Fisheries Management* (S.J. Smith, J.J. Hunt, & D. Rivard, eds), pp. 31–57. Canadian Special Publication in Fisheries and Aquatic Sciences No 120.

Roberts, C.M. (1993) Coral reefs: health, hazards and history. *Trends in Ecology and Evolution* **6**, 83–87.

Roberts, C.M. (1995) Effects of fishing on the ecosystem structure of coral reefs. *Conservation Biology* **9**, 988–95.

Rothschild, B.J. (1991) Multispecies interactions on Georges Bank. *ICES Marine Science Symposium* **193**, 86–92.

Rounsevell, D.E. & Copson, G.R. (1982) Growth rate and recovery of a King penguin *Aptenodytes patagonicus* population after exploitation. *Australian Wildlife Research* **9**, 519–25.

Rowe, G.T. (1981) The benthic processes of coastal upwelling ecosystems. In *Coastal Upwelling* (F.A. Richards, eds.). American Geophysical Union, Washington DC.

Ruddle, K. (1996) Traditional management of reef fishing. In *Reef Fisheries* (N.V.C. Polunin & C.M. Roberts, eds), pp. 315–35. Chapman & Hall, London.

Rudstam, I., Aneer, G. & Hilden, M. (1994) Top-down control in the pelagic Baltic ecosystem. *Dana* **10**, 105–29.

Ruiz, X., Oro, D., Martinez-Vilalta, A. & Jover, L. (1996) The feeding ecology of Audouin's Gull in the Ebro Delta. In *Ecology, Conservation and Management of Colonial Water Birds in the Mediterranean Region*. Colonial Waterbirds 19. Special Publication (A.J. Crivelli, H. Hafner, M. Fasola, R.M. Erwin, & D.A. McCrimmon, eds), pp. 68–74.

Russ, G.R. (1991) Coral reef fisheries: effects and yields. In *The Ecology of Fishes on Coral Reefs* (P.F. Sale, ed.), pp. 601–35. Academic Press, San Diego.

Russ, G.R. & Alcala, A.C. (1996a) Do marine reserves export adult fish biomass? Evidence from Apo Island, central Philippines. *Marine Ecology Progress Series* **132**, 1–9.

Russ, G.R. & Alcala, A.C. (1996b) Marine reserves: rates and patterns of recovery and decline of large predatory fish. *Ecological Applications* **6**, 947–61.

Safina, C. (1995) The world's imperiled fish. *Scientific American* **273**, 46–53.

Saila, S. (1983) Importance and assessment of discards in commercial fisheries. FAO Circular 765, UN/FAO, Rome, Italy.

Saila, S.B., Kokic, V.L. & McMAnus, J.W. (1993) Modelling the effects of destructive fishing practices on tropical coral reefs. *Marine Ecology Progress Series* **94**, 51–60.

Sainsbury, K.J. (1987) Assessment and management of the demersal fishery on the continental shelf of northwestern Australia. In *Tropical Snappers and Groupers: Biology and Fisheries Management* (J.J. Polovina & S. Ralston, eds), pp. 465–503. Westview Press, Boulder, CO.

Sainsbury, K.J. (1988) The ecological basis of multispecies fisheries management of a demersal fishery in tropical Australia. In *Fish Population Dynamics* (J.A. Gulland, ed.), pp. 349–82. John Wiley, Chichester.

Sainsbury, K.J., Campbell, R.A., Lindholm, R. & Whitelaw, A.W. (1997) Experimental management of an Australian multispecies fishery: examining the possibility of trawl-induced habitat modification. In *Global Trends: Fisheries Management* (K. Pikitch, D.D. Huppert & M.P. Sissenwine, eds), pp. 107–12. American Fisheries Society, Bethesda, Maryland.

Scandol, J.P. & Walters, C.J. (1997) Converting research results into improved management using Bayesian statistics: the example of Pacific salmon gauntlet fisheries. In *Developing and Sustaining World Fisheries Resources: The State of Science and Management* (D.A. Hancock, D.C. Smith, A. Grant, & J.P. Beumer, eds), pp. 233–37. CSIRO Publishing, Collingwood, Victoria.

Schnack-Schiel, S.B. & Muijica, A. (1994) The zooplankton of the Antarctic Peninsula region. In *Southern Ocean Ecology: The Biomass Perspective* (S.Z. El-Sayed, ed.), pp. 79–92. Cambridge University Press, Cambridge.

Schoener, T.W. (1989) Food webs from the small to the large. *Ecology* **70**, 1559–89.

Schraeder-Frechette, K.S. & McCoy, E.D. (1993) *Method in Ecology: Strategies for Conservation.* Cambridge University Press, Cambridge.

Schwinghamer, P., Guigne, J.Y. & Siu, W.C. (1996) Quantifying the impact of trawling on benthic habitat structure using high resolution acoustics and chaos theory. *Canadian Journal of Fisheries and Aquatic Science* **53**, 288–96.

Sebens, K.P. (1994) Biodiversity of coral reefs: What are we losing and why? *American Zoologist* **34**, 115–34.

Sheldon, R.W., Prakash, A. & Sutcliffe, W.H.J. (1972) The size distribution of particles in the ocean. *Limnology and Oceanography* **17**, 327–40.

Shepard, A.N. & Auster, P.J. (1991) Incidental (non-capture) damage to scallops caused by dragging on rock and sand substrates. In *An International Compendium of Scallop Biology and Culture* (S.E. Shumway & P.A. Sandifer, eds), pp. 219–30. World Aquaculture Society, Baton Rouge.

Sherman, K. (1991) The large marine ecosystem concept: A research and management strategy for living marine resources. *Ecological Applications* **1**, 349–60.

Sherman, K., Jones, C., Sullivan, L., Smith, W., Berrien, P. & Ejsymont, L. (1981) Congruent shifts in sand eel abundance in western and eastern North Atlantic ecosystems. *Nature* **291**, 486–89.

Siemann, E., Tilman, D. & Haarstad, J. (1996) Insect species diversity, abundance and body size relationships. *Nature* **380**, 704–706.

Simberlof, D. (1987) Simplification, danger, and ethics in conservation biology. *Bulletin of the Ecological Society of America* **68**, 156–57.

Sissenwine, M.P. & Cohen, E.B. (1991) Resource productivity and fisheries management of the Northeast shelf ecosystem. In *Food Chains, Yields, Models, and Management of large Marine Ecosystems* (K. Sherman, L.M. Alexander & B.D. Gold, eds), pp. 107–123. Westview Press, Oxford.

Sissenwine, M.P., Brown, B.E., Palmer, J.E., Essig, R.W. & Smith, W. (1982) Empirical examination of population interactions for the fishery resources off the northeastern USA. *Canadian Special Publication in Fisheries and Aquatic Sciences* **59**, 82–94.

Sladen, W.J.L. (1964) The distribution of the Adelie and Chinstrap penguins. In *Biologie Antarctique* (R. Carrick, M.W. Holdgate & J. Prevost, eds), pp. 359–65. Hermann, Paris.

Smith, C.R. (1992) Whale falls. *Oceanus* **35**, 74–78.

Smith, C.R. & Brumsickle, S.J. (1989) The effects of patch size and substrate isolation on colonization modes and rates in an intertidal sediment. *Limnology and Oceanography* **34**, 1263–77.

Smith, P.E. (1972) The increase in spawning biomass of northern anchovy *Engraulis mordax*. *Fishery Bulletin* **70**(3), 849–74.

Snelgrove, P., Blackburn, T.H., Hutchings, P.A., Alongi, D.M. *et al.* (1997) The importance of marine sediment biodiversity in ecosystem processes. *AMBIO* **26**, 578–83.

Solow, A.R. (1994) Detecting change in the composition of a multispecies community. *Biometrics* **50**, 556–65.

Soutar, A. & Isaacs, J.D. (1974) Abundance of pelagic fish during the 19th and 20th centuries as recorded in anaerobic sediment off the Californias. *Fishery Bulletin* **72**(2), 257–73.

Sparholt, H. (1990) An estimate of the total biomass of fish in the North Sea. *Journal du Conseil. Conseil International pour l'exploration de la mer* **46**, 200–10.

Sprules, W.G. & Goyke, A.P. (1994) Size-based structure and production in the Pelagia of Lakes Ontario and Michigan. *Canadian Journal of Fisheries and Aquatic Science* **51**, 2603–11.

Sprules, W.G. & Stockwell, J.D. (1995) Size-based biomass and the production models in the St Lawrence Great Lakes. *ICES Journal of Marine Science* **52**, 705–710.

Steele, J.H. (1996) Regime shifts in fisheries management. *Fisheries Research* **25**, 19–23.

Steele, J.H. (1998) Regime shifts in marine ecosystems. *Ecological Applications* **8**, 33–36.

Steele, J.H. & Henderson, E.W. (1984) Modelling long-term fluctuations in fish stocks. *Science* **224**, 985–87.

Stevens, J. (in press). The population status of highly migratory oceanic sharks in the Pacific Ocean. National Coalition for Marine Conservation.

Stewart-Oaten, A., Murdoch, W. & Parker, K. (1986) Environmental impact assessment: 'Pseudoreplication' in time? *Ecology* **67**, 929–40.

Stone, M. & Bublitz, C.G. (1995) Cod trawl separator panel: potential for reducing halibut bycatch. In *Solving Bycatch: considerations for today and tomorrow*. Proceedings of the Solving Bycatch Workshop, 25–27 September, 1995, Seattle.

Svendsen, E. & Magnusson, A.K. (1991) Climatic variability in the North Sea. ICES CM 1991/Variability Symposium, Paper No 10

Takekawa, J.E., Carter, H.R. & Harvey, T.E. (1990) Decline of the common murre in central California, 1980–1986. *Studies in Avian Biology* **14**, 149–63.

Taylor, R.H., Wilson, P.R. & Thomas, B.W. (1990) Status and trends of Adelie penguin populations in the Ross Sea region. *Polar Record* **26**, 293–304.

Testa, J.W., Oehlert, G., Ainley, D.G., Bengtson, J.L., Siniff, D.B., Laws, R.M. & Rounsevell, D. (1991) Temporal variability in Antarctic marine ecosystems: periodic fluctuations in the phocid seals. *Canadian Journal of Fisheries and Aquatic Science* **48**, 631–39.

Thiebaux, M.L. & Dickie, L.M. (1993) Structure of the body size spectrum of the biomass in aquatic ecosystems: a consequence of allometry in predator-prey interactions. *Canadian Journal of Fisheries and Aquatic Science* **50**, 1308–17.

Thouzeau, G., Robert, G. & Ugarte, R. (1991) Faunal assemblages of benthic mega-invertebrates inhabiting sea scallop grounds from eastern Georges Bank, in relation to environmental factors. *Marine Ecology Progress Series* **74**, 61–82.

Thrush, S.F., Whitlach, R.B., Pridmore, R.D., Hewitt, J.E., Cummings, V.J. & Wilkinson, M.R. (1996) Scale-dependent recolonization: the role of sediment stability in a dynamic sandflat habitat. *Ecology* **77**, 2472–87.

Tilman, D. & Downing, J.A. (1994) Biodiversity and stability in grasslands. *Nature* **367**, 363–65.

Tuck, I.D., Hall, S.J., Robertson, M.R., Armstrong, E. & Basford, D.J. (1998) Effects of physical trawling disturbance in a previously unfished sheltered Scottish sealoch. *Marine Ecology Progress Series* **162**, 227–42.

Turrell, W.R. (1992) New hypotheses concerning the circulation of the northern North Sea and its relation to North Sea fish stock recruitment. *ICES Journal of Marine Science* **49**, 107–23.

Underwood, A.J. (1984) The analysis of competition by field experiments. In *Community Ecology – Pattern and Process* (D.J. Anderson & J. Kikkawa, eds), pp. 240–68. Blackwell, Oxford.

Underwood, A.J. (1992) Beyond BACI: the detection of environmental impacts on populations in the real, but variable, world. *Journal of Experimental Marine Biology and Ecology* **161**, 145–78.

Valdemarsen, J.W., Engas, A. & Isaksen, B. (1985) Vertical entrance into a trawl of Barents Sea gadoids as studied with a two-level fish trawl. ICES CM 1985/B:46.

Van Beek, F.A., Van Leeuwen, P.I. & Rijnsdorp, A.D. (1990) On the survival of plaice and sole discards in the otter-trawl and beam-trawl fisheries in the North Sea. *Netherlands Journal of Sea Research* **26**, 151–61.

Van den Heiligenberg, T. (1987) Effects of mechanical and manual harvesting of lugworms *Arenicola marina* L. on the benthic fauna of tidal flats in the Dutch Wadden Sea. *Biological Conservation* **39**, 165–77.

Van der Elst, R.P. (1979) A proliferation of small sharks in the shore-based Natal sports fishery. *Environmental Biology of Fishes* **4**(4), 349–62.

Verity, P.G. & Smetacek, V. (1996) Organism life cycles, predation, and the structure of marine pelagic ecosystems. *Marine Ecology Progress Series* **130**, 277–93.

Vermeij, G.J. (1977) The Mesozoic marine revolution: evidence from snails, predators and grazers. *Paleobiology* **3**, 245–58.

Vermeij, G.J. (1987) *Evolution and Escalation: an Ecological History of Life.* Princeton University Press, Princeton, New Jersey.

Vitousek, P.M., Ehrlich, P.R., Erlich, A.H. & Matson, P.A. (1986) Human appropriation of the producers of photosynthesis. *BioScience*, **36**, 368–73.

Wainright, S.C., Fogarty, M.J., Greenfield, R.C. & Fry, B. (1993) Long-term changes in the Georges Bank food web: trends in stable isotopic compositions of fish scales. *Marine Biology* **115**, 481–93.

Walker, P. (1996) Sensitive skates or resilient rays? Spatial and temporal shifts in ray species composition in the central and northwestern North Sea. ICES CM 1996/Mini:4.

Walters, C.J. (1986) *Adaptive Management of Renewable Resources.* Macmillan, New York.

Walters, C.J. & Collie, J.S. (1988) Is research on environmental factors useful to fisheries management? *Canadian Journal of Fisheries and Aquatic Science* **45**, 1848–54.

Walters, C.J. & Hilborn, R. (1978) Ecological optimization and adaptive management. *Annual Review of Ecology and Systematics* **9**, 157–88.

Walters, C.J. & Pearse, P.H. (1996) Stock information requirements for quota management systems in commercial fisheries. *Reviews in Fish Biology and Fisheries* **6**, 21–42.

Warwick, R.M. & Clarke, K.R. (1995) New 'biodiversity' measures reveal a decrease in taxonomic distinctness with increasing stress. *Marine Ecology Progress Series* **129**, 301–305.

Wassenberg, T.J. & Hill, B.J. (1987) Feeding by the sand crab *Portunus pelagicus* on material discarded from prawn trawlers in Moreton Bay, Australia. *Marine Biology* **95**, 387–93.

Wassenberg, T.J. & Hill, B.J. (1990) Partitioning of material discarded from prawn trawlers in Moreton Bay. *Australian Journal of Marine and Freshwater Reseach* **41**, 27–36.

Watson, M. & Ormond, R.F.G. (1994) Effects of an artisanal fishery on the fish and urchin populations of a Kenyan coral reef. *Marine Ecology Progress Series* **109**, 115–29.

Watson, M., Righton, D., Austin, T. & Ormond, R. (1996) The effects of fishing on coral reef fish abundance and diversity. *Journal of the Marine Biological Association (U.K.)* **76**, 229–33.

Weimerskirch, H., Clobert, J. & Jouventin, P. (1987) Survival in five southern albatrosses and its relationship with their life history. *Journal of Animal Ecology* **56**, 1043–55.

Wells, S. (1988) *Coral Reefs of the World.* UN Environment Program and the International Union for the Conservation of Nature and Natural Resources., Cambridge, UK.

Westra, L. (1996) Ecosystem integrity and the 'fish wars'. *Journal of Aquatic Ecosystem Health* **5**, 275–82.

Wetherall, J. & Seki, M. (1992) Assessing impacts of North Pacific high seas driftnet fisheries on Pacific pomfret and sharks: progress and problems. Compendium of documents submitted to the scientific review of North Pacific highseas driftnet fisheries, Sidney, B.C., Canada, 11–14 June, 1991, Vol. 2.

Wilson, J.A., Acheson, J.M., Metcalfe, M. & Kleban, P. (1994) Chaos, complexity and community management of fisheries. *Marine Policy* **18**, 291–305.

Wilson, J.A., French, J., Kleban, P., McKay, S. & Townsend, R. (1991a) Chaotic dynamics in a multiple species fishery: a model of community predation. *Ecological Modelling* **58**, 303–22.

Wilson, J.A., Kleban, P., McKay, S. & Townsend, R. (1991b) Management of multiple species fisheries with chaotic population dynamics. *ICES Marine Science Symposium* **193**, 287–300.

Witbaard, R. & Klein, R. (1994) Long-term trends on the effects of the southern North Sea beamtrawl fishery on the bivalve mollusc *Arctica islandica* L. (Mollusca, Bivalvia). *ICES Journal of Marine Science* **51**, 99–105.

Witman, J.D. & Sebens, K.P. (1992) Regional variation in fish predation intensity: a historical perspective in the Gulf of Maine. *Oecologia (Berlin)* **90**, 305–315.

Wright, P.J. (1996) Is there a conflict between sandeel fisheries and seabirds? A case study at Shetland. In *Aquatic Predators and their Prey* (S.P.R. Greenstreet & M.L. Tasker, eds), pp. 154–65. Fishing News Books, Oxford.

Wright, P. & Bailey, M. (1993) Biology of sandeels in the vicinity of seabird colonies at Shetland. Fisheries Research Report No 15/93. SOAFD Marine Laboratory, Aberdeen.

Yamasaki, A. & Kuwahara, A. (1990) Preserved area to effect recovery of overfished Zuwai crab stocks off Kyoto Prefecture. Proceedings of the International Symposium on King and Tanner Crabs, Alaska Sea Grant College Program, University of Alaska, Fairbanks. pp 575–78

Yang, J. (1982) An estimate of the fish biomass in the North Sea. *Journal du Conseil. Conseil International pour l'exploration de la mer* **40**, 161–72.

Index